S0-BWX-132

PHYSICS POTENTIAL AND DEVELOPMENT OF $\mu^+\mu^-$ COLLIDERS SECOND WORKSHOP

AIP CONFERENCE PROCEEDINGS 352

PHYSICS POTENTIAL AND DEVELOPMENT OF $\mu^+\mu^-$ COLLIDERS SECOND WORKSHOP

SAUSALITO, CA NOVEMBER 1994

EDITOR: DAVID B. CLINE
UNIVERSITY OF CALIFORNIA
AT LOS ANGELES

American Institute of Physics **Woodbury, New York**

L.C. Catalog Card No. 95–81413
ISBN 1-56396-506-2
DOE CONF- 9411210

Printed in the United States of America.

Contents

PREFACE

With the demise of the Superconducting Super Collider, the U.S. particle physics community is left without a frontier accelerator for the first time in 50 or more years. Even in 1992, when the SSC seemed assured to be constructed, a small group of physicists started discussions of a possible $\mu^+\mu^-$ collider. This was largely due to the fact that e^+e^- linear colliders in the multi-TeV energy range are not likely to be practical due to the enormous radiative energy losses. However, $\mu^+\mu^-$ colliders provide for less radiative losses and can still use circular machines up to the multi-TeV range.

The first $\mu^+\mu^-$ collider workshop was held in Napa California in 1992. Later a small workshop was held in early 1993 at LANL. In 1994 a much larger workshop was held in Sausalito California, which provides most of the papers in this book. We begin this book with a brief historical perspective on the early ideas for $\mu^+\mu^-$ colliders and $\mu^\pm$ storage rings. We then provide the salient papers from the Napa meeting. The essentials of the conclusions of the Sausalito workshop are covered in the later two thirds of the book. I believe these efforts to develop a new powerful machine to study elementary particles will someday lead to the actual construction of such a collider. I note that similar workshops were held between 1976 and 1982 as the $p\bar{p}$ colliders at CERN and FNAL were being designed. The $p\bar{p}$ colliders succeeded beyond anyone's imagination in the 1970's, 1980's, and 1990's.

A brief list of the possible advantages of a $\mu^+\mu^-$ colliders, as presented in the papers in this book, includes:

1) The synchrotron radiation of muons compared with electrons is reduced by a factor of $(m_e/m_\mu)^2$ allowing $\mu^+\mu^-$ to use circular storage rings even for extremely high energies, $E \gg 10$ TeV. (Circular machines larger than LEP are not practical.)
2) The next major goal of particle physics is to detect and study fundamental Scalar particles (i.e., the Higgs Boson). Leptons couple to those Scalar particles as m_l^2; therefore, the relative coupling between muons and electrons is m_μ^2/m_e^2, for example. In Supersymmetric models there can be several low-mass Scalars — $\mu^+\mu^-$ colliders may provide a unique possibility to help determine those states.
3) $\mu^\pm$ are produced with great abundance at hadron machines through $\pi^\pm$ production and decay — a strong interaction process.
4) Polarized $\mu^+\mu^-$ colliders may be feasible, greatly enhancing the physics reach.

Of course all of these possibilities, in fact, lead to technical challenges; including the use of μ cooling, rapid acceleration, high intensity hadron sources, and the particle detectors themselves.

I wish to thank M. Laraneta, J. Kolonko, and D. Sanders for their help in these meetings and with these proceedings, and the advisory committee for both the Napa and Sausalito workshops. Finally, I wish to thank the authors who took the time to write up their papers for these proceedings, and the Department of Energy for its support.

D. B. Cline
Editor

ACKNOWLEDGMENTS

Acknowledgment is made to the following for permission to reprint copyright material:

Nuclear Instruments and Methods in Physics, A350, David B. Cline, "Physics potential of a few hundred GeV $\mu^+\mu^-$ collider," 24–26 (1994), with kind permission from Elsevier Science B. V., Amsterdam, The Netherlands.

Nuclear Instruments and Methods in Physics, A350, David V. Neuffer, "$\mu^+ - \mu^-$ colliders: possibilities and challenges," 27–35 (1994), with kind permission from Elsevier Science B. V., Amsterdam, The Netherlands.

Nuclear Instruments and Methods in Physics, A350, William A. Barletta and Andrew M. Sessler, "Characteristics of a high energy $\mu^+\mu^-$ collider based on electro-production of muons," 36–44 (1994), with kind permission from Elsevier Science B. V., Amsterdam, The Netherlands.

Nuclear Instruments and Methods in Physics, A350, Alessandro G. Ruggiero, "A muon collider scenario based on stochastic cooling," 45–52 (1994), with kind permission from Elsevier Science B. V., Amsterdam, The Netherlands.

Nuclear Instruments and Methods in Physics, A350, S. Chattopadhyay, W. Barletta, S. Maury, D. Neuffer, A. Ruggiero, and A. Sessler, "Critical issues in low energy muon colliders — a summary," 53–56 (1994), with kind permission from Elsevier Science B. V., Amsterdam, The Netherlands.

I. EARLY CONCEPTS FOR $\mu^+\mu^-$ COLLIDERS AND HIGH ENERGY μ STORAGE RINGS

Some Comments on the Early History of the $\mu^+\mu^-$ Concept and the High Intensity μ Storage Ring

D. B. Cline

University of California at Los Angeles
Center for Advanced Accelerators, Department of Physics and Astronomy
Los Angeles, CA 90095-1547

It is always difficult to pin down the origin of a good idea in particle physics. Take the case of the $p\bar{p}$ collider. The concept of a colliding $p\bar{p}$ was discussed a few years after the discovery of the anti-proton but a realistic collider could not be considered until ~1976[1] when electron and stochastic cooling had been invented.

The $\mu^+\mu^-$ collider story will likely be the same. There are two aspects of the development:

1. High Energy $\mu^\pm$ Storage Ring
2. Collection and Cooling of $\mu^\pm$ and Concept of the $\mu^+\mu^-$ Collider.

There is a long history to (1). For example, the various μ(g-2) experiments store the μ's for a short time. One of the earliest concepts of a high energy μ storage ring for a μ-p collider seems to have started with Rochester[2]. A more detailed discussion of a μ storage ring - this time to produce specific neutrino beams - was proposed by Cline and Neuffer in 1980 (see following paper reproduced here)[3].

The development of μ ionization cooling is also an old idea (two other early papers on this are reproduced here: Neuffer[4]; Parkhomchuk & Skrinsky[5]). We have obtained some information concerning even earlier discussions at Novosibirsk from Skrinsky. We provide those extracted statements here. We also provide reference to the early work by Neuffer on $\mu^+\mu^-$ colliders[6].

REFERENCES:

1. C. Rubbia, P. McIntyre and D. Cline, Proc. of 1976 Aachen Neutrino Conf., (Eds, H. Faissner, H. Reithler and P. Zerwas)(Vieway Braunschweig), (1977) 683.
2. A. Melissinos, private communication. He and others at Rochester proposed a μp collider in the 60's.
3. D. Cline and D. Neuffer, AIP Conf. Proc. no. 68 pt 2, (1980) 846.
4. D. Neuffer, Proc. 12th Int. Conf. on High Energy Accelerators, (Eds. F. T. Cole and R. Donaldson), (1983) 481.
5. A. N. Skrinsky and V. V. Parkhomchuk, Sov. J. Part. Nucl. 12(3) (1981) 223.
6. D. Neuffer, Part. Accel. 14 (1983) 75; D. Neuffer, in: Adv. Accel. Concepts [AIP Conf. Proc. 156 (1987) 201.

Extracts:

The First INP Publications on Muon Colliders

From the Budker's talk at the International Accelerator Conference, Erevan, 1969:

"On the other hand, of obvious interest are the lepton collisions at energies of a few hundred GeV (in the center of mass system). For this, the most convenient particles are mu-mesons. One can consider two variants:

1. The mu-mesons injected into a storage ring make some certain number of turns during their lifetime independent on energy: 400 turns in a 20 kG field and 4000 turns in a field of 200 kG. Using accelerators of very high intensity as the injectors of mu-mesons one can attempt to realize my-meson collisions in a special accelerator. The small cross-section of the event and low factor of conversion for protons into mu-mesons with a low spread of energy and angles make this problem very complex even with the presence of a proton accelerator at very high energy.
2. Another method currently under development in Novosibirsk suggests the long time storing of a very large number of protons of rather low energy (25 GeV), slamming them into a target, obtaining a large number of mu-mesons of both signs of low energy and in a good phase volume, and their further acceleration in a special coreless pulsed accelerator up to high energies in the range where the reactions with mu-meson interactions are being studied. Because of an increase in the mu-meson lifetimes with an increase in energy, the acceleration can be performed practically with no loss."

From the Budker's talk at the International High Energy Physics Conference, Kiev, 1970:

"Therefore we are seriously considering the somewhat exotic but quite real project of mu-meson accelerators and mu-meson colliding beams. The relativistic effects enable one to accelerate mu-mesons in cyclic accelerators with practically no loss at a sufficiently high growth rate of the magnetic filed. Therefore, for mu-meson accelerators with fixed targets and colliding beams, we plan to use the short-focus ironless magnets developed at INP. The magnets accurately form fields due to their thin skin current distribution. For obtaining a large current of mu-mesons, which is very important for colliding beams, we expect to use the proton accelerator VAPP as well as the whole megagauss focusing instrumentation being developed at the VAPP-NAP installation for focusing of antiprotons.

In order to decrease the phase volume of mu-mesons from the decay of pi-mesons generated prior to their entrance into the storage ring we expect to build an intermediate pulsed storage ring with a field of 200 kG with the damping of the mu-meson beam due to the energy losses on a target installed inside the ring.

At present, two stages are under consideration: the installation of a 12-15 m radius ring with an energy of 7-100 GeV and the installation of a 160 m radius ring at 1000 GeV. It should be noted, that these installations can be used both as the proton and antiproton accelerators up to these energies and as the installations with the proton-anti-proton colliding beams. The low duty factor of the pulse accelerators is practically entirely compensated for by the high particle current and this gives them experimental possibilities compatible with the iron magnet at the corresponding energy. Their much lower cost makes their construction feasible for small laboratory having reasonably low funds such as the Novosibirsk Institute of Nuclear Physics. One should note, that the possibility of using these accelerators for protons provides us with the decisiveness to start their construction. Though we hope to get a lot of experimental data with the mu-meson colliding beams, it does not exclude the possibility that it might be an installation of a single experiments, namely, the discovery of the form-factor at the lepton interactions. If the cross-section of the weak interaction stops to grow with energy, the installation's luminosity is not sufficient for any observation. Whatever interesting result could it be, however, this does not justify the decisiveness for such a complex project."

Morges Seminar 1971 - Intersecting Storage Rings at Novosibirsk
A.N. Skrinsky

$\mu^+\mu^-$ Possibilities

These experiments at hundreds GeV energy region will be available, only when several very difficult things will be discovered:

1. To have a very large number of protons with 10 GeV energy in rather short bunches. It is necessary to have about 10^{14} or even 10^{15} protons in about 10 sec in several meters long bunch.
2. To produce with maximum efficiency muons with 1 GeV or less energy, using nuclear cascade, strong focusing in the target and in pion decay channel. It seems possible to have 0.1 or even more useful muon per proton.
3. To cool muons in special 100 kilogauss pulsed storage ring, using ionization energy losses. If the targets are in places with very small β-function, the finishing emittance of muon beam should be small enough to be injected into the main muon accelerator with small aperture and to be well compressed in interaction points.
4. To accelerate muons rapidly in some accelerators. If the muons are accelerated to their rest energy in a time, several times less than their life time at rest, most of the muons will be accelerated up to the required energy. It is possible to use a linear accelerator, or to use a synchrotron with more than a 100 kilogauss and magnetic field with a short rise time. In the last case, the accelerator will be at the same time the colliding beams ring. In the ring with such a magnetic field it is possible to have several thousands of useful turns of muon beams.

If all of these conditions are satisfied, it seems to be possible to have an average luminosity $10^{31} cm^{-2} sec^{-1}$ and may be a bit more, which should be sufficient.

Accelerator and Detector Prospects of Elementary Particle Physics - A.N. Skrinsky

Institute of Nuclear Physics of the Siberian Division of the Academy of Sciences of the USSR, Novosibirsk, Usp. Fiz. Nauk 138, I-43 (September 1982)

This review treats the progressive changes in the filed of physics and technology of accelerators, and in part, of detectors, which have exerted and will in the near future exert a fundamental influence on the development of elementary-particle physics. In particular, it discusses the possibilities of generation of beams of elementary particles and the prospects of performing experiments with colliding beams involving the development of methods of cooling charged-particle beams, designing superconducting systems, and developing Superlinacs.

PACS numbers: 29.15Dt.29.20 - c. 29.40. -n. 29.25.Fb

IONIZATION COOLING: PHYSICS AND APPLICATIONS

V.V.Parkhomchuk and A.N.Skrinsky

Institute of Nuclear Physics,
630090, Novosibirsk 90, USSR

Ionization cooling is based on use of the ionization energy losses of charged particles, which travel in matter, with the external source compensation for the average energy losses of equilibrium particles /1-3/. When the motion in matter occurs at the average electron density N_e, the power of losses $\mathcal{P}_{fr}$ and the frictional force F_{fr} are, correspondingly, equal to

$$\mathcal{P}_{fr} = 4\pi \frac{N_e e^4 L}{m_e v}\ ; \quad \vec{F}_{fr} = -\frac{\mathcal{P}_{fr}}{v^2}\vec{v} \qquad (1)$$

where e and m_e are the charge and mass of an electron, $L_i = \ln\left(\frac{2m_e\gamma^2 v^2}{Z_i I}\right) \approx 8 \div 12$, Z_i is the atomic number of matter, I is the effective ionization potential, and $\vec{v}$ is the velocity of particle motion.

1. For simplicity, let us first consider the case of a straight-line motion of the beam under cooling in matter with ionization losses. The specific features of the ionization cooling manifest themselves in this case as well, but the sum of the decrements is completely in dependent of the specific features of the particle motion, including the presence of focusing. Assuming that the deviations of the longitudinal momentum and the transverse momentum (with respect to the direction of the force which recovers the energy losses) are small, $|\Delta P_\parallel|$, $|\Delta P_x|, |\Delta P_z| \ll P$, we immediately obtain, in a linear approximation, the damping decrements in the form

$$\delta_{x,z} = \frac{F_{fr}}{p} = \frac{\partial F_{fr}^{x,z}}{\partial P_{x,z}}\ ; \quad \delta_\parallel = \frac{\partial F_{fr}}{\partial P_\parallel} \qquad (2)$$

and an expression for the sum of the decrements:

$$\sum\delta = \delta_x + \delta_z + \delta_\parallel = \mathrm{div}_p \vec{F}_{fr} = \frac{2\mathcal{P}_{fr}}{pv}\left(1 - \frac{p_\parallel}{2v}\frac{\partial v_\parallel}{\partial p_\parallel}\right) + \frac{1}{v}\frac{\partial \mathcal{P}_{fr}}{\partial P_\parallel} =$$

$$= \begin{cases} \dfrac{\mathcal{P}_{fr}}{2E_k} + \dfrac{\partial \mathcal{P}_{fr}}{\partial E_k}, & v \ll c \\ \dfrac{2\mathcal{P}_{fr}}{E} + \dfrac{\partial \mathcal{P}_{fr}}{\partial E}, & v \to c \end{cases} \qquad (3)$$

In the non-relativistic case, a rapid fall in the frictional force with increasing of the kinetic energy E_k leads to a negative value of the longitudinal decrement and to a substantial decrease of the sum of the decrements compared with $\delta_x + \delta_z$:

$$\sum\delta = \frac{\mathcal{P}_{fr}}{E_k}\cdot\frac{2}{L_i} = 16\pi r_e^2 c\,\frac{m_e}{M} N_e \left(\frac{v}{c}\right)^{-3}, \ v \ll c. \qquad (3a)$$

This is 3÷5 times less than the sum of the transverse decrements only.

In the relativistic region, where the ionization losses become, in practice, energy-independent, the sum of the decrement is equal to

$$\sum\delta = \frac{2\mathcal{P}_{fr}}{E} = 8\pi r_e^2 c L_i N_e \gamma^{-1}, \ v \to c. \qquad (3b)$$

The magnetic structure of a cyclic ionization cooler and the shape of a target should be such that the decrements from transverse degrees of freedom are 'transferred' to a longitudinal one. To ensure the energy spread damping, it is sufficient to place the targets on a section where the position of the closed orbit of a particle depends on its energy and to make the target thickness variable in order that the larger thickness t of the target correspond to the higher energy:

$$\frac{1}{t}\frac{dt}{dx} = \left(\frac{\delta_\parallel}{\delta_x}\psi\right)^{-1}, \qquad (4)$$

where ψ is the dispersion function in the target region ($\psi = p\,dx/dp$). If $v \ll c$ it is necessary to use the coupling with both transversal degrees simultaneously.

2. The multiple scattering on the nuclei and electrons of the target itself and the fluctuations of ionization losses are referred to the main 'heating' factors determining the equilibrium of the momentum spreads. To a sufficient accuracy, the direct diffusion of transverse and longitudinal momenta can be written down as follows:

$$\frac{d(\overline{\Delta P_{x,z}^2})}{d\tau} = \frac{4\pi N_i Z_i^2 e^4 L_c}{v} = Z_i m_e \frac{L_c}{L_i}\mathcal{P}_{fr}\ ;$$

$$\frac{d\overline{(\Delta E)^2}}{d\tau} = \Delta E_{max}\cdot\frac{2\pi N_e e^4}{m_e v} = \Delta E_{max}\cdot\mathcal{P}_{fr}/L_i\ ; \qquad (5)$$

where $L_c = \ln\left(\frac{\theta_{max}}{\theta_{min}}\right) \approx 15$ and ΔE_{max} is the maximum energy transfer to a target electron:

$$\Delta E_{max} = \frac{2m_e\left(\frac{E^2}{c^4} - M^2\right)c^2}{2m_e E/c^2 + M^2 + m_e^2} = \begin{cases} 4\frac{m_e}{M}E_k, & v \ll c \\ 2\frac{m_e}{M}\gamma E, & 1 \ll \gamma \ll \frac{M}{2m_e} \\ E, & \gamma \gg \frac{M}{2m_e} \end{cases}$$

(no aperture limitation is assumed). If each decrement is regarded to be made equal to one third of the sum of decrements (3) as a result of the choice of the magnetic structure, the equilibrium values of the transverse angles and of the energy spread are then determined by the following expressions (with the excitation of transverse oscillations by the fluctuations of energy losses taken into account):

$$\left.\begin{aligned} \left(\frac{\Delta P_{x,z}}{p}\right)^2 &= \frac{3}{2}L_c Z_i \frac{m_e}{M} + \frac{3}{4}\frac{\psi_{x,z}^2}{\beta_{x,z}^2}\frac{m_e}{M} \\ \left(\frac{\Delta P_\parallel}{p}\right)^2 &= \left(\frac{1}{2}\frac{\sigma_E}{E_k}\right)^2 = \frac{3}{4}\frac{m_e}{M} \end{aligned}\right\} v \ll c\ ;$$

$$\left.\begin{aligned} \left(\frac{\Delta P_{x,z}}{P}\right)^2 &= \frac{3}{2}\frac{L_c}{L_i} Z_i \frac{m_e}{\gamma M} + \frac{3}{2}\frac{m_e}{M}\frac{\gamma}{L_i}\frac{\psi_{x,z}^2}{\beta_{x,z}^2} \\ \left(\frac{\Delta P_{\parallel}}{P}\right)^2 &= \left(\frac{\sigma_E}{E}\right)^2 = \frac{3}{2}\frac{m_e}{M}\cdot\frac{\gamma}{L_i} \end{aligned}\right\} 1 \ll \gamma \ll \frac{M}{2m_e} \quad (6)$$

where $\beta_{x,z}$ are the values of the beta-function in the target region and σ_E is the energy spread in the beam.

Thus, one can obtain a small value of the equilibrium transverse emittance

$$\Omega_{x,z} = \left(\frac{\Delta P_{x,z}}{P}\right)^2 \cdot \beta_{x,z}$$

by choosing the magnetic structure so that the minimum of the beta-function take place in the target region (assuming $\psi_{x,z} \leq \beta_{x,z}\sqrt{z L_i/\gamma}$)

The equilibrium geometrical longitudinal emittance of the beam will be equal to

$$\Omega_{\parallel} = \sigma_E \cdot \frac{\ell_\varphi}{v} \cdot \frac{1}{\gamma M v} \quad (7)$$

$$\Omega_{\parallel} \approx \frac{\ell_\varphi \sigma_E}{E}, \quad v \to c$$

where ℓ_φ is the mean azimuthal deviation of the particles in the bunch from the equilibrium particle, which is proportional to σ_E. Let us assume that compensation of the ionization losses takes place on a q-harmonic of the revolution frequency ω_s in the phase stability regime. The relation between ℓ_φ and σ_E is then given by the expressions

$$\ell_\varphi = \frac{R}{\omega_\gamma}\frac{d\omega}{dE}\sigma_E$$

where R is the mean radius of the accelerator, $\omega_\gamma = \sqrt{eU_0 q \sin\varphi_s \omega_s \frac{d\omega}{dE}/2\pi}$ is the frequency of phase oscillations, eU_0 is the amplitude of energy gain from the accelerating element per turn, φ_s is the equilibrium phase (note that $eU_0\cos\varphi_s$ equals the ionization losses per turn) and, finally, $d\omega/dE$ characterizes the variation rate of the particle revolution frequency in the accelerator as the energy varies and tends to zero at the so-called 'critical energy', dependent on the accelerator magnetic structure. Hence, we obtain

$$\Omega_{\parallel} = \left(\frac{\sigma_E}{E}\right)^2 \cdot R\frac{E}{\omega_\gamma}\cdot\frac{d\omega}{dE} \sim \frac{3}{2}\frac{m_e}{M}\cdot\frac{\gamma}{L_i} R\frac{E}{\omega_\gamma}\frac{d\omega}{dE}, \quad 1 \ll \gamma \ll \frac{M}{2m_e} \cdot \quad (8)$$

Thus, in order to obtain as small equilibrium longitudinal emittance of the beam as possible at a given energy of cooling, it is necessary to maximally decrease the effective 'longitudinal focal length' $R\frac{E}{\omega_\gamma}\cdot\frac{d\omega}{dE}$ of the accelerator, in particular, trying to increase the harmonic number of the applied compensating RF voltage in the single-bunch regime, it is necessary to strive for working at an energy close to a critical one.

After cooling one can take advantage of the smallness of the equilibrium longitudinal emittance (due to smallness of the equilibrium length of the bunch) for a sharp monochromatization of the beam. To do this, it is sufficient, for example, to eliminate first the ionization losses adiabatically slowly, smoothly reducing the thickness of targets, and then to decrease the accelerating voltage. This results in lengthening the bunch, thereby gaining, proportionally, in monochromaticity. The adiabatic lengthening of the bunch can also be made in an additional matched accelerating track, or using a bending expander with further compensation of the energy gradient along the bunch.

3. Let us now consider for what particles the ionization cooling is reasonable to use /3/.

For electrons and positrons, the ionization cooling is not applicable at all. At low energies the multiple scattering takes place more rapidly than the ionization deceleration, and at high energies the main energy losses are the radiation ones which occur by large 'portions' because the bremsstrahlung spectrum is uniform up to the quanta energies of the order of the initial energy of electrons.

When cooling the protons and antiprotons, the major obstacle at not too low energies is a strong (nuclear) interaction with the target nuclei. (It is worth emphasizing that the necessity for confining the particles, scattered in the target at an angle several times larger compared with the equilibrium angular spread, offers the possibility of neglecting the particle losses because of the single Coulomb scattering). In this case, the particle loss cross section is nearly equal to the total nuclear cross section. The estimates show that the proton beams can be cooled only at 100 MeV even for the hydrogen target /2/.

For antiprotons, the nuclear cross section at low energies is much larger and, hence, the ionization cooling is inapplicable for them, except, possibly, the region of very low energies where the cross section for antiprotons is unknown.

The most interesting and promising is the application of ionization cooling to muon beams since there are no, in practice, radiation losses and nuclear interaction (except the Coulomb scattering) for these beams. The lifetime of the muon beam is limited by the muon decay with the time $\gamma\tau_\mu$ even under the conditions when the accelerator confines the beam with the equilibrium emittance with a sufficient reserve. Therefore, the time of cooling has to be several times shorter compared with the decay time:

$$(\Sigma\delta)^{-1} \ll \gamma\tau_\mu$$

In our further consideration we will restrict ourselves to the case $1 \ll \gamma \ll \mu/2m_e$, where μ is the mass of a muon, since at nonrelativistic energies the equilibrium angle proves to be too large - of the order of unity, whereas at higher energies the equilibrium energy spread becomes too high. The condition for a sufficiently high rate of cooling takes the form

$$\frac{1}{3}\cdot\frac{2P_{tr}}{E} > \frac{\mu c^2}{E\tau_\mu} \to \left\langle\frac{dE}{dx}\right\rangle_{orbit} > \frac{3}{2}\frac{\mu c}{\tau_\mu} \approx 1.5\ keV/cm.$$

If we use, for example, lithium targets, then an admissible fraction ξ of the orbit, used for a target, should be as follows:

$$\xi \gg \frac{1.5 \cdot 10^{3}}{2 \cdot 10^{6}} \simeq 0.7 \cdot 10^{-3},$$

i.e. the target substance has to occupy of the order of 1% of the accelerator orbit.

Importance is of the fact that the several phase oscillations must occur during the damping time, and, hence, the frequency of phase oscillations should be substantially higher than

It is worth noting that the cooling can be made both in the cyclic (the above formulae are addressed just to this case) and quasi-linear accelerators with a total energy gain several times higher compared with the energy of cooling; the accelerator has to include sections with bending magnetic field. In this case, the energy dispersion function in the target region needs to be chosen correctly in order to carry out the longitudinal cooling as well. The general estimates of the equilibrium emittances remain valid.

4. Let us briefly discuss, according mainly to Refs /4-6/, possible applications of ionization-cooled muon beams. To produce intense, completely pure and deeply-cooled muon beams, the following steps are necessary to make:

1) to produce a pion beam with as low emittance as possible at an energy E_c of about 1 GeV, by using the most intense proton beams with an energy of hundreds of GeV and higher and the nuclear cascade in the conversion target;

2) to let the pions decay in a strongly-focusing straight-line channel; this ensures the minimum increase of the transverse emittance;

3) to perform the ionization cooling of the produced muon beam;

4) to accelerate, to the necessary energy, the produced muon beam in a linear or cyclic accelerator with a rate of cooling several times higher, per unit length, than $\mu c/\tau_\mu$; this provides the smallness of intensity losses because of the muon decay.

These muon beams is possible to use either for a direct study of the interaction of muons with nucleons and nuclei or for obtaining, after the injections of muons into a special magnetic track, a generator of electron and muon neutrinos and antineutrinos up to the total energy with a very small angular spread. The latter can be made close to $\mu c^2/E_\mu$, where E_μ is the energy of accelerated muons. With E_μ = = 1 TeV and a 300-m-thick shield, this will enable one to have the transverse sizes of the neutrino beam of the order of 3 cm. This circumstance will permit the setting up of neutrino experiments to be simplified to a considerable extent.

However, the most interesting possibility is to make the colliding muon beams experiments. If the aberrations do not increase, during acceleration, the emittance of muon beams gathered in two very short bunches with the number of particles $N_{\mu^+} = N_{\mu^-} = N_\mu$, then having injected them into the ring with a strong magnetic field H (to increase the number of collisions N_τ during the lifetime of accelerated muons), one can obtain the luminosity

$$\mathcal{L} = \frac{N_\mu^2}{4\pi \Omega_\perp \beta_\upsilon \left(\frac{E_i}{E_\mu}\right)} \cdot N_\tau \cdot f = N_\mu^2 \frac{E_\mu}{6\pi Z_i m_e c^2 \beta_i \beta_\upsilon} \cdot \frac{eH\tau_\mu}{2\pi \mu c}$$

where β_i and β_υ are the values of the beta-functions in the region of ionization targets and in the interaction point, respectively; and f is the repetition frequency of injection cycles.

If we set N_μ = 10^{11}, E_μ = 1 TeV, H = 100 kG, f = 10 Hz and $\beta_i = \beta_\upsilon$ = 1 cm, we obtain the luminosity exceeding 10^{31} cm^{-2} s^{-1}. As shown in Ref. /6/, close parameters can be achieved with the help of proton klystrons, by utilizing the intense proton beams of the modern and future accelerators at ultimately high energies.

References

1. A.A.Kolomensky, Atomnaya Energiya 19 (1965) 534.
2. Yu.M.Ado and V.I.Balbekov, Atomnaya Energiya 39 (1971) 40.
3. A.N.Skrinsky and V.V.Parkhomchuk, Physics of Elementary Particles and Atomic Nucleus (Sov) 12, vyp. 3 (1981) 557.
4. A.N.Skrinsky, Proc. of International Seminar on Prospects of High-Energy Physics, March 1971.
5. G.I.Budker and A.N.Skrinsky, Usp. Fiz. Nauk 124, No. 4 (1978).
6. E.A.Perevedentsev and A.N.Skrinsky, Proc. of the 6th National Conf. on Charged Particle Accelerators, 1978. Dubna, 1979, p. 272; INP preprint No. 79-80, 1979, Novosibirsk; ICFA-II Proc., 1979; paper presented at this Conference.

A MUON STORAGE RING FOR NEUTRINO OSCILLATIONS EXPERIMENTS

David Cline
University of Wisconsin, Madison, WI 53706

David Neuffer
Fermilab,* Batavia, IL 60510

ABSTRACT

$\mu^{\pm}$ decay in a $\mu^{\pm}$ Storage Ring can provide ν_e, ν_μ beams uniquely suitable for the study of ν oscillations. The Fermilab $\bar{p}$ precooler is studied as a possible first μ storage ring.

INTRODUCTION

Recent experimental reports[1,2] of a non-zero ν_e mass and of possible $\bar{\nu}_e$ oscillations reveal the need for more complete study of neutrino properties. Previously, accelerator ν beams have been muon neutrino (ν_μ) beams from $\pi \to \mu\nu_\mu$. In this paper we note that a muon storage ring (see Figure 1) can provide ν_e and $\bar{\nu}_\mu$ beams from $\mu \to e \nu_e \bar{\nu}_\mu$ as earlier suggested by Wojicicki and Collins.[3] We further note that a μ storage ring provides clean ν beams of precisely knowable flux, and therefore an excellent tool for the study of ν_e and ν_μ oscillations.

DESCRIPTION OF A μ STORAGE RING

We also note that the Tevatron $\bar{p}$ precooler (see Figure 2) inescapably functions as a 4.5 GeV/c μ storage ring during the first ms of its cycle, and that its large acceptance designed for $\bar{p}$ acceptance make it a very good storage ring, and therefore a candidate for use in the first experiment of this type.

The 80 GeV proton line, the production target, the transport line and the pre-cooler are shown in Figure 2. Pulses of 1.8×10^{13} protons are focussed on the target producing many secondary particles (π, k, $\bar{p}$, etc.) which follow the transport line to insertion in the ring. The production is dominated by π's which decay ($\pi \to \mu\nu$) and a substantial number of the decay muons will circulate in the ring, a first estimate indicates 10^{10} μ.[4] The decay of these muons in precooler straight sections will provide collimated ν_e and $\bar{\nu}_\mu$ beams with $\sim 8 \times 10^8$ ν per beam per p pulse.

---

*Operated by the Universities Research Association, Inc., under contract with the U.S. Department of Energy.

0094-243X/81/680856-02$1.50

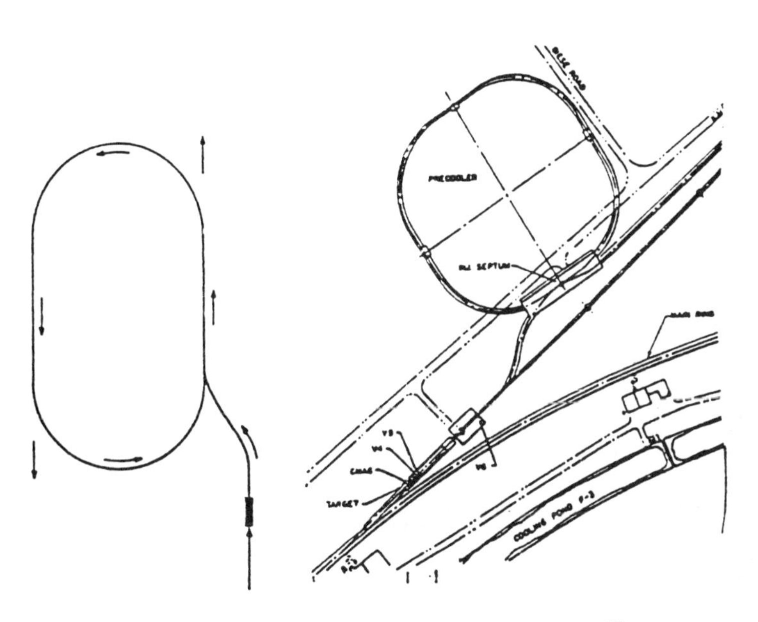

Figure 1: A μ Storage Ring.

Figure 2: The $\bar{p}$ Precooler/μ Storage Ring.

Modifications of the precooler to increase   and   acceptance and to increase the decay straight section length could increase this flux by a factor of ∿10 and the proton pulse period of 10 seconds can be reduced from 10 seconds with $\bar{p}$ cooling (parisitic ν beam) to two seconds (dedicated mode). These intensities and designs are discussed in Reference 4, and will be improved in future work.

EXPERIMENTAL COMMENTS

The precooler μ storage ring can provide adequate event rate for a variety of experiments. A 100 ton detector, 0.5 km away will receive ∿4-400 events/day with 5×10^8 - 5×10^9 $\nu_e \bar{\nu}_\mu$/pulse, 10^4-10^5 pulses/day. The Fermilab 15' bubble chamber could also observe events. A suitable compromise between detector size, sensitivity, and cost is left as a challenge to interested experimenters. Since the ν flux can be precisely known from monitoring the decaying muon current, the μ storage ring can provide a unique tool for future ν experiments.

REFERENCES

1. F.H. Reines, et al., this proceedings.
2. V.A. Lyubimov, et al., ITEP-62, submitted to Yadernaya Fizika (1980).
3. S. Wojcicki, unpublished (1974), T. Collins, unpublished (1975).
4. D. Cline and D. Neuffer, paper contributed to XX International Conference on High Energy Physics (1980).

PRINCIPLES AND APPLICATIONS OF MUON COOLING

David Neuffer
Los Alamos National Laboratory, Los Alamos, New Mexico 87545

Summary

The basic principles of the application of "ionization cooling" to obtain high phase-space density muon beams are described, and its limitations are outlined. Sample cooling scenarios are presented. Applications of cooled muon beams in high-energy accelerators are suggested; high-luminosity μ^+-μ^- and μ-p colliders at $\gtrsim$1-TeV energy are possible.

Introduction

Electron-positron (e^+-e^-) colliders have been essential tools in gaining an understanding of particle physics. However, their future use at higher energies is severely limited by radiation processes. Synchrotron radiation in storage rings causes electrons to lose energy at a rate proportional to the fourth power of the electron energy, and this radiation effectively prevents construction of $e^{\pm}$ storage rings at energies greater than 100 GeV (LEP). e^+-e^- linear colliders are proposed to circumvent this problem. However, they have substantial practical difficulties in obtaining adequate luminosity, are very expensive, and also have particle radiation problems that prevent practical implementation at particle energies $\gtrsim$300 GeV. Another approach, $\bar{p}p$ and pp colliders, can indeed reach multi-TeV energies, but hadron-hadron interactions lack the simplicity of lepton-lepton collisions; lepton-lepton and lepton-hadron colliders are necessary to provide a complete picture of high-energy processes.

Synchrotron radiation varies inversely as the fourth power of the mass, so the radiation difficulties of e^+-e^- machines can be avoided by the use of "heavy electrons," muons (μ^+-μ^-), and that possibility is the subject of this paper.

The principal liabilities of muons are their short lifetimes and the large initial phase-space area of a muon beam as produced in π decay. The lifetime τ is given by

$$\tau = 2.197 \times 10^{-6} \frac{E_\mu}{m_\mu} \text{ sec} \quad , \tag{1}$$

where E_μ, m_μ are the muon energy and mass. However, this is ~0.02 s for 1-TeV $\mu^{\pm}$ and is adequate for any linac and for high-energy storage rings and rapid cycling synchrotrons (see below). The large phase-space area of a muon beam can be damped using "ionization cooling"[1] (as described below) to a small value suitable for high-luminosity colliders.

In the following sections, we will describe the principles of muon cooling, discuss cooling scenarios and experiments, and then high-energy collider applications.

Muon Cooling

The basic mechanism of μ cooling is displayed in Fig. 1. The muon beam is passed through a material medium in which it loses energy, principally through

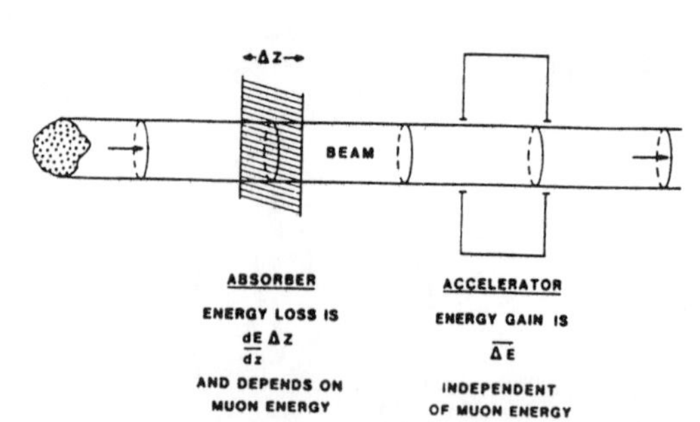

Fig. 1. Sketch of "ionization cooling" principle.

interactions with atomic elections. Following this, it passes through an accelerating cavity where the average longitudinal energy loss is restored. Energy cooling occurs following

$$\frac{d(\Delta E_\mu)}{dn} \equiv - \frac{\partial \Delta_\mu}{\partial E_\mu} \Delta E_\mu \quad , \tag{2}$$

where ΔE_μ is the muon energy deviation from the central value, n is the number, of cooling cycles, Δ_μ is the muon energy loss in the absorber, and the derivative is taken at the central value E_μ. Cooling occurs if the derivative is positive. For $E_\mu \lesssim 0.3$ GeV, this energy-loss rate derivative is steeply negative for all absorbing materials, but for $E_\mu \gtrsim 0.5$ GeV, it is positive with

$$\frac{\partial \Delta_\mu}{\partial E_\mu} \cong 0.2 \frac{\Delta_\mu}{E_\mu} \quad . \tag{3}$$

The precise value has a weak dependence on the absorber material, and the energies $0.5 \lesssim E_\mu \lesssim 2$ GeV are reasonable energies for muon collection. The μ beam is recirculated through many absorber/accelerator cycles either by a return path (cooling ring) or repeated structure (linac) to obtain the desired final distribution.

Transverse damping also occurs because energy loss is parallel to the particle trajectory, whereas energy gain is longitudinal. Expressing this transverse energy loss in terms of rms emittance, one obtains[2]

$$\frac{d\,\epsilon_\perp}{dn} = - \frac{\Delta_\mu}{E_\mu} \epsilon_\perp \tag{4}$$

for both transverse degrees of freedom.

An exchange in cooling rate between the longitudinal (Eq. 3) and a transverse (Eq. 4) dimension can be obtained if a "wedge" absorber in a nonzero Courant-Snyder[3] dispersion region is used[2] (see Fig. 2). Enhanced energy damping with this method implies decreased transverse damping, whereas the sum of ε_x, ε_y, ΔE damping rates is constant:

$$\sum_{x,y,\Delta E} \frac{E_\mu}{E_{cool,i}} \cong 2.2 \text{ is constant} \quad , \tag{5}$$

where $E_{cool,i}$ is the e^{-1} damping energy in each dimension, $E_{cool} = n_e{-1}\,\Delta_\mu$.

The process is basically similar to radiation damping in e± storage rings,[4] where energy loss in bending sections by synchrotron radiation is recovered in rf cavities. Radiation damping is limited by quantum fluctuations; similarly, muon cooling is limited by statistical fluctuation in muon-atom interactions in the absorber.

The important difference is that muons decay, and cooling must be completed before decay occurs. The muon lifetime (Eq. 1) can be translated to a path length ($\beta_\mu \cong 1$)

$$L_\mu = 6.59 \times 10^2 \frac{E_\mu}{m_\mu} \text{ meters} \quad , \tag{6}$$

which can be translated into a number of turns of beam storage,

$$N = \frac{L_\mu}{2\pi \bar{R}} = \frac{L_\mu \bar{B}}{2\pi\, B_\rho} = 297\ \bar{B}\ (T) \text{ turns} \quad , \tag{7}$$

where $\bar{B}$ is the ring-averaged bending field and B_ρ the magnetic rigidity $[B_\rho(T\text{-}m) \cong 3.3\ E_\mu\ (GeV)]$. N is independent of E_μ.

Muon cooling is limited by heating due to statistical fluctuations in the number and energy exchange in the muon-electron collisions in the absorber. An estimate of this heating in energy cooling can be obtained by noting that the mean energy, exchange is the mean electron ionization energy I[5]

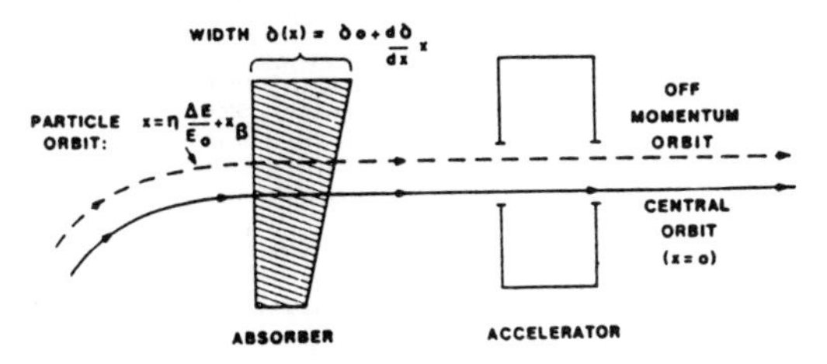

Fig. 2. Use of varying thickness absorber to enhance energy dependence of energy loss.

$$I \cong 10\ Z_{abs}\ eV \quad . \tag{8}$$

The number of collisions per cooling cycle is $N \sim \Delta_\mu/I$ and the rms energy spread is $\sim\sqrt{N}\ I$ or $\sqrt{\Delta_\mu I}$. Combining cooling with heating, we obtain

$$\frac{d}{dn} \langle(\Delta E)^2\rangle \cong - \frac{2\Delta_\mu}{E_{cool,z}} \langle(\Delta E)^2\rangle + \Delta_\mu I \quad , \tag{9}$$

which has an equilibrium solution indicating the limits of muon energy cooling,

$$\langle(\Delta E)^2\rangle_{rms} \approx \frac{I\ E_{cool,z}}{2} \quad . \tag{10}$$

For typical values ($E_{cool,z} = 2$ GeV),

$$\overline{\Delta E_f} = \sqrt{\langle(\Delta E)^2\rangle_{rms}} \cong 10^5 \sqrt{Z_{abs}}\ eV \quad . \tag{11}$$

Transverse cooling is severely limited by multiple small-angle elastic scattering, mostly Coulomb scattering from the nuclei. The mean scattering angle in passing through an absorber of thickness δ can be estimated by the following equation:[5]

$$\theta_{rms} \cong \frac{14\ (MeV)}{E_\mu\ (MeV)} \sqrt{\frac{\delta}{L_R}} \quad , \tag{12}$$

where L_R is the radiation length of the absorber material. The cooling equation for transverse emittance can then be written as

$$\frac{d\varepsilon_x}{dn} = - \frac{\Delta_\mu}{E_{cool,x}} \varepsilon_x + \frac{\beta_x}{2} \left(\frac{14}{E_\mu}\right)^2 \frac{\delta}{L_R} \quad , \tag{13}$$

where β_x is the C-S betatron function at the absorber. The equilibrium emittance is

$$\varepsilon_0 = \frac{\beta_x}{2} \left(\frac{14}{E_\mu}\right)^2 \frac{E_{cool,x}}{\left(\frac{dE}{dz} L_R\right)} \quad . \tag{14}$$

The product $\left(\frac{dE}{dz} L_R\right)$ depends upon the absorber and is largest for light elements[5] (~100 MeV for Be or C but ~7 MeV for Pb or W). The appearance of β_x in Eq. (14) indicates optimum emittance cooling requiring very strong focusing to a low value of β_x at the absorber. Note that the length of the absorber must be less than

$\sim 2\ \beta_x$, and obtaining maximum absorption in minimum length requires heavy elements, opposing the previous constraint.

We note here that the constraint $L_{abs} \lesssim 2\ \beta_x$ can be relaxed if the absorber is an active focusing element (such as a lithium lens). Optimum μ coolers will probably include such elements in some portion of their structure.

Muon Cooler-Design Outlines and Experiments

In this section we outline some feasible μ cooler designs and experiments. Two basic approaches are suggested: storage ring and linacs. We expect that optimum designs to obtain minimum phase space will combine these in a multistage system.

We first consider a storage-ring system. Figure 3 shows the basic components: a rapid-cycling p synchrotron for π production, a π-decay line, and a storage ring for 1-GeV μ. The muon storage ring is a relatively modest device with conventional magnets (<2T) and modest cooling goals suitable for a μ-p collider.

A second stage would probably be necessary to achieve the lower transverse phase-space densities necessary for a μ^+-μ^- collider; it is difficult to cool transverse emittance by more than a factor of ~30 in a single dc storage ring because of the focusing required in the absorber. A second stage using superconducting magnets ($B \gtrsim 10T$) will obtain $\varepsilon_\perp(1\ \text{GeV}) \lesssim 2.0$ mm-mrad. Higher fields and optimized designs will obtain smaller $\varepsilon_\perp$, but $\varepsilon_\perp \lesssim 0.5$ mm-mrad appears impractical. We note that μ cooling requirements are ideal for use of maximum field superconducting magnets (dc operation, low-particle flux, modest sizes).

Focusing requirements are relaxed in a μ linac where magnet apertures can be reduced, providing stronger focusing, as the beam is cooled. Beam loss from decay is also reduced. Figure 4 shows a modest first-stage μ cooling ($100\ \pi \rightarrow 10\ \pi$) linac using conventional magnets. A second stage with superconducting magnets can obtain $\varepsilon_\perp \lesssim 2\ \pi$ mm-mrad.

Existing ~1-GeV storage rings may be modified with low-beta insertions and additional rf for experiments testing μ cooling concepts. Acceleration on the order of 10 MeV/turn is required to obtain cooling before decay. One candidate is the Fermilab 600-MeV/c "electron cooling ring," outfitted with ~5 to 10 MeV of rf borrowed from the future "Debuncher." The $\bar{p}$ production target can be used to provide π's to be transported in a decay line to provide μ's. Another candidate is a SLC damping ring with additional rf and some μ source (instead of e^+).

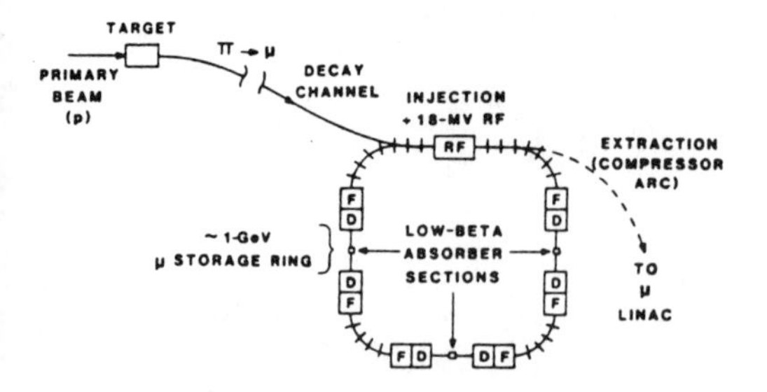

Fig. 3. Muon cooling ring.

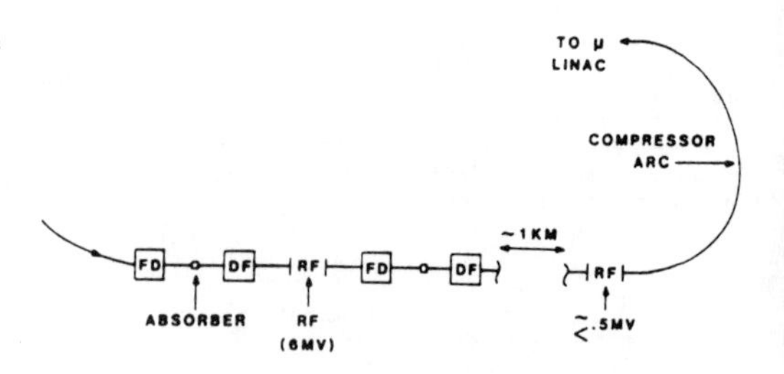

Fig. 4. Sketch of muon cooling linac.

Application of Cooled Muons in High-Energy Accelerators

Cooled muon beams have many possible uses in high-energy accelerators. In this section, we will emphasize applications not accessible to $e^\pm$ machines, such as colliders at ≳0.5-TeV energies.

A. μ^+-μ^- Rapid-Cycling Collider

Most collider applications will require a high-intensity muon source at a frequency matched to the muon lifetime. At 1 TeV, $\tau_\mu \cong 0.02$ s, this is reasonably well matched to a high-intensity 30- to 60-Hz rapid-cycling proton synchrotron.

In Fig. 5 we outline the major components of a 1-TeV μ collider: a rapid-cycling proton synchrotron with target to produce π's; a decay channel (or "stochastic injection" into a storage ring)[6] for $\pi \rightarrow \mu$ decay; a storage-ring/linac system for μ cooling; and a μ linac (or "booster") for injection into a rapid-cycling synchrotron with period matched to the proton synchrotron. In this example the μ synchrotron is simply a conventional larger version of the proton synchrotron. We assume from previous calculations that $\sim 5 \times 10^{-3}$ stored muons are obtained from each primary proton.

The μ^+-μ^- collider luminosity L may be estimated using

$$L \cong \frac{f_o\ n_t\ n_B\ N^+\ N^-}{4\pi\ \beta^*\ \varepsilon^*} \quad . \tag{15}$$

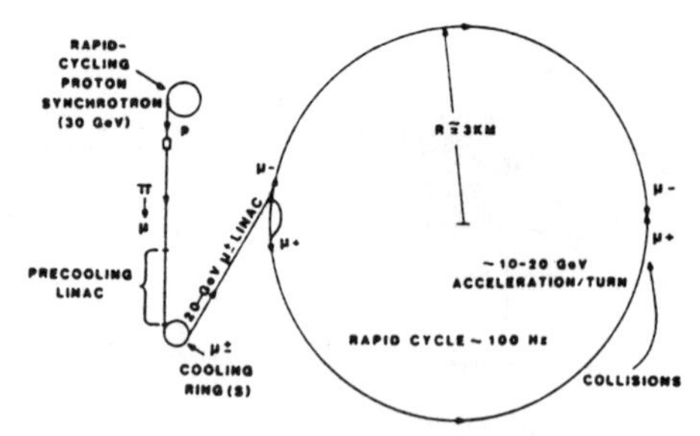

Fig. 5. 1-TeV μ rapid-cycling synchrotron.

With 2×10^{13} protons/pulse, we obtain $\sim 10^{11}$ stored $\mu^{\pm}$ which may be organized into $n_B \cong 2$ bunches with $N^+ = N^- = 5 \times 10^{10}$ μ/bunch. The cycling frequency (30 Hz) is f_o; n_t is the mean number of storage turns (300); and we may estimate $\beta^* \cong 0.3$ cm and $\varepsilon^* \approx 2 \times 10^{-7}$ cm-R at 1 TeV. We obtain $L \gtrsim 10^{32}$ cm^{-2} s^{-1}, an adequately high luminosity. The same synchrotron may be used as a μ^--p^+ collider at high luminosity, with more relaxed requirements on N^-, β^*, and ε^*. The scenario is, in principle, easier but more expensive at higher energies.

B. Linac/Storage-Ring Scenario

Cooled muons may be suitable for injection in a high-gradient linac. Skrinsky suggested acceleration of μ's in his "proton klystron."[1] Other linac ideas such as "surfatrons," "wake fields," etc., may be more readily adaptable to μ acceleration than e because of the μ immunity to synchrotron radiation, bremsstrahlung and particle-medium interaction.

Assuming a suitable high-gradient linac, μ^+-μ^- (and μ-p) collisions may be obtained in a dc superconducting ring which accepts the high-energy output beam (see Fig. 6). Luminosity can, in principle, be higher than in the previous example ($L \gtrsim 10^{33}$), since stronger fields will increase n_t (the number of beam-storage turns) and decrease β^*, and beam loss in acceleration is reduced.

C. μ-p Colliders

An important advantage of μ^+-μ^- colliders over e^+-e^- is that μ-p collisions may also occur in the same ring. In the rapid-cycling synchrotron protons may be injected with μ^-, and in the storage ring scenario they may be stored before μ^- injection. The revolution frequencies are naturally mismatched because of the different velocities at equal energies. They can be rematched[2] by displacing the beams in energy under the condition (high energy),

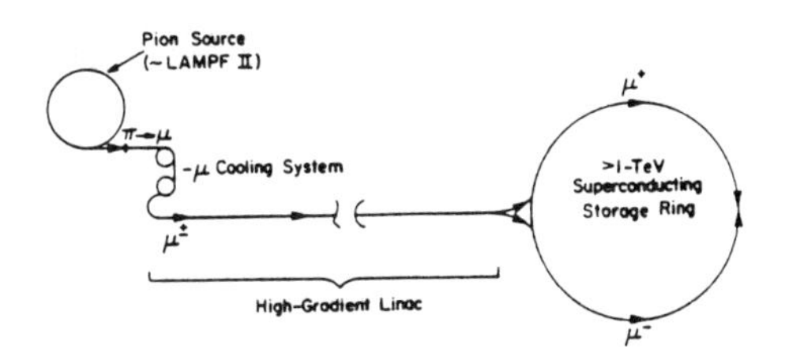

Fig. 6. μ linac/storage-ring system.

$$\frac{\Delta p}{p}\left[\frac{1}{\gamma_t^2}\right] \cong \frac{1}{2\gamma_p^2} - \frac{1}{2\gamma_\mu^2} , \qquad (16)$$

where $\Delta p/p$ is the momentum offset, γ_t the transition energy of the ring, and γ_μ, γ_p are the kinetic factors. At 1 TeV with $\Delta p/p = 10^{-3}$ we obtain $\gamma_t \cong 45$, a reasonable value.

References

1. A. N. Skrinsky, Proc. XXth Int. Conf. on HEP, AIP Conf. Proc. 68 (1980), p. 1056.

2. D. Neuffer, Fermilab FN-378, January 1983, to be published in Particle Accelerators.

3. E. D. Courant and H. S. Snyder, Ann. Phys. 3, 1 (1958).

4. M. Sands, SLAC-121, Proc. Int. School of Physics, Varenna, 1969 (Academic Press, New York, 1971).

II. RESULTS FROM THE NAPA WORKSHOP (NIM PAPERS)

Physics potential of a few hundred GeV $\mu^+\mu^-$ collider

David B. Cline

University of California Los Angeles, Center for Advanced Accelerators, Physics Department, 405 Hilgard Ave., Los Angeles, CA 90024–1547, USA

There is growing evidence that both a Standard Model and a SUSY Model Higgs should exhibit one resonance at a mass less than $2M_Z$. This is precisely in the mass range that is very difficult (but not impossible) to detect at the LHC and possibly beyond the reach of LEP II. At a $\bar{\mu}\mu$ collider the direct channel $\mu^+\mu^- \to h^0 \to b\bar{b}$ can be used to search for the Higgs. We discuss the collider requirements for this search on the $\mu\bar{\mu}$ collider luminosity and the detector. These results arose from a $\mu^+\mu^-$ Collider Workshop held in Napa Valley, California, November 1992.

1. Introduction

At the Port Jefferson Advanced Accelerator Workshop in the Summer of 1992, a group investigating new concepts of colliders studied anew the possibility of a $\mu^+\mu^-$ collider since e^+e^- colliders will be very difficult in the several TeV range [1]. A small group also discussed the possibility of a $\mu^+\mu^-$ collider. A special workshop was then held in Napa, California in the fall of 1992 for this study. There are new accelerator possibilities for the development of such a machine, possibly at an existing or soon to exist storage ring [2, 3]. For the purpose of the discussion here, a $\mu^+\mu^-$ collider is schematically shown in Fig. 1. In this brief note we study one of the most interesting goals of a $\mu^+\mu^-$ collider: the discovery of a Higgs boson in the mass range beyond that to be covered by LEP I & II ($\sim$ 80–90 GeV) and the natural range of the supercolliders $\gtrsim 2M_Z$ [4]. In this mass range, as far as we know, the dominant decay mode of the h^0 will be

$$h^0 \to b\bar{b}, \tag{1}$$

whereas the Higgs will be produced by the direct channel

$$\mu^+\mu^- \to h^0 \tag{2}$$

which has a cross section enhanced by

$$(M_\mu/M_e)^2 \sim (200)^2 = 4 \times 10^4 \tag{3}$$

larger than the corresponding direct product at an e^+e^- collider. However, we will see that the narrow width of the Higgs partially reduces this enhancement.

There is growing evidence that the Higgs should exist in this low mass range from

1) the original paper of Cabibbo et al., which shows that, when $M_t > M_Z$ and assuming a Grand Unification of Forces, $M_h < 2M_Z$ [4] (Fig. 2),

2) fits to LEP data, which imply a low mass h^0 could be consistent with $M_t > 150$ GeV [4],

3) the extrapolation to the GUT Scale that is consistent with SUSY also implies that one of the Higgs should have a low mass, perhaps below 130–150 GeV [4].

This evidence implies the exciting possibility that the Higgs mass is just beyond the reach of LEP II and in a range that is very difficult for the super-colliders to extract the signal from background, ie. either $h^0 \to \gamma\gamma$ or the very rare $h^0 \to \mu\mu\mu\mu$ in this mass range, since $h \to b\bar{b}$ is swamped by hadronic background. However, detectors for the LHC are designed to extract this signal [5].

In this low mass region the Higgs is also expected to be a fairly marrow resonance and thus the signal should stand out clearly from the background from

$$\begin{aligned} \mu^+\mu^- &\to \gamma \to b\bar{b} \\ &\to Z_{\text{tail}} \to b\bar{b}. \end{aligned} \tag{4}$$

In Fig. 3 we plot the Higgs resonance signals and various types of background for this mass range. While the cross sections are fairly large, the requirements on the beam energy resolution of the $\mu^+\mu^-$ collider are very constraining. In Fig. 4 we show the relationship between the Higgs width and the required machine energy resolution. If the resolution requirements can be met, the machine luminosity of $\sim 10^{31}$ cm^{-2} s^{-1} could be adequate to facilitate the discovery of the Higgs in the mass range of 100–180 GeV.

For masses above 180 GeV, the dominant Higgs decay is

$$H^0 \to W^+W^- \quad \text{or} \quad Z^0Z^0 \tag{5}$$

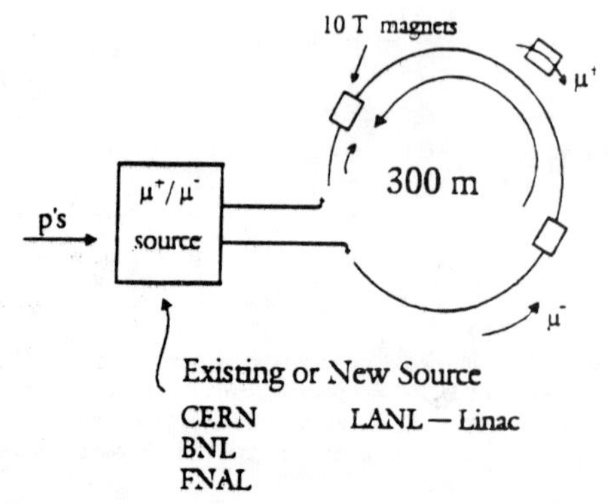

Fig. 1. Schematic of a possible $\mu\mu$ collider scheme (few hundred GeV).

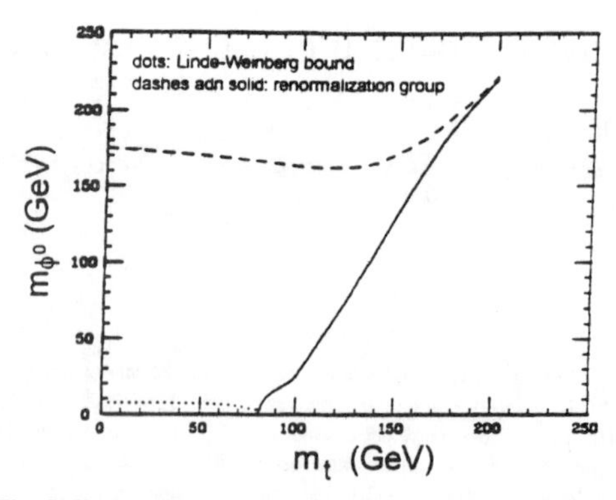

Fig. 2. Upper and lower bounds on $m_{\bar{\phi}}\,0$ as a function of m_t, coming from the requirement of a perturbative theory [4].

giving a clear signal and a larger width; the machine energy resolution requirements could be relaxed somewhat!

Another possibility for the intermediate Higgs mass range is to search for

$$\mu^+\mu^- \to Z^0 + H^0 \tag{6}$$

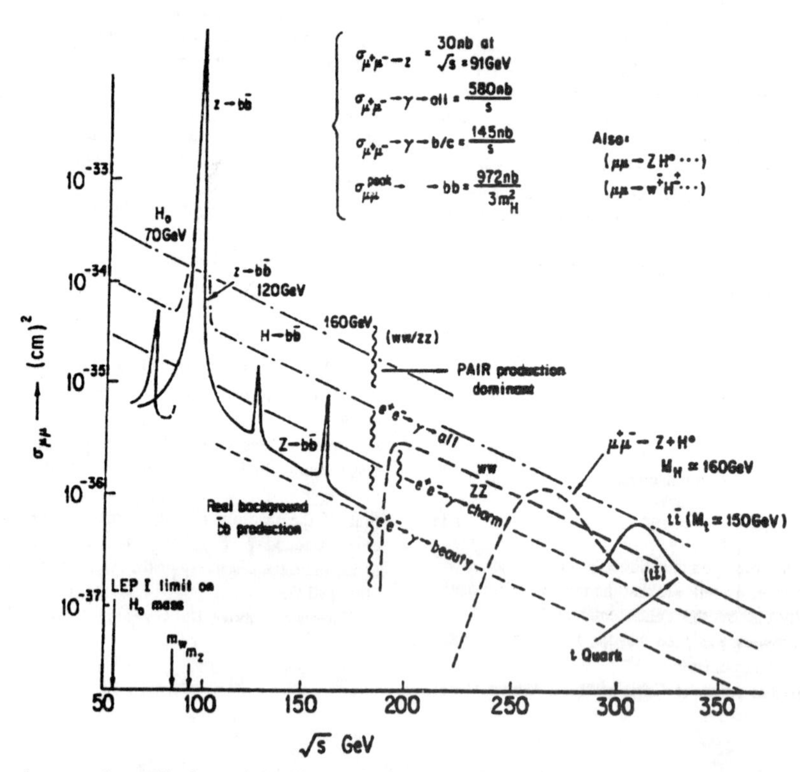

Fig. 3. Estimated cross sections and backgrounds for Higgs production in the direct channel or by associated production for various Higgs mass.

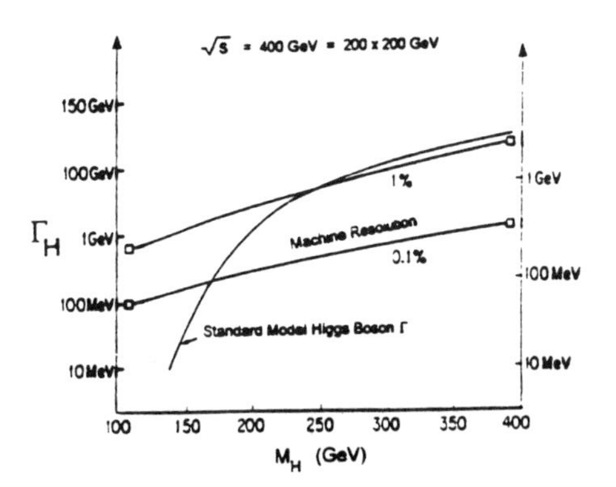

Fig. 4. Higgs search at a $\mu^+\mu^-$ collider. Required machine resolution and the expected Higgs width.

using a broad energy sweep. The corresponding cross section is small (see Figs. 3 and 5). Once an approximate mass is determined, a strategy for the energy sweep through the resonance can be devised. The study of the t quark through $t\bar{t}$ production would also be interesting.

Finally, another possibility is to use the polarization of the $\mu^+\mu^-$ particles orientated so that only scalar interactions are possible (thus eliminating the background from single photon intermediate states as shown in Fig. 3) [6]. However, there would be a trade-off with luminosity and thus a strategy would have to be devised to maximize the possibility of success in the energy sweep through the resonance.

At the Napa workshop the possibility of developing a $\mu^+\mu^-$ collider in the 10^{31} cm^{-2} s^{-1} region was considered and appears feasible. It is less certain that the high energy resolution required for the Higgs sweep can be

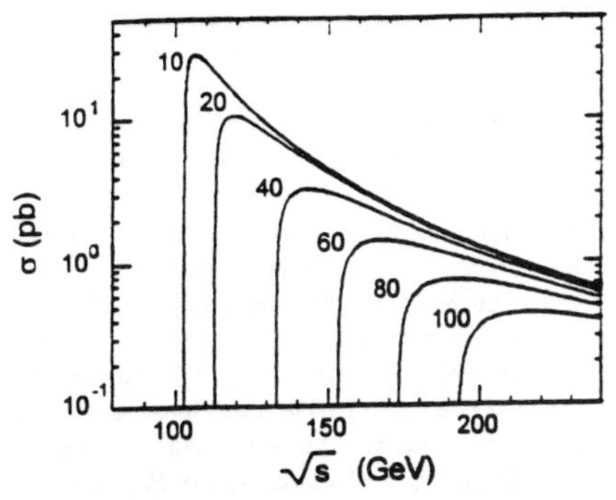

Fig. 5. Total cross section for $\mu^+\mu^- \to \phi^0$ Z as a function of $\sqrt{s}$ for the fixed $m\phi_0$ values indicated by the numbers (in GeV) beside each line. (Adapted from Ref. [4].)

Table 1
Key issues in the Higgs search

There is growing evidence that one Higgs particle is below 2 M_Z
SUSY Higgs – 3 Higgs – one near M_Z (possibly up to ~130 GeV); extremely hard to detect
Hadron machines can search for these Higgs provided: i $\int \mathscr{L}\, dt \geq 10^5$ pb (LHC); ii the background for $H \to \gamma\gamma$ is small enough
A $\mu^-\mu^-$ collider with $\mathscr{L} \geq 10^{31}$ cm^{-2} s^{-1} operating between 100–180 GeV could discover the Higgs in 2000+ provided sufficient energy resolution is achieved

obtained. We summarize in Table 1 some of the key issues in the Higgs search.

$\mu^+\mu^-$ colliders could also be very important in the TeV energy range; however, since the cross sections for new particle production are much smaller, the luminosity requirements would be $\mathscr{L}_{\mu^+\mu^-} \geq 10^{33}$ cm^{-2} s^{-1}. This is the energy range where e^+e^- linear colliders are extremely difficult to develop [1]. This possibility will be the subject of a second Napa workshop to be held in 1994.

Acknowledgement

I wish to thank the participants of the workshop, and in particular Dave Neuffer and Andy Sessler, Bill Barletta and Kirk McDonald for discussions.

References

[1] M. Tigner, presented at the 3rd Int. Workshop on Advanced Accelerator Concepts, Port Jefferson, NY, (1992); other paper to be published in AIP, 1992, ed. J. Wurtle.

[2] Early references for $\mu\mu$ colliders are: E.A. Perevedentsev and A.N. Skrinsky, Proc. 12th Int. Conf. on High Energy Accelerators eds. F.T. Cole and R. Donaldson, 1983, p. 485; D. Neuffer, Proc. 12th Int. Conf. on High Energy Accelerators, eds F.T. Cole and R. Donaldson, 1983, p. 481; D. Neuffer, Particle Accel. 14 (1983) 75; D. Neuffer, in: Advanced Accelerator Concepts, AIP Conf. Proc. 156 (1987) 201.

[3] Reports from the Napa Workshop: D. Neuffer, $\mu^+\mu^-$ Colliders: Possibilities and Challenges; W. Barletta and A. Sessler, Characteristics of a high energy $\mu^+\mu^-$ collider based on electro-production of muons.

[4] See for primary references to the theoretical estimates here S. Dawson, J.F. Gunion, H.E. Haber and G.L. Kane, The Physics of the Higgs Bosons: Higgs Hunter's Guide (Addison Wesley, Menlo Park, 1989).

[5] For example the Compact Muon Solenoid detector at the LHC, CERN reports.

[6] This possibility came up in discussion with K. McDonald, Princeton.

μ^+–μ^- colliders: possibilities and challenges ☆

David V. Neuffer

Continuous Electron Beam Accelerator Facility, 12000 Jefferson Avenue, Newport News, VA 23606, USA

The current status of the $\mu^+-\mu^-$ collider concept is reviewed and discussed. In a reference scenario, a high-intensity pulsed proton accelerator (of K-factory class) produces large numbers of secondary π's in a nuclear target, which produce muons by decay. The muons are collected and cooled (by "ionization cooling") to form high-intensity bunches that are accelerated to high-energy collisions. High-luminosity $\mu^+-\mu^-$ and μ^-–p colliders at TeV or higher energy scales may be possible. Challenges in implementing the scenario are described. Possible variations in muon production, accumulation, and collisions are discussed; further innovations and improvements are encouraged.

1. Introduction

Current and planned highest-energy colliders are hadronic proton–(anti)proton (p–p or p–$\bar{p}$) colliders or electron–positron (e^+–e^-) colliders, and both approaches have significant difficulties in extension to higher energies. Hadrons are composite objects; so only a small fraction of the total energy participates in a collision, and this fraction decreases with increasing energy. Also, production of new particle states is masked by a large background of nonresonant hadronic events; identification of new physics becomes increasingly difficult with increasing energy. Leptonic (e^+–e^-) colliders have had the advantage of providing simple, single-particle interactions with little background. However, extension to higher energies is limited in energy, luminosity, and resolution by radiative effects (synchrotron radiation in circular colliders, beam–beam radiation and pair creation in linear colliders). At very high energies, the collisions are no longer point-like, because of the radiative background.

However, this radiation scales inversely as the fourth power of the lepton mass. Thus, we can extend the high-quality features of e^+–e^- colliders to much higher energies by colliding higher-mass leptons such as muons. The μ^+–μ^- collider concept has been suggested, and is described in some detail in Refs. [1–4], and an example is displayed graphically in Fig. 1. In those initial concepts, a high-intensity multi-GeV hadron accelerator beam produces pions from a hadronic target, and muons are obtained from π-decay. The μ^+'s and μ^-'s are accumulated and cooled by ionization cooling, and then accelerated (in linacs and/or synchrotrons) to high energies for high-energy collisions in a storage ring. The process is repeated at a rate matched to the high-energy muon lifetime to obtain potentially high luminosities.

Since the initiation of the muon collider concept, some subsequent developments have increased interest in the possibility of muon colliders, and recent progress in related fields may increase their potential capabilities.

In high-energy physics (HEP), plans for the current generation of high-energy facilities are now reasonably well established. The next HEP devices are to be high-energy (8–20 TeV) p–p colliders (SSC and/or LHC), to be followed by an e^+–e^- linear collider (up to 0.5 TeV per beam). It is now possible to begin serious consideration of projects to follow these, such as a μ^+–μ^- collider. The current generation of HEP devices includes large components (injectors, tunnels, storage rings, linacs) and technology (high-gradient linacs, low-β^* optics, large-bandwidth stochastic cooling, etc.) which may be incorporated into a muon collider.

Development of high-intensity accelerator concepts (for kaon factories, or accelerator transmutation of waste, or tritium production, or for μ-catalyzed fusion) can also provide methods and facilities for improved μ production, collection and cooling. Some of the possible extensions are suggested below.

Recent studies of e^+–e^- linear colliders show that that approach appears to be limited in energy or luminosity at the TeV scale by beamstrahlung radiation effects. This calls for a "new paradigm" in extension to higher energies [5], possibly including a μ^+–μ^- collider option. Also, the next frontier in HEP appears to be in understanding the "Higgs sector", the generation of masses. The greater direct coupling of muons to the Higgs sector (by $(m_\mu/m_e)^2$) may provide an important incentive for devel-

☆ Work supported by Department of Energy contract #DE-AC05-84ER40150.

oping μ^+–μ^- colliders, possibly in the energy-specific form of a Higgs-resonance "factory".

In this paper we present an overview of possible high-energy μ^+–μ^- colliders for the μ^+–μ^- workshop (Napa, CA, 12/1992), following the concepts previously presented in the Refs. [1–4]. In this overview, we will identify some of the critical problems in implementing the scenario, suggest some possible variations and improvements, and, hopefully, inspire some further directions for innovation and invention from the participants and other readers.

2. μ production

A critical problem in μ^+–μ^- colliders is the production and collection of adequate numbers of muons. The critical difficulties are the large initial phase-space volume of the muons (since they are produced as secondary or tertiary products of high-energy collisions) and the fact that muons decay, with a lifetime of 2.2×10^{-6} γ decay time. (γ is the usual kinetic factor (E_μ/m_μ).) The problem is to obtain, compress, and use sufficient muons before decay.

We will first consider as a baseline reference a high-energy hadronic (HEH) muon source of the type described in Refs. [1,2,4]: a medium-energy high-intensity hadron accelerator beam is transported onto a high-density target to produce GeV-energy π's, and the π's are confined in a transport channel where they decay to produce μ's, which are collected, compressed, and accelerated. In this section, the baseline HEH source is described, and guidelines for optimization and improvement are suggested. Variations and alternative approaches are then mentioned; these include a low-energy (GeV-scale) "π-factory" beam, possibly producing surface muon beams, or an e^- beam producing μ^+–μ^- pairs by photoproduction.

Production of large numbers of muons in the baseline scenario is not difficult. Hadronic interactions of the beam with the target produce large numbers of π's, and almost all of these π's decay by producing a muon plus a neutrino. The difficult problem of optimizing production and collection for maximal μ intensity is not yet solved; however, some guidelines may be obtained from approximate calculations. A first estimate of $\pi^\pm$ production in proton–hadron collisions may be obtained using known and calculated particle spectra, such as the empirical formulae of Wang [6]:

$$\frac{d^2N}{dP\, d\Omega} \cong AP_p X(1-X) e^{-BX^C - DP_t} \frac{\text{pions}}{\text{sr-GeV}/c} \text{ /interacting proton} \qquad (1)$$

where P_p is the incident proton momentum, $X = P_\pi/P_p$ is the pion/proton momentum ratio, P_t is the pion transverse momentum and $A = 2.385$ (1.572), $B = 3.558$ (5.732), $C = 1.333$ (1.333), and $D = 4.727$ (4.247) for positive (negative) pions. In this formula, pions are produced with a mean transverse momentum of $\sim D^{-1}$ or ~ 0.2 GeV. Also, if $P_\pi/P_p \ll 1$, pion production is nearly independent of proton energy, and pion production within a given momentum bite ($\Delta P_\pi/P_\pi$) is nearly constant. If the Wang model is accurate, an optimal π-source may be obtained from a medium-energy proton beam (20–50 GeV) which collides into a nuclear target, followed by strong-focusing optics which collects secondaries (P_t acceptance ~ 0.3 GeV) and a strong-focusing transport line for $\pi \to \mu\nu$ decay. Small spot sizes on the production target and small beam sizes in the transport line are desired to minimize π and μ emittance. High momentum acceptance ($\delta P/P >$ $\pm 10\%$) is desired for maximal production. Economy would favor lower energies.

Fig. 2 displays a possible configuration for μ^+ or μ^-

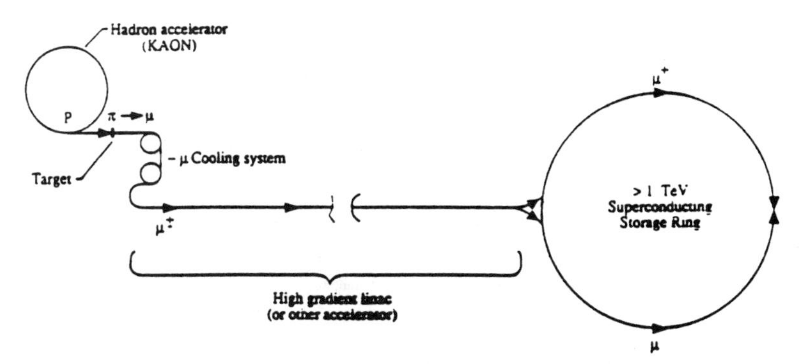

Fig. 1. Overview of linac-based μ^+–μ^- collider, showing a hadronic accelerator, which produces π's on a target, followed by a π-decay channel ($\pi \to \mu\nu$) and μ-cooling system, followed by a μ-accelerating linac, feeding into a high-energy storage ring for μ^+–μ^- collisions. (The linac could be replaced by a rapid-cycling ring; see Fig. 6.)

production. A 40 GeV proton beam is focused (possibly with a Li lens) onto a ~ 5 cm W target to a mm-scale spot, producing π's, which are captured by the following optics, which is designed to accept 2 GeV π's at angles up to $\Theta_T = 150$ mrad ($P_t < 0.3$ GeV/c). This optics is approximated by a 2.0 cm radius Li lens of 10 cm length centered 15 cm downstream from the target center. The capture optics is to be followed by a strong-focusing transport of ~ 1 π decay length (~ 110 m at 2 GeV), which maintains a mean betatron function of 1 m. The resulting muons are inserted into a muon accumulator system for cooling and acceleration.

According to the Wang model, this outline system would produce and accept 0.12 (0.08) π's /(GeV/c)/interacting proton. This must be multiplied by a target efficiency factor $\eta_T \approx 0.4$ (the probability that an incident proton produces an exiting π), and by the momentum acceptance width (0.4 GeV/c for $\pm 10\%$ acceptance), to obtain the number of π's accepted in the decay channel per primary proton (0.019/0.012 $\pi^+\pi^-$).

In the decay channel, π-decay ($\pi \to \mu\nu$) produces μ's within a uniform energy distribution (between 0.57 and $1.0E_\pi$) and with a maximum transverse momentum of 29.8 MeV/c. With 63%(1/e) of the π's decaying and $\sim 40\%$ of the product μ's within the transport acceptance, we find ~ 0.005 (0.0033) $\mu^+(\mu^-)$ per primary proton are delivered to the μ-collector.

The transverse emittance is determined by the π-production phase space, and the π-decay phase space. The transverse emittance from π-production from a thin target is of order $r_T\Theta_T$ or 3×10^{-4} m rad at 2 GeV, if the primary beam size on the target, r_T, is 0.002 m. The π-decay increases emittance by $\sim \beta\theta_d^2/2$, where β is the mean decay-line betatron function and θ_d is the maximum decay angle. With $\theta_d = P_t/P_\mu \approx 0.02$ and $\beta = 1$ m in our reference case, an emittance ϵ_t of $\sim 2 \times 10^{-4}$ m rad at 1.5 GeV is obtained ($\sim 3 \times 10^{-3}$ m rad normalized). (Total emittance from target size, target length, and decay effects should be less than twice this value, and could be reduced somewhat by further optimization.)

This reference case is oversimplified. The problems of separating primary p's from secondary π^+ and π^- beams, and separating the π beams from each other are not addressed; the actual optics will be more interesting. The parameters are not optimized, and the production estimates are not very accurate. The calculations do not include π's (and μ's) from secondary interactions. For $E_\pi \ll E_p$, secondary and cascade production could be large. However, μ-decay from source to collider will reduce the final number (by $\sim 2\times$). The calculations do show that an HEH source can obtain $\geq 10^{-3}$ μ's per primary proton, and that is adequate for a high-luminosity collider. This results is in agreement with an independent analysis of Noble [7]. It is possible that improvements, acceptance increases, and optimization could increase this to the 10^{-2} level, but probably not much greater.

The HEH scenario has been motivated from a high-energy physics bias, and imitates $\bar{p}$ source methods. π-production also occurs in low-energy hadronic (LEH) "π-factories", from GeV/nucleon protons or deuterons, at a level of ~ 0.5 π/p. For μ-collider use, a compressor ring would be needed to combine the proton (deuteron) linac beams into sub-μs pulses. The π's are produced at the 100 MeV energy level, where they can be stopped in an absorber to produce 29 MeV/c μ's, which will lose further energy in the absorber. (Stopping times are at ns levels; the μ-lifetime at rest is 2.2 μs.) (Nagamine has proposed extracting such slow muons in a high-intensity surface muon source [8].) An LEH source can produce large numbers of μ's with low energy and momentum spread, which may require little or no further cooling, and the LEH source could be preferable to an HEH source. The difficulty is in extracting sufficient μ's in a small phase-space volume which are suitable for acceleration in a μ^+–μ^- collider. The problem of calculating and optimizing μ production in an LEH configuration is unsolved; it is an important challenge for the reader.

Another possible μ-source can be obtained by colliding multi-GeV e^- beams into a hadronic target; bremsstrahlung would produce μ^+–μ^- pairs by photoproduction, but not as frequently as e^+–e^- pairs. Obtaining the μ's through pair production avoids the phase-space dilution of π-decay; however, μ-production does not seem to be as copious as in a hadronic source. An optimized calculated comparison has not yet been made.

The best possible μ-source is not yet identified or developed. It may follow some of the ideas suggested in this section, with added improvements and innovations, or it may be dramatically different. This is an important challenge for the workshop.

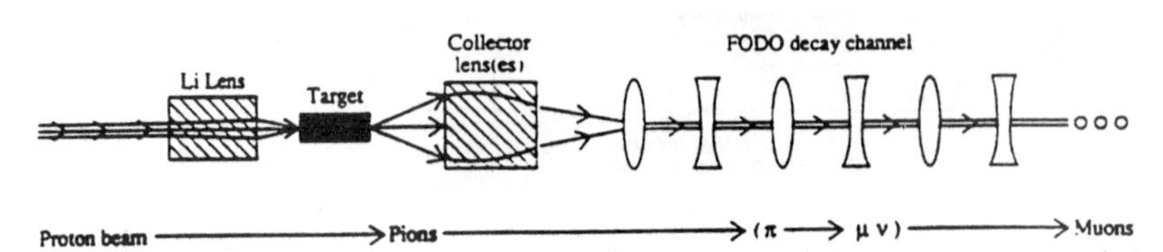

Fig. 2. Schematic view of μ-production from π-decay, with π's produced from hadrons. A high-energy hadronic beam is focused onto a target; a collector lens(es) collects the resulting π's into a strong-focusing "FODO" channel, where π-decay produces μ's for collider use.

3. μ-cooling, and combination

The μ's are produced in a relatively large phase-space volume which must be compressed to obtain high-luminosity collisions. Most of the needed compression is obtained from adiabatic damping; acceleration from GeV-scale μ collection to TeV-scale collisions reduces phase-space by $\sim 10^9$ (10^3 per dimension). Additional phase-space reduction can be obtained by "ionization cooling" of muons ("μ cooling"), which is described in some detail in Refs. [1–3], and is conceptually similar to radiation damping. In this section we first describe transverse μ cooling. Longitudinal cooling and bunch combination, and muon survival and acceleration are then discussed.

The basic mechanism of transverse μ cooling is quite simple, and is shown graphically in Fig. 3. Muons passing through a material medium lose energy (and momentum) through ionization interactions. The losses are parallel to the particle motion, and therefore include transverse and longitudinal momentum losses; the transverse energy losses reduce (normalized) emittance. Reacceleration of the beam (in rf cavities) restores only longitudinal energy. The combined process of ionization energy loss plus rf reacceleration reduces transverse momentum and hence reduces transverse emittance. However, the random process of multiple scattering in the material medium increases the emittance.

The equation for transverse cooling can be written in a differential-equation form as:

$$\frac{\mathrm{d}\epsilon_\perp}{\mathrm{d}z} = -\frac{\frac{\mathrm{d}E_\mu}{\mathrm{d}x}}{E_\mu}\epsilon_\perp + \frac{\beta_0}{2}\frac{\mathrm{d}\langle\theta^2_{\mathrm{rms}}\rangle}{\mathrm{d}z}, \tag{2}$$

where $\epsilon_\perp$ is the (unnormalized) transverse emittance, $\mathrm{d}E_\mu/\mathrm{d}z$ is the absorber energy loss per cooler transport length z, β_0 is the betatron function in the absorber and Θ_{rms} is the mean accumulated multiple scattering angle in the absorber. Note that $\mathrm{d}E_\mu/\mathrm{d}z = f_A \mathrm{d}E_\mu/\mathrm{d}s$, where f_A is the fraction of the transport length occupied by the absorber, which has an energy absorption coefficient of $\mathrm{d}E_\mu/\mathrm{d}s$. Also the multiple scattering can be estimated from:

$$\frac{\mathrm{d}\langle\Theta^2_{\mathrm{rms}}\rangle}{\mathrm{d}z} \cong \frac{f_A}{L_R}\left(\frac{0.014}{E_\mu}\right)^2, \tag{3}$$

where L_R is the material radiation length and E_μ is in GeV. (The differential-equation form assumes the cooling system is formed from small alternating absorber and reaccelerator sections; a similar difference equation would be appropriate if individual sections are long.)

If the parameters are constant, Eqs. (2) and (3) may be combined to find a minimum cooled (unnormalized) emittance of

$$\epsilon_\perp \to \frac{(0.014)^2}{2E_\mu}\frac{\beta_0}{L_R\frac{\mathrm{d}E_\mu}{\mathrm{d}z}} \tag{4}$$

or, when normalized

$$\epsilon_N = \epsilon_\perp\gamma \Rightarrow \frac{(0.014)^2}{2m_\mu c^2}\frac{\beta_0}{L_R\frac{\mathrm{d}E_\mu}{\mathrm{d}z}} \tag{5}$$

(all energies are in GeV).

Avoiding longitudinal phase-space dilution implies cooling at $E_\mu \geq 0.3$ GeV, and economy implies cooling at relatively low energies (since cooling by e^{-1} requires E_μ energy loss and recovery). $E_\mu \approx 0.5$–1.5 GeV seems reasonable. Cooling can be obtained in either linacs (possibly using recirculation lines) or storage rings. Multiple stages

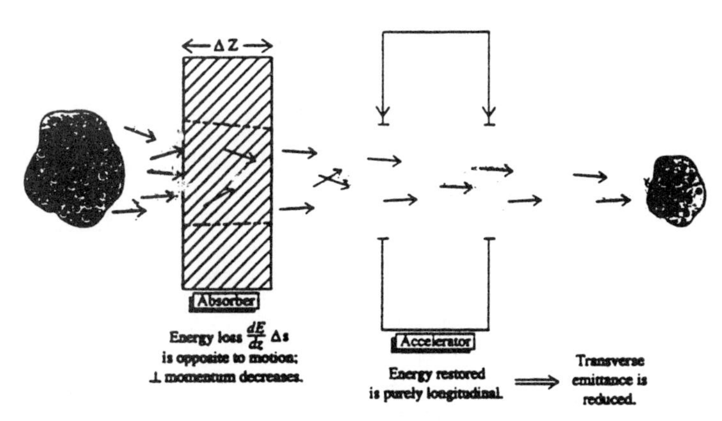
ΔZ
Absorber
Energy loss $\frac{dE}{dz}$ Δs
is opposite to motion;
⊥ momentum decreases.
Accelerator
Energy restored
is purely longitudinal.
Transverse
emittance is
reduced.

Fig. 3. Schematic view of transverse "ionization cooling". Energy loss in an absorber occurs parallel to the motion; therefore transverse momentum is lost with the longitudinal energy loss. Energy gain is longitudinal only; the net result is a decrease in transverse phase-space area.

can be used to optimize cooling scenarios. The important constraint is that cooling must be completed within a muon lifetime, which can be expressed as $\sim 300\overline{B}$ (T) turns in a storage ring, where $\overline{B}$ is the mean bending field, or as a length $L_{\mu} = 660\gamma$ m of path length. This constraint may be surmountable.

Some guidelines for optimal cooling may be obtained from Eqs. (2)–(5). They indicate that it is desirable to obtain small β_0 (strong focusing) at the absorber. It is also desirable to have materials with large values of the product $L_R\, dE/ds$. $L_R\, dE/ds$ is largest for light elements (0.1 GeV for Li, Be but ~ 0.01 GeV for W, Pb), indicating the desirability of light absorbers. However, the need for small β_0 and the depth of focus constraint that a (non-focusing) absorber section must be less than $2\beta_0$ would favor large dE/ds (heavy) absorbers. A conducting light-metal absorber (Li, Be, Al) can also be a continuously focusing lens, which could then be arbitrarily long while maintaining small β_0. With current technology, lenses maintaining $\beta_0 < 1$ cm appear possible.

With $\beta_0 \approx 1$ cm, and $L_R\, dE/ds \approx 0.1$ GeV, a normalized emittance of $\epsilon_N \approx 10^{-2}\beta_0 \approx 10^{-4}$ m rad is obtained as a reasonable goal for transverse cooling. Some improvements may be possible; the reader may develop ideas for optimal implementation.

Longitudinal (energy-spread) cooling is also possible, if the energy loss increases with increasing energy. The energy loss function for muons, dE/ds, is rapidly decreasing (heating) with energy for $E_{\mu} < 0.3$ GeV, but is slightly increasing (cooling) for $E_{\mu} > 0.3$ GeV. This natural dependence can be enhanced by placing a wedge-shaped absorber at a "non-zero dispersion" region where position is energy-dependent (see Fig. 4). Longitudinal cooling is limited by statistical fluctuations in the number and energy of muon–atom interactions. An equation for energy cooling is:

$$\frac{d\langle(\Delta E)^2\rangle}{dz} \approx -2\frac{\partial \frac{dE_{\mu}}{dz}}{\partial E_{\mu}}\langle(\Delta E)^2\rangle + \frac{d\Delta E_{\mathrm{rms}}^2}{ds}, \tag{6}$$

where the derivative with energy combines natural energy dependence with dispersion-enhanced dependence. An expression for this enhanced cooling derivative is:

$$\frac{\partial \frac{dE_{\mu}}{dz}}{\partial E_{\mu}} = f_A\left(\frac{\partial \frac{dE_{\mu}}{ds}}{\partial E_{\mu}}\right)_0 + f_A\frac{dE_{\mu}}{ds}\frac{d\delta}{dx}\frac{\eta}{E_{\mu}\delta_0}, \tag{7}$$

where η is the dispersion at the absorber, and δ and $d\delta/dx$ are the thickness and tilt of the absorber. Note that using a wedge absorber for energy cooling will reduce transverse cooling; the sum of transverse and longitudinal cooling rates is invariant. In the long-pathlength Gaussian-distribution limit, the heating term or energy straggling term is given by [9]:

$$\frac{d(\Delta E_{\mathrm{rms}})^2}{ds} \cong 4\pi\left(r_e m_e c^2\right)^2 N_0 \frac{Z}{A}\rho\gamma^2(1-\beta^2/2),$$

where N_0 is Avogadro's number and ρ is the density. Since this increases as γ^2, cooling at low energies is desired. From balancing heating and cooling terms, we find that cooling of $\Delta E_{\mu}/E_{\mu}$ to ≤ 0.02 at $E_{\mu} = 0.25$ GeV is possible, which adiabatically damps to a 1 GeV energy spread of $\Delta E_{\mu}/E_{\mu} < 0.005$. An rf buncher plus compressor arc (or synchrotron oscillations) can use this reduced energy spread to obtain reduced bunch length.

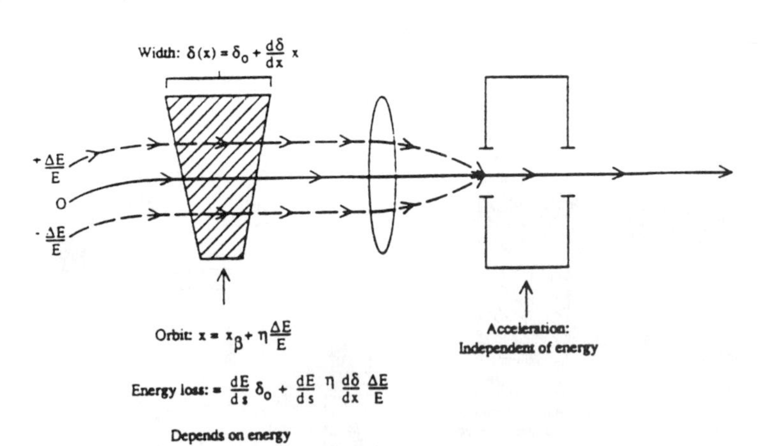

Fig. 4. Enhancement of energy cooling by using a wedge absorber placed in a non-zero dispersion region. The thickness of the absorber depends on transverse position ($\Delta = \delta_0 + (d\delta/dx)x$), and the position at the absorber depends on the energy ($x = \eta(\Delta E/E)$), producing an enhanced energy dependence of energy loss, decreasing energy spread. Energy recovery in the accelerator is independent of energy. (Transverse cooling decreases with enhanced energy cooling.)

The major problem in longitudinal space derives from the mismatch between the initial bunch structure of the μ-source and the desired μ-collider bunch configuration. The primary proton beam from a rapid-cycling synchrotron (RCS) would consist of ~ 100–200 bunches, while compression to one (or a few) μ-collider bunches is desired. Phase-space manipulations will be needed.

Bunch combination procedures could include:

1) Proton-bunch overlap: Before extraction from the RCS, the proton bunches can be compressed with a lower harmonic (or sideband) rf system to provide spatially overlapping bunches (with different momenta) on the target. Since target spot sizes need not be very small and π-production is not energy-dependent, a broad primary energy spread can be accepted at the target, with no degradation of π-production. Combination by at least a factor of 10 should be obtainable. A separate extraction-energy proton compressor ring for the rf bunch manipulations may be desired, and may be capable of combining the number of bunches to ~ 2–4.

2) Non-Liouvillian "stochastic injection" [10]: Bunch combination without phase-space dilution can occur during π decay, as in "stochastic injection" into a μ-storage ring, as shown in Fig. 5. In this process a train of π bunches from a hadronic target is injected into a decay channel, which is also a zero-dispersion straight-section of a μ-storage ring. The π-bunch spacing is matched to the storage ring period (or a low harmonic). The injected π's are not in the acceptance of the ring; the ring accepts lower-energy particles, in particular, those μ's from π decay in the straight section that are within that acceptance. Successive π-bunch arrivals are timed to overlap an accumulated μ bunch. The π lifetime is ≈ 1% of the μ lifetime, and is naturally matched for decay within the first-turn decay channel, while permitting multiple turn accumulation. At reasonable parameters, μ's from 10–30 π bunches can be accumulated in a single bunch, without large μ-decay losses. (The storage ring can also be used for cooling the accumulated μ's.)

3) Beam cooling with bunch combination: Transverse or energy cooling of μ bunches can compress beams to a degree where bunches can be stacked together, using conventional Liouvillian bunch-combination optics. The stacked bunch can then be further cooled to a phase-space volume ≤ the previous cooled single-bunch size. The process can continue through further stacking and cooling steps. The process adds the complications of multiple beam-combination transport lines and optics; however, combinations of many bunches (10–30) could be obtained.

Some combination of these three (plus other to-be-developed) methods can be used to reduce the number of μ bunches to a few enhanced-intensity bunches. (The first two procedures are complementary: proton bunch stacking naturally combines nearby bunches while stochastic injection more naturally combines widely spaced bunches.) Ionization cooling can then be used for further compression, in bunch length or energy spread, for optimal collider use.

4. Acceleration and collider scenarios

Cooled and compressed muon bunches can be used in high-energy high-luminosity colliders. In this section, we describe some potential scenarios. We use as a reference case a 1 TeV per beam μ^+–μ^- collider (2 TeV in the center-of-mass), as this is the energy scale where the μ^+–μ^- collider may begin to be preferable to e^+–e^- colliders. Table 1 shows a reference case, with relatively conservative choices of parameters.

1) Linac–storage ring. This scenario is displayed graphically in Figure 1, and this is probably the highest-luminosity case. μ^+–μ^- bunches from the collector/cooler are both accelerated to full energy in a high-gradient linac to 1 TeV, where both bunches are injected into a superconducting storage ring for high-energy collisions at low-β_0 interaction points. The μ beam lifetime is $\sim 300B$ turns, where B is the mean bending field in T. $B \approx 8$ T, imply-

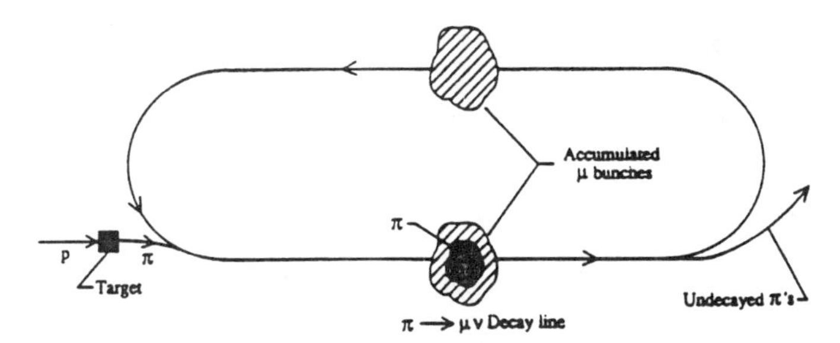

Fig. 5. Schematic view of "stochastic injection" into a storage ring. A train of p bunches produces π bunches, which are injected into a storage ring for multi-turn stacking. The initial spacing is matched to a ring harmonic ($h = 2$ in the figure), so that following π bunches overlap accumulated μ's from previous bunches. π decays in the straight section which produce μ's within the ring acceptance add to the accumulation. (Note that the short but finite π lifetime is nearly optimally matched to make this scheme practical.)

Table 1
Parameter list for TeV μ^-–μ^- collider (2 TeV collisions)

Parameter	Symbol	Value
Energy	$E_{\mu\pm}$	1 TeV
Luminosity	L	3×10^{32} cm^{-2} s^{-1}
HEH-source parameters		
Proton energy	E_p	40 GeV
P/pulse	N_p	10^{14}
Pulse rate	f_0	30 Hz
μ production efficiency	μ/p	10^{-3}
Collider parameters		
$\#\mu^+/\mu^-$ per bunch	$N^{\pm}$	10^{11}
#bunches	n_B	1
Storage turns	n_s	1200
μ emittance	$\epsilon_\perp = \frac{\epsilon_N}{\gamma}$	10^{-8} m rad
Interaction focus	β^*	1 mm
Beam size	σ	3 μm

ing 2400 turns, is currently achievable. The main difficulty is the relatively large cost of the full-energy TeV linac.

2) Linac–linac collider. As in e^+–e^- linear colliders, μ bunches from opposing linacs can collide. This scenario loses the luminosity magnification obtained from multiple collisions in a storage ring, which is permitted by the long μ lifetime. It also requires two full-energy linacs, and a TeV storage ring is cheaper than a TeV linac. However, an existing e^+–e^- linear collider could be modified to obtain μ^+–μ^- collisions, with the addition of a μ source.

3) Rapid-cycling synchrotron collider. At 1 TeV, the μ lifetime has increased to 0.021 s, and the lifetime increase with energy is sufficient to permit acceleration in a rapid-cycling synchrotron with acceptable losses. Fig. 6 shows the basic components: a μ source injected into a $\sim$20–50 GeV linac followed by a rapid cycling synchrotron with 20 km circumference ($B \approx 1$ T). Acceleration from injection to full energy in $\sim$50–100 turns could follow a 60–120 Hz waveform followed by $\sim$0.02 s at fixed field for collisions or, for higher luminosity, transfer to a fixed-field (8 T) collider ring. Luminosity would be naively expected to be about an order of magnitude smaller than in the linac–storage ring scenario.

4) The μ–p collider. A μ^+–μ^- collider can also be operated as a μ^-–p collider with both μ and p beams at full energy. High luminosity would be relatively easily obtained because only one beam (μ) is unstable and diffuse. This is a probable initial and debugging operating mode for a storage ring μ^+–μ^- collider. Revolution frequencies of equal energy μ and p beams would be naturally mismatched because of unequal speeds. They can be rematched by displacing the beams in energy and using the ring nonisochronicity [2]. The required energy displacement is

$$\frac{\delta E}{E} = \gamma_T^2\left(\frac{1}{2\gamma_p^2} - \frac{1}{2\gamma_\mu^2}\right) \tag{8}$$

where γ_T^2 is the ring transition-gamma. At reference parameters ($E_\mu = 1$ TeV, $\gamma_T = 30$) $\delta E/E = 0.0005$ is required.

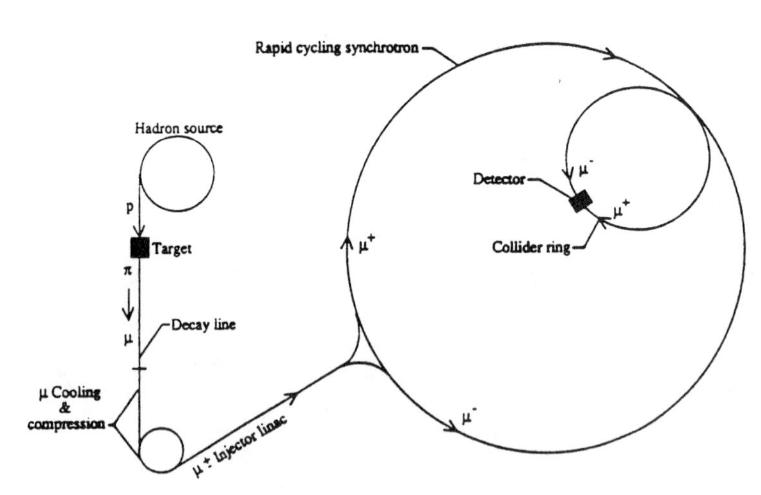

Fig. 6. Overview of a μ^+–μ^- collider, with a rapid-cycling synchrotron for the primary accelerator. The figure shows a primary proton source, producing π's on a target, which decay to μ's in a decay channel. After cooling and compression, μ^+–μ^- bunches are accelerated in a linac and in a rapid-cycling synchrotron to full energy, where they are injected into a high-field storage ring for multi-turn collisions. For example, a 20 GeV linac feeding into a 20 GeV/turn (up to 1.2 T) rapid-cycling synchrotron would produce 1 TeV μ^+–μ^- beams with acceptable decay losses.

5) Physics–opportunity colliders. A resonance, such as a new Z particle or a Higgs particle, may exist or be predicted in μ^+–μ^- collisions. At such a resonance, a lower-luminosity and/or lower-energy collider could still provide extremely important physics. Such a collider would be a simplified form of the baseline high-luminosity models, possibly with μ sources using existing accelerators, omitting μ cooling, and/or using existing storage rings for collisions. Such a facility would, of course, provide excellent training for a high-energy high-luminosity collider.

5. Luminosity possibilities and constraints

Using the previously discussed techniques, high luminosity can be obtained in a μ^+–μ^- collider. The luminosity is given by the equation:

$$L = \frac{f_C N^+ N^-}{4\pi\sigma^2} = \frac{f_C N^+ N^-}{4\pi\beta^* \epsilon_\perp}, \tag{9}$$

where N^+,N^- are the number of μ^+, μ^- per colliding bunch, f_C is the frequency of bunch collisions, σ^2 is the colliding beam size, β^* is the betatron function at the collision point and $\epsilon_\perp = \epsilon_N/\gamma$ is the transverse emittance. In a storage-ring collider, $f_C = f_0 n_B n_S$, where f_0 is the system cycling rate, n_B is the number of bunches, and n_S is the number of turns of storage per cycle.

This formula is applied to a reference TeV μ^+–μ^- collider (Table 1). The parameters we use include $N^+ = N^- = 10^{11}$ (which can be obtained assuming a modest production rate of 10^{-3} μ/p from 10^{14} proton high-energy pulses), $n_B = 1$, $f_0 = 30$ Hz, and $n_S = 1200$ turns storage. With $\epsilon_N = 10^{-4}$ m rad ($\epsilon_\perp = 10^{-8}$) from $\beta_0 = 1$ cm and $\beta^* = 1$ mm ($\sigma \approx 3$ μm), we obtain a respectable baseline luminosity of $L \approx 3 \times 10^{32}$.

Note that the above parameter set is relatively modest, and improvements in some of the parameters (i.e., N^+, N^-, σ) by up to an order of magnitude are conceivable. However, reliable accomplishment of high luminosity in a novel and complicated facility which uses unstable particles and has several difficult design components will still not be easy.

Luminosity in a μ^+–μ^- collider ring may be expected to be limited by the beam–beam interaction. Since long-term stability is not needed, the allowable beam–beam tune shift should be somewhat greater than the e^+–e^- storage limit of $\Delta\nu \lesssim 0.05$. The tune shift is given by

$$\Delta\nu = \frac{N r_\mu \beta^*}{4\pi\gamma\sigma^2}. \tag{10}$$

where $r_\mu = 1.363 \times 10^{-17}$ m. At the baseline parameters $\Delta\nu \approx 0.001$, luminosity would have to be increased dramatically for the beam–beam limit to be significant.

A μ^+–μ^- collider has substantial beam power requirements, particularly in the primary proton beam. In the reference case, the requirements are 10^{14} 40 GeV protons at 30 Hz, which implies 20 MW beam power. This is an order of magnitude above present facilities, but is the same magnitude as proposed K-factories. (The high-energy μ beams themselves require only 0.32 MW.) More efficient μ-production may be desirable (obtaining more μ/p, using lower-energy p's or a LEH source). However, a high-luminosity high-energy μ^+–μ^- collider is a successor to SSC- or TLC-size facilities, and on that scale, a K-factory type source is small. An even higher intensity source could be affordable.

A significant difficulty in a storage ring is that μ's decay ($\mu \to e\nu\nu$), and decay electrons at ~ 0.3 TeV will hit the walls of the storage ring. At the reference parameters, with half the μ's decaying during storage, we find 50 kW of ~ 0.3 TeV electrons will be deposited evenly within a narrow strip on the inner wall of the ring. The ring must be designed to accept this.

The obtainable luminosity L is expected to increase with increasing end point energy E_μ, as the μ lifetime increases and the emittance and energy spreads are adiabatically damped. As discussed in Ref. (4), the beam size (σ^2) at collision should decrease as E_μ^{-2}, as both β^* and $\epsilon_\perp$ can decrease. Cycle time would increase; however, the longer cycle time could permit accumulation of successive rapid-cycling proton pulses to obtain magnified-intensity μ-bunches. The net effect is that N^+ and N^- increase as E_μ and f_C is reduced by E_μ^{-1}. In all, L naturally increases as E_μ^3. (Costs increase linearly with E_μ.) This scaling would be expected to dominate until the $E_\mu \approx 100$–1000 TeV region, where μ synchrotron radiation excludes μ storage-ring colliders.

6. Summary

In this paper, we have introduced concepts which show the promise of the development of high-luminosity TeV-scale μ^+–μ^- colliders. These initial concepts need considerable practical development. While key ingredients of a future facility have been introduced, further innovations and improvements are greatly desired. These concepts plus further developments must be integrated into a fully self-consistent design for a μ^+–μ^- facility. The discussions at the first μ^+–μ^- collider workshop at Napa, California, should elucidate the possibilities and set a basis for further development. Contributions and improvements from the workshop participants and other readers are encouraged. A functional μ^+–μ^- collider will be obtained only through further innovation and development.

Acknowledgements

I thank D. Cline, A. Ruggiero, J.D. Bjorken, S. Glashow, J. MacLachlan, H.A. Thiessen, R. Palmer and R.

Noble (and any others I forgot) for important conversations contributing to the ideas expressed in this paper.

Appendix

Since this report is appearing subsequent to the 1992 Napa μ^+–μ^- Collider workshop, in this section I am adding some initial impressions of the proceedings of the workshop.

D. Cline presented an interesting immediate application for a μ^+–μ^- collider [11]. There are theoretical reasons to believe that the Higgs boson might exist in the 90–180 GeV region, and that is an energy region at which only a μ^+–μ^- collider could obtain a clean observation of a Higgs boson. (μ^+–$\mu^- \rightarrow$ H is favored because of the relatively large muon mass.) The observation would require luminosity greater than $\sim 10^{29}$ cm^{-2} s^{-1}. Sample parameters with this luminosity goal are shown in Table 2. This relatively low-energy collider could be developed relatively inexpensively, possibly using existing facilities for major components, although it is unclear if any existing accelerator could deliver sufficient muon intensity. An important future goal is developing an optimal short-term path to this extremely important physics goal.

At the workshop, it was speculated that maximal muon production could be obtained from a hadronic (or leptonic) cascade source, possibly at a beam dump, rather than the single-interaction source outlined above. H. Thiessen suggested that the most efficient source could be a ~ 5 GeV hadronic source. Serious target problems were noted for any high-intensity source. Further study/optimization is needed.

Table 2
Parameter list for 100 GeV μ^+–μ^- collider

Parameter	Symbol	Value
Energy	$E_{\mu\pm}$	100 GeV
Luminosity	L	10^{29} cm^{-2} s^{-1}
Pulse rate	f_0	10 Hz
Storage turns	n_s	1000
#bunches	n_B	10
#μ^+/μ^- per bunch	$N^{\pm}$	10^{10}
μ-emittance	$\epsilon_\perp = \dfrac{\epsilon_N}{\gamma}$	10^{-7} m rad
Interaction β	β^*	1 cm
Beam size (at IR)	σ	30 μm

In beam cooling, the limitations on ionization cooling due to multiple scattering (described above) were discussed. It may be possible to have a muon source with initial emittance smaller than the multiple scattering limit, and therefore to avoid cooling.

The bunch combination/compression (see above) was identified as a key problem, particularly bunch length reduction to match small β^* optics. At multi-GeV energies, the muons will be relativistic and have no longitudinal motion within a linac. However, relativistic muon bunches can be compressed with an rf-induced energy tilt and transported through a bending arc (or ring). Scenario design/optimization is needed.

Although the workshop did not identify a clear path to a sufficient luminosity design, it did show important (and perhaps obtainable) physics goals, particularly the Higgs discovery opportunity. Future study goals, particularly in source design and scenario development, were identified, and the need for future workshop(s) after further development was suggested.

References

[1] D. Neuffer, Particle Accelerators 14 (1983) 75.
[2] D. Neuffer, Proc. 12th Int. Conf. on High Energy Accelerators, eds. F.T. Cole and R. Donaldson ibid., 1983, p. 481.
[3] E.A. Perevedentsev and A.N. Skrinsky, ibid., 1983, p. 485.
[4] D. Neuffer, in: Advanced Accelerator Concepts, AIP Conf. Proc. 156 (1987) 201.
[5] M. Tigner, Proc. 3rd Int. Workshop on Advanced Accelerator Concepts, Port Jefferson, NY, AIP Conf. Proc. 279 (1993) p. 1.
[6] C.L. Wang, Phys. Rev. D 9 (1973) 2609 and Phys. Rev. D 10 (1974) 3876.
[7] R. Noble, Proc. 3rd Int. Workshop on Advanced Accelerator Concepts, Port Jefferson, NY, AIP Conf. Proc. 279 (1993) p. 949.
[8] K. Nagamine, unpublished communication (1986).
[9] U. Fano, Ann. Rev. Nucl. Sci. 13 (1963) 1.
[10] D. Neuffer, IEEE Trans. Nucl. Sci. NS-28, (1981) 2034.
[11] D. Cline, unpublished communication (1992).

Characteristics of a high energy $\mu^+\mu^-$ collider based on electro-production of muons

William A. Barletta [*,+], Andrew M. Sessler [++]
Lawrence Berkeley Laboratory, Berkeley, CA, USA

We analyze the design of a high energy $\mu^+\mu^-$ collider based on electro-production of muons. We derive an expression for the luminosity in terms of analytic formulae for the electron-to-muon conversion efficiency and the electron beam power on the production target. On the basis of studies of self-consistent sets of collider parameters under "realistic" ("optimistic") assumptions about available technology with beam cooling, we find the luminosity limited to 10^{27} cm^{-2} s^{-1} (10^{28} cm^{-2} s^{-1}). We also identify major technological innovations that will be required before $\mu^+\mu^-$ colliders can offer sufficient luminosity (10^{30} cm^{-2} s^{-1}) for high energy physics research.

1. Introduction

Many physicists consider that the recent determinations of lower bounds for the mass of the top meson reinforce arguments that a Standard Model Higgs should have a mass less than twice the mass of the Z. This consideration has led to renewed interest in muon colliders as an ideal means of probing the mass range from m_Z to $2m_Z$. More generally, a muon collider with center-of-mass energy in the range of 200 to 400 GeV has the potential to produce very large numbers of Higgs particles because of the enhanced (vis a vis electrons) muon coupling to the Higgs. For such a collider to have maximum discovery potential the luminosity should be $\geq 10^{30}$ cm^{-2} s^{-1} [1]. As the muon is an unstable particle, the muons must be generated as secondary beams from either a proton beam or an electron beam striking a production target. The muons that emerge from the target must then be gathered and accelerated rapidly to high energy, at which point they can be injected into a storage ring collider with superconducting magnets.

This paper analyses the possibility of using electro-production to generate the muon beams. The chief advantage of producing the muons with an electron beam from a high energy, linear accelerator is that the bunches of muons are naturally formed with a short bunch length (< 1 cm) for acceleration to the desired high energy in a subsequent linear accelerator. As the muons will retain their short bunch length in the collider, a low β interaction region can be employed. This scheme is illustrated in Fig. 1.

* Corresponding author.

+ Work performed under the auspices of the Lawrence Livermore National Laboratory under contract W-7405-eng-48.

++ Work performed under the auspices of the Lawrence Berkeley Laboratory under contract DE-AC03-76SF00098.

2. Electro-production

Muons can be produced by an electron beam via two classes of processes, 1) $\mu^+\mu^-$ pair production and 2) photo-production of π's and K's, which subsequently decay into muons. It is known experimentally [2] that the cross-section for pair production is much more than an order of magnitude greater than that for process (2). Consequently, in the discussion that follows we will consider only pair production.

To estimate the muon pair production from an electron beam of energy, E_e, incident upon a thick target of atomic number Z, one can use the expression from Nelson [3] based on approximation A of shower theory. F is number of muons per electron produced at an angle $\geq \phi$ with respect to the incident electron beam:

$$\frac{dF}{dE}(E_e, E, \phi) = \frac{1}{(2\pi)^2}\frac{m^2}{\mu^2}\frac{0.572E_e\eta}{\mu^2\ln(183Z^{-1/3})}2\ln(\gamma_\mu) \times\{(1-\nu^2)-0.33[1-4\nu^3(1-0.75\nu)] \times\eta[1+\gamma^2]\}, \quad (1)$$

where m = electron mass, μ = muon mass, E = energy of

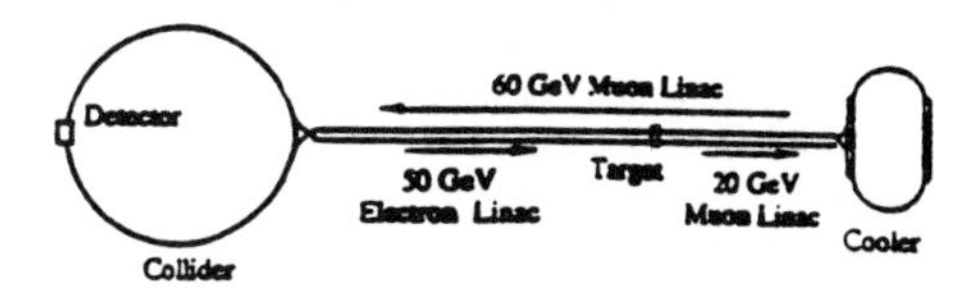

Fig. 1. The scheme for a μ^+/μ^- collider using electro-production.

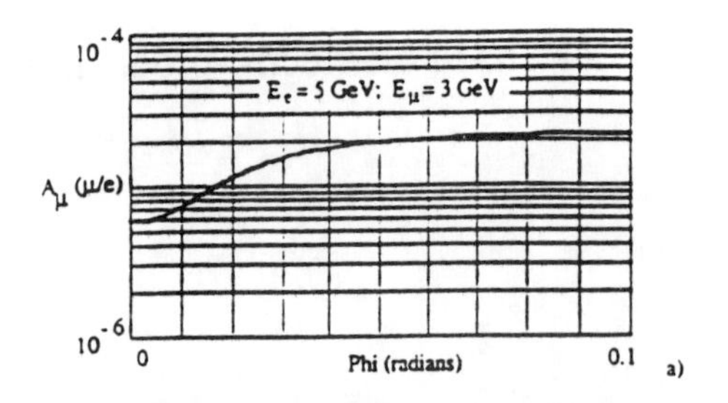

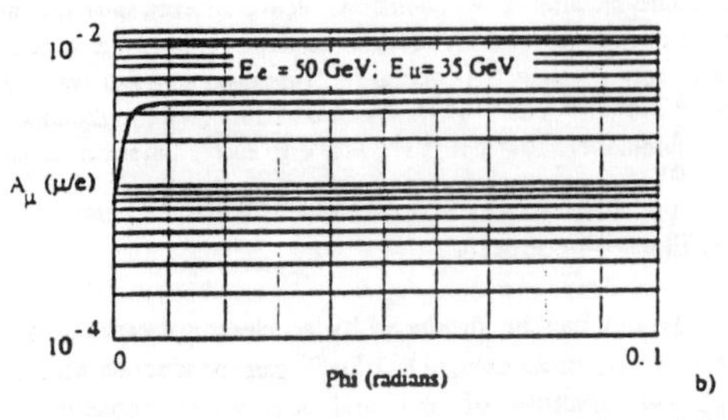

Fig. 2. Number of μ pairs per e^- accepted at an angle $\leq \phi$ for (a) a 5 GeV electron beam with $E_\mu = 3$ GeV and (b) a 50 GeV electron beam with $E_\mu = 35$ GeV.

the muon at the production target, $\gamma_\mu = E/\mu$ at the production target, $\nu = E/E_e$, $\lambda = \gamma^2\phi^2$, and $\eta = (1+\lambda)^{-2}$. Eq. (1) is known to overestimate the muon pair production by a factor of two.

The number of muons per electron accepted in an angle $\leq \phi$, in a momentum bite of $\pm\Delta p/p$ at a muon energy E is

$$A_\mu = \left[\frac{dF}{dE}(E_e, E, 0) - \frac{dF}{dE}(E_e, E, \phi)\right] E \frac{2\Delta p}{p}. \quad (2)$$

From Eq. (2) it is immediately obvious that one will prefer to accept muons with a large value of E/E_e rather than with a small E/E_e as long as the function dF/dE is relatively flat in energy. For small muon production angles this condition obtains for the energy range, $0.2 < E/E_e < 0.8$. The same consideration also argues that one should choose a large initial electron beam energy. Figs. 2a and 2b display plots of Eq. (2) for a low energy and a high energy production option respectively.

At the front surface of the production target the electron beam can be focused to a spot of radius, $r_b \approx 1$ mm. The muons will, however, appear to originate from a somewhat larger spot with a size given by the radial extent of the electromagnetic shower at a depth corresponding to the shower maximum, which occurs approximately six radiation lengths ($6X_0$) inside the target. The radiation length, X_0, for tungsten is 3 mm; hence the shower maximum will occur at ≈ -20 mm, and a tungsten target 30 mm long will yield almost the entire thick target conversion to muons.

At the depth corresponding to the shower maximum the primary electron beam will have suffered a mean scattering angle of

$$\Theta^2 = \left(\frac{0.028 \text{ GeV}}{E_e}\right)^2 \left(\frac{6X_0}{X_0}\right), \quad (3)$$

which will induce a radial spread of $6X_0\Theta$ in the primary beam. Actually in a high Z target, the shower will spread by an amount roughly double this value. Hence, the shower radius can be approximated by

$$r_{sh} = \left(r_b^2 + (12\Theta X_0)^2\right)^{1/2}. \quad (4)$$

At production the geometrical emittance, $\epsilon(E)$, of the muon beam of energy, E, accepted into an angle ϕ_{accept} will be

$$\epsilon(E) = \frac{\epsilon_{n,prod}}{\gamma_{prod}} = r_{sh}\phi_{accept}, \quad (5)$$

where $\epsilon_{n,prod}$ is the normalized emittance at production.

To increase the muon production efficiency one might consider alternate techniques of photo-production. The production process consists of two steps: 1) conversion of the electron energy into photons and 2) muon pair production from the photons. Rather than using bremsstrahlung, one might employ synchrotron radiation as the conversion process. Synchrotron radiation conversion could either take place in a crystal or in a plasma [4], which has an obvious advantage of being more amenable to high average power operation.

The choice of synchrotron radiation conversion is unlikely to increase the rate of muon production as the mean photon energy is lower for the synchrotron radiation photons than for the bremsstrahlung photons. The synchrotron radiation photons are more numerous, but only at low energies for which muon pair production is not energetically allowed. The angular distribution of the muons produced will be dominated by the spread of angles of the electrons in the primary beam as the average electron angle will be significantly larger than γ^{-1}.

The pair production rate in crystals is known experimentally [5] to be larger than in amorphous materials due to the coherent field effects. For photons of 100 GeV, the coherent production is a few times the Bethe–Heitler rate; however, for 20 GeV photons this effect increases pair production by only 10%. As the mean energy of bremsstrahlung photons is $\approx 20\%$ of the incident beam energy, pair production in a crystal will not significantly enhance the muon yield for a 100 GeV per beam collider. Hence, in the analysis that follows we restrict our attention to the use of a conversion bremsstrahlung production target.

3. Ionization cooling

In designing a collider one will inevitably seek a means of having as low an emittance as possible for the beams. One suggested means of cooling the muons (Fig. 3) is to pass the beam through a succession of alternating slabs of material (ionization cells) and rf-accelerating sections. In the ionization cells each of the muons gives up momentum along its particular trajectory, thereby losing transverse and longitudinal momentum. In the accelerating sections the longitudinal momentum is restored to the beam. Thus the transverse emittance of the beam is reduced in a manner analogous to radiation damping.

Neuffer [6] has shown that the ionization cooling of the transverse emittance is limited by beam heating due to multiple Coulomb scattering. If the transverse cooling is performed at an energy, E_c, using a medium for which the radiation length is X_R and the ionization loss rate is dE/dx, then the equilibrium, normalized emittance will be

$$\epsilon_{eq,n} = \frac{\beta_{cool}}{2} \frac{(14 \text{ MeV})^2}{m_\mu c^2 \left(X_R \frac{dE}{dx} \right)}, \tag{6}$$

where β_{cool} is the value of the beta function in the scattering medium. From Eq. (6) it follows that efficient cooling requires that one employ a very strong focusing system that brings the beam to a symmetric waste of small radius in the ionization medium. For high energy muons traversing a medium of density ρ (g/cm^3), of atomic number Z, and of atomic weight, A, the ionization loss rate can be approximated [7] by

$$\frac{dE}{dx} = \frac{DZ\rho}{A\beta^2} \left\{ \ln\left(\frac{2m_e\gamma^2\beta^2c^2}{I} - \beta^2 \right) \right\}, \tag{7}$$

where $\beta = v/c$, $D = 0.307$ and $I = 16Z^{0.9}$ eV. For materials with $Z \geq 6$, the radiation length may be approximated by

$$X_R(\text{cm}) = \frac{716.4A}{\rho Z(Z+1)\ln(287Z^{-1/2})}. \tag{8}$$

Multiplying Eqs. (7) and (8), one observes that the product $(X_R \, dE/dx)$ is independent of the density of the ionization medium and is greatest for small values of Z. Hence,

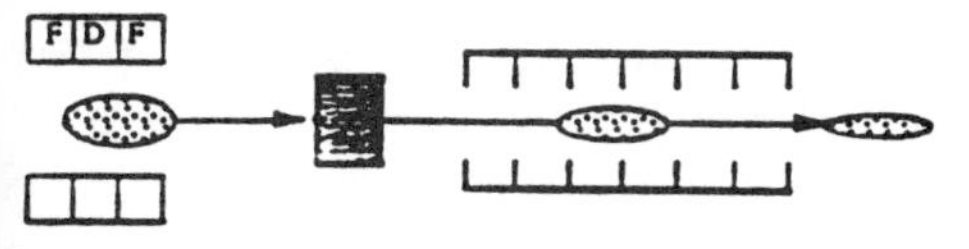

Fig. 3. Schematic of the basic components of an ionization cooling array: a strong lens to focus the beam, the ionization medium in which the particles lose both transverse and longitudinal momentum, and an accelerating structure to restore the longitudinal momentum of the beam.

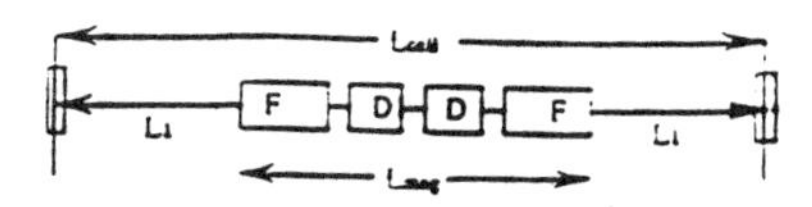

Fig. 4. Schematic of the triplet optics of an ionization cooling cell: the disks of the ionizing medium are shaded and have a half width of $\beta_{cool}/4$.

low Z media will be preferred over high Z media for ionization cooling. The length of the scattering medium in any individual ionization cell will have to be limited to $\beta_{cool}/2$.

As the momentum bite of the selected muons will be relatively large, one should consider using optics with second order chromatic corrections to focus the beam onto the ionization targets as otherwise the spot size will be unacceptably large. The focusing system may be a strong quadrupole triplet. Brown [8] has suggested a focal system that is suitable for scaling calculations. In this triplet transverse dimensions are scaled by a factor a_q, which is the aperture (radius) of the first quadrupole of the triplet; longitudinal dimensions are scaled by the "ideal" focal length, f,

$$f = \left(\frac{a_q}{B_q} (B\rho) \right)^{1/2}, \tag{9}$$

where B_q is the pole tip field in the first quadrupole, and $(B\rho)$ is the magnetic rigidity of the beam. For a beam of momentum p,

$$B(\text{T})\rho(\text{m}) = 3.3p(\text{GeV}/c)c. \tag{10}$$

The optical invariants of this particular triplet design are incorporated into the scaling equations that follow; the geometry of the design is illustrated in Fig. 4.

The free space from the focus to the first quadrupole, L_1, is $1.36f$; the length of the triplet, L_{mag}, is $3.13f$ and the length of the cooling cell, l_{cell} is $5.85f$. Without chromatic correction the value of β_{cool} for a beam with fractional momentum spread $\sigma_p(=\Delta p/p)$ is given by

$$\beta_{cool} = 5.92\sigma_p f. \tag{11}$$

With second order chromatic correction of the focusing optics the beta function can be reduced to

$$\beta_{cool} = 74.0(B\rho)\left(\frac{a_q}{fB_q} \right)\sigma_p^2. \tag{12}$$

In the analysis that follows we chose the corrected optics described by Eq. (12). As the cooling disks have a length, $\beta_{cool}/2$, each cooling cell produces an energy loss of e_{cell}, limited to

$$e_{cell} = \frac{\beta_{cool}}{2} \frac{dE}{dx}. \tag{13}$$

In optimizing the production/cooling scenario for the muon collider, one can now choose both the energy of muon production, E_μ, and the energy at which the cooling is performed, E_c. Note that although the equilibrium emittance of Eq. (6) does not depend explicitly on the cooling energy, the choice of E_μ, E_c, and the momentum acceptance will determine σ_p and thereby β_{cool} in the cooling lattice. Thus the choice of E_c will determine the transverse cooling coefficient, C_μ, via

$$C_\mu = \frac{\epsilon_{n,prod}}{\epsilon_{eq,n}} = \frac{r_{sh}\phi_{accept}\gamma_{prod}}{\epsilon_{eq,n}}. \tag{14}$$

The choice of E_μ will also influence the number of muons per bunch that are available to be injected into the collider as some of the muons will decay as they traverse the cooling lattice. If energy is replaced during the cooling process by accelerator cells with an average accelerating gradient G, the total path length in the cooling lattice, L_{cool}, will be

$$L_{cool} = \frac{E_c}{F_c}\left(\frac{L_{cell}}{e_{cool}} + \frac{1}{G}\right)\ln(C_\mu). \tag{15}$$

In Eq. (15) F_c is the overall packing fraction of the ionization and acceleration cells in the cooler lattice. F_c accounts for pumping ports, flanges, diagnostics, bending magnets, and sextupoles in the cooling lattice. If the number of muons per bunch that is injected into the cooling lattice is N_μ, then the number of muons per bunch available for injection into the collider will be

$$N_\mu^* = N_\mu \exp\left(-\frac{L_{cool}}{c\tau_\mu\gamma_c}\right), \tag{16}$$

where τ_μ is the muon lifetime at rest, and γ_c is $E_c/m_\mu c^2$.

Longitudinal cooling of the beam would allow smaller values of β_{cool} and consequently lower equilibrium emittances. Such reduction of the momentum spread can be accomplished by two means: 1) adiabatic damping by accelerating the muons prior to transverse cooling and 2) ionization cooling either in the transverse damper or in a separate damping structure. If the longitudinal cooling were limited to the ionization damping in the zero-dispersion cells of the transverse damper described above, the amount of acceleration, A_μ needed [4] to reduce the momentum spread by a factor $1/e$ would be

$$A_\mu = E_c\left(\frac{dE/dx}{E_c\dfrac{\partial^2 E}{\partial E_\mu\,\partial x}}\right) \approx 5E_c. \tag{17}$$

If the longitudinal cooling is done in a dispersive section, the energy spread might be reduced by $1/e$ with as little as $2E_c$ of total energy exchange.

As computed from Eq. (16), the path length of the muons in the cooler will typically be tens of kilometers, even if the packing fraction of the cooling lattice is large. A large packing fraction in conjunction with a high muon energy implies that the transverse emittance cooler should be constructed in the form of a recirculating linac such as CEBAF with high field bending magnets in the arcs and with as much as a few GeV per turn of acceleration in the straight, cooling sections.

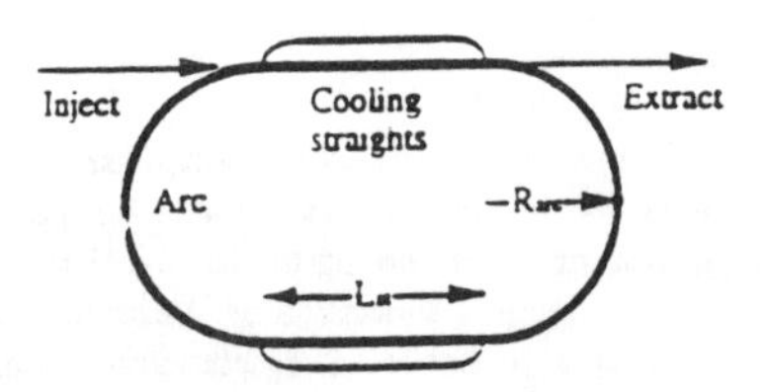

Fig. 5. A 2-in-1 muon cooling ring. Transverse coolers are in each of the straight sections. The gray sections have large aperture quadrupoles for the first stage of cooling; the black straights have stronger, small aperture quadrupoles.

At injection into the cooler the transverse emittance and the momentum spread of the muon beam will be large. Consequently the apertures of the quadrupoles in the cooling straights must be relatively large. One may envision a more effective form of cooler in which the emittance is reduced by an order of magnitude before injection into a final cooler which can have stronger, smaller aperture quadrupoles. As the value of β_{cool} can be much smaller in the second cooler, the equilibrium emittance could be much smaller than achievable in a single cooling ring. In such a scheme there is no need to duplicate the cost of the high field, dipole arcs. Rather the two coolers can share common arcs in a 2-in-1 arrangement illustrated in Fig. 5. The choice of straight-through or by-pass paths for the cooling stages can be selected to minimize the total path length of the muons in the coolers.

In the cooling ring the total length of the cooling cells plus re-acceleration cavities is $2L_{st}P_c$ where P_c is the packing fraction of ionization cells plus accelerator cells in the straight sections. As the overall packing fraction, F_c, is just $[2L_{st}P_c(2\pi\ R_{arc} + 2L_{st})^{-1}]$, the number of cooling cells, N_c is related to the average dipole field in the bends, $\langle B_d\rangle$ and the accelerating field, G, by

$$N_c = F_c\frac{2\pi(B\rho)}{\langle B_d\rangle}\left[1 - \frac{F_c}{P_c}\right]^{-1}\left[l_{cell} + \frac{e_{cell}}{G}\right]^{-1}. \tag{18}$$

Hence, the rf-system of the cooling ring must supply $N_c e_{cell}$ volts per turn. In damping the emittance of the muons by a factor C_μ, the muons must execute $[E_c N_c e_{cell}^{-1} \ln C_\mu]$ turns.

4. Collider considerations

The number of muons per bunch, N_μ^*, that circulate in the collider will be determined by the production efficiency, A_μ, by the charge, N_e in the electron bunch that strikes the production target, and by the path length through the cooling lattice. The number of electrons per bunch will be limited by the beam loading in the linac and by the design of the electron gun. The present SLAC gun (thermionic) produces bunches of 10 nC. If the electron beam emittance is not critical, the charge in the electron bunch can be raised to 20–30 nC. Bunches with as much as 50 nC may be produced with a photocathode gun, but such a large charge would lead to large beam loading and complications form the beam-breakup instability in an S-band linear accelerator.

The electron bunches will be produced in a macropulse of duration, τ_e, which is chosen to match the circulation period of the muons in the storage ring collider (Fig. 1). If the average dipole field in the ring is 3 T and if the muon energy is 100 GeV, then the circulation period will be 2 μs:

$$\tau_e = 2\mu s \times \left(\frac{3\,T}{B_{ave}}\right)\left(\frac{E_\mu}{100\,GeV}\right). \qquad (19)$$

If the number of bunches per macropulse is N_b, then the frequency of collisions in the collider will be

$$f_{coll} = N_b/\tau_e. \qquad (20)$$

To maintain the muon population in the collider the linac must be pulsed at a frequency of τ_μ^{-1}, where τ_μ is the muon lifetime as seen in the laboratory; at 100 GeV. $\tau_\mu = 2$ ms. Hence, the duty factor of the linac will be τ_e/τ_μ. The average power of the electron beam on the muon production target is, therefore,

$$P_{beam} = q\frac{N_b N_e}{\tau_e}\frac{\tau_e}{\tau_\mu}E_e = q\frac{N_b N_e}{\tau_\mu}E_e, \qquad (21)$$

where q is the electron charge.

The peak luminosity of the collider with muons with a geometrical emittance, ϵ, can be written as

$$L = \frac{N_\mu^{*} f_{coll}}{4\pi\epsilon\beta^{*}}, \qquad (22)$$

where $\gamma = E_\mu/\mu$ and β^* is the value of the beta function at the collision point. Combining Eqs. (5), (16), (20) and (22), and evaluating the average luminosity of a collider of repetition rate, R, we obtain the following expression for the average luminosity of the collider,

$$\langle L\rangle = \frac{A_\mu^2 N_e^2 N_b \gamma C_\mu}{4\pi r_{sh}\phi_{accept}\beta^*\tau_e}\left(\frac{\gamma}{\gamma_{prod}}\right)\exp\left(-\frac{2L_{cool}}{c\tau_\mu\gamma_c}\right) \times\left[1-\exp\left(\frac{2}{\tau_\mu\gamma R}\right)\right]\left(\frac{\tau_\mu\gamma R}{2}\right). \qquad (21)$$

The factor, γ/γ_{prod} implies that maximizing the luminosity argues for accepting the muons into the muon linac at an energy somewhat lower than the energy which maximizes A_μ. The factor, C_μ, accounts for the possibility of cooling the muons; if no transverse cooling is used, $C_\mu = 1$ and $L_{cool} = 0$. At 100 GeV a reasonable value of β^* can be assumed to be 1 cm, although smaller values are possible, limited by the muon bunch length and by the design of the detector. Hence, the length of the muon bunch should be less than 1 cm. Such a short pulse is assured, if the length of the electron beam pulses are ≈ 0.5 cm.

5. Examples and parametric dependences

One now has a complete set of equations with which to maximize the luminosity of the muon collider as a function of the electron beam power incident on the production target and other system characteristics. As a first step in examining parametric dependences, we formulate a "realistic", baseline scenario that does not employ cooling of the muon beam.

The CLIC group at CERN [9] has developed a design concept for a high power positron production target to operate at 500 to 750 kW, more than an order of magnitude greater than presently operating designs. For the "realistic", baseline scenario assume that this target design can be realized at 0.5 MW. Using a 50 GeV electron beam with 20 nC per bunch and one bunch per macropulse, one can produce muon bunches of ≈ 0.1 nC at 29 GeV with an acceptance of ±3% in the capture section of the muon linac. The geometrical emittance of the muon beam at 29 GeV will be 5 π mm-mrad. If the average dipole field in the collider is 3 T, the revolution period will be 2 μs. Hence, the collision frequency will be ≈ 0.5 MHz. Then for β^* equal to 1 cm, the luminosity of the muon collider at 100 GeV will be ≈ 2×10^{26} $cm^{-2}s^{-1}$. This scenario, which we will use as a base case for parametric studies, is summarized as column 1 in Table 1 along with a more optimistic case without cooling (column 3).

The improvements to the "realistic" and "optimistic" cases that would obtain from damping the transverse emittance of the muons via ionization cooling are shown in columns 2 and 4 respectively. A far more optimistic scenario (column 5), which also requires several technological inventions including considerable cooling of the muon beam, is discussed in Section 6. In all the examples with beam cooling the ionization media are beryllium disks of thickness of $\beta_{cool}/2$. The beam is re-accelerated in rf-cavities with rf-cavities operating with an average accelerating gradient of 17 MeV/m.

The effect of the choice of the electron beam energy on the production efficiency can be seen in Fig. 6, which displays the maximum luminosity versus the electron beam energy for the realistic scenario. In this calculation the number of electron bunches is varied to keep the beam power on the muon production target fixed at 0.5 MW. The momentum acceptance is fixed at ±3%; however, the value of muon energy accepted and the angular spread of muons accepted is varied so as to maximize the luminos-

Table 1
Characteristics of a 100 GeV × 100 GeV muon collider using electro-production. The repetition rate in all cases is 500 Hz. For multiple rings, B_q, a_q, β_{cool} refer to the second ring. The quantities with daggers require technological inventions

	"Realistic" no cooling	"Realistic" with cooling	"Optimistic" no cooling	"Optimistic" with cooling	Needs inventions
Production					
E_e (GeV)	50	50	50	50	50
P_{beam} (MW)	0.5	0.5	2	2	5 †
N_e (nC)	20	20	30	30	50 †
E_{accept} (GeV)	29	21	29	22	25
$(\Delta p/p)_\mu$(%)	±3	±3	±4	±4	±8
N_μ (nC)	0.1	0.1	0.2	0.18	0.6
ϵ_n (π m rad)	1.95×10^{-3}	2.2×10^{-3}	1.95×10^{-3}	2.2×10^{-3}	3.0×10^{-3}
Cooler					
E_{cool} (GeV)	-	40	-	45	100 †
Number of rings	0	1	0	1	2
F_{cool}	-	0.5	-	0.5	0.6
$\langle B_d \rangle$ in (arc) (T)	-	4.5	-	4.5	4.5
V_{ring} (GeV/turn)	-	0.95	-	1.2	3.2
C_{ring} (m)	-	491	-	553	1840
(B_q (T), a_q (cm))	-	(4, 1.5)	-	(6, 1.2)	(8†, 0.5)
β_{cool} (cm)	-	1.3	-	1.5	0.4
ϵ_{med} (π m rad)	1.7×10^{-3}	5.7×10^{-5}	1.9×10^{-3}	7.4×10^{-5}	1.6×10^{-5}
C_μ	1	38	1	28	136
Collider					
N_μ (nC)	0.1	0.068	0.2	0.14	0.35
N_{bunch}	1	1	2	2	2
B_{ave} (T)	3	3	4.5	4.5	6 †
$C_{collider}$ (m)	690	690	460	460	345
f_{coll} (MHz)	0.5	0.5	1.33	1.33	2
β (cm)	1	1	1	1	0.4
$(\Delta E/E)_{collider}$ (%)	±0.9	±0.8	±1.3	±1.0	±0.6
$\langle L \rangle$ (cm^{-2} s^{-1})	1.5×10^{26}	9.5×10^{26}	7.1×10^{26}	8.6×10^{27}	1.0×10^{30}

ity. As can be seen from Eqs. (2) and (22), the optimum acceptance energy will be a large fraction of the beam energy as the luminosity is quadratic in the conversion efficiency.

The scenarios employing production of muons at high initial energy (20–30 GeV) achieve relatively high luminosity at the expense of producing a muon beam with a relatively large (≈ 1%) momentum spread at the interaction point. If a much lower spread, say ±0.1%, were required for physics reasons, then the accepted muon energy, E_{accept}, would have to be reduced to ≈ 5 GeV. The luminosity is still maximized by maximizing the electron beam energy. Making this change in E_{accept} to the "realistic" scenario reduces the luminosity to ≈ 6×10^{24} cm^{-2}s^{-1}. As transverse cooling is accompanied by damping of the momentum spread, this consideration is not as severe in the scenarios with beam cooling.

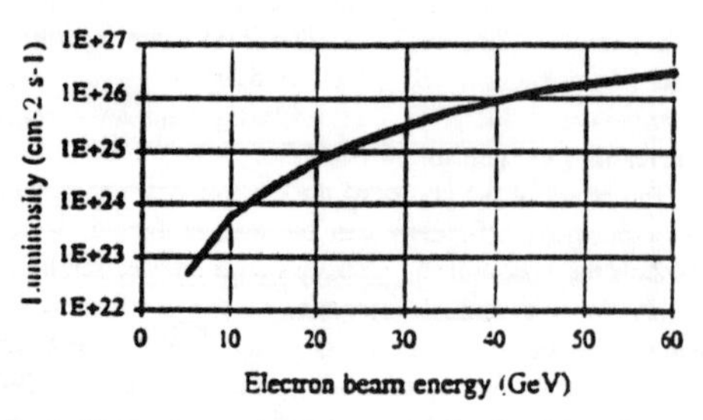

Fig. 6. The variation of collider luminosity with energy of the electron beam at the production target in the "realistic" scenario. The beam power is fixed at 0.5 MW.

The optimum energy for accepting the muons in the absence of cooling is 29 GeV. If instead we employ an ionization cooler, the optimum acceptance energy would be reduced to 21 GeV; a curve of the luminosity versus muon acceptance energy for the "realistic scenario" is given in Fig. 7. In this calculation cooling energy has been optimized but limited to ≤ 40 GeV.

Somewhat surprisingly, the higher the energy at which the muons are cooled, the higher the final luminosity. The reason is that the adiabatic damping of the energy spread permits a much smaller value of β_c. If an initial stage of

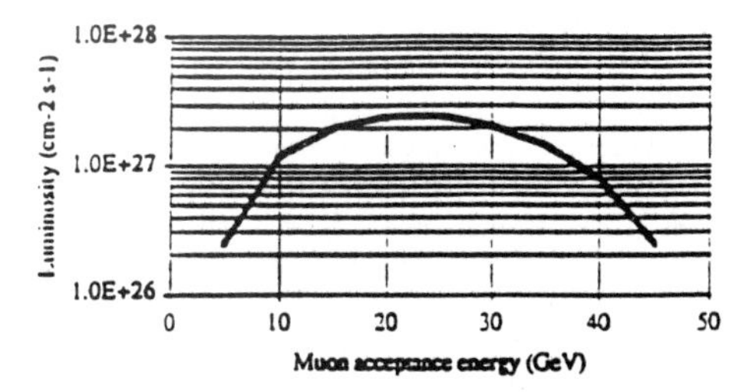

Fig. 7. Luminosity as a function of muon acceptance energy for the "realistic" scenario with ionization cooling.

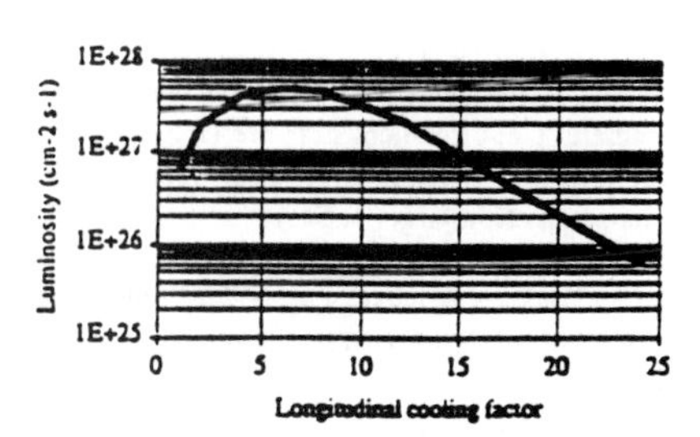

Fig. 8. Luminosity variation with longitudinal cooling for the "realistic" scenario.

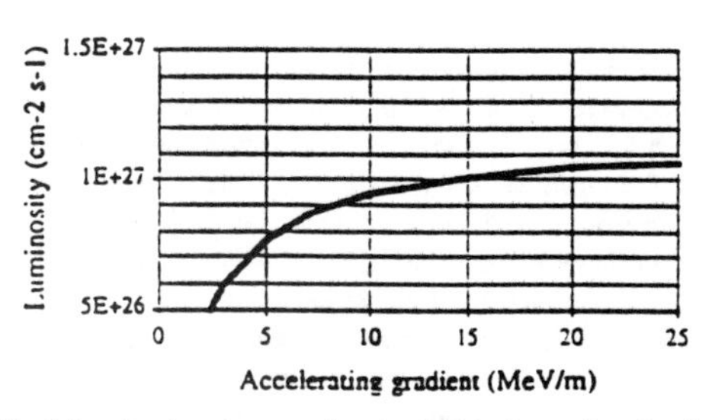

Fig. 9. Luminosity versus accelerating field in the cooling ring for the "realistic" scenario of Table 1.

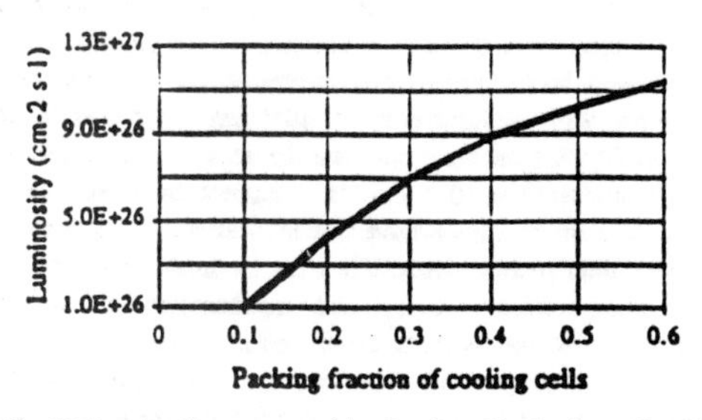

Fig. 10. Luminosity versus packing fraction of ionization cells and re-acceleration cavities in the cooling ring for the "realistic" scenario of Table 1.

ionization cooling were employed to reduce the energy spread, the optimum energy at which transverse cooling is performed could shift to a lower value. In the "realistic" example, the traverse cooling by a factor of 48 at 40 GeV requires an energy exchange of only $3.8 E_c$. From Eq. (17) one expects a slight improvement in C_μ from the damping of the energy spread in the zero-dispersion cells. Adding ionization cells in the dispersive sections of the ring as suggested in Ref. [6] could improve the luminosity significantly.

Fig. 8 illustrates the variation in luminosity for the "realistic" scenario with longitudinal cooling accompanying the transverse emittance damping. In this calculation cooling is done at the muon acceptance energy, 21 GeV so that no additional adiabatic reduction in energy spread is included. The field strength and aperture of the cooling channel optics is kept fixed. The decrease in luminosity as the cooling factor increases beyond six comes from the decay of the muon population as the transverse cooling path length increases to allow the beam to reach the equilibrium emittance.

As the muons must remain in the cooling lattice for hundreds of microseconds, it may not be possible to maintain an accelerating gradient of 17 MeV/m as assumed in the examples of Table 1. The consequence of reducing the gradient to allow for a lower power accelerating system in the cooling ring is displayed in Fig. 9. The degradation of the luminosity becomes especially large as the gradient falls below 10 MeV/m. As the number of muons in the ring is small, the beam loading in the cooling ring will be very small. One might consider the use of superconducting rf-cavities to keep rf-power requirements relatively small. Whether the superconducting cavities can function in the presence of radiation from the muon decay is uncertain.

A second characteristic of the ionization cooling lattice that can have a strong effect on the final luminosity of the collider is the packing fraction, F_c, of the ionization cells plus the rf-cavities in the cooling ring. Fig. 10 illustrates the variation of luminosity with F_c for the "realistic" case with cooling.

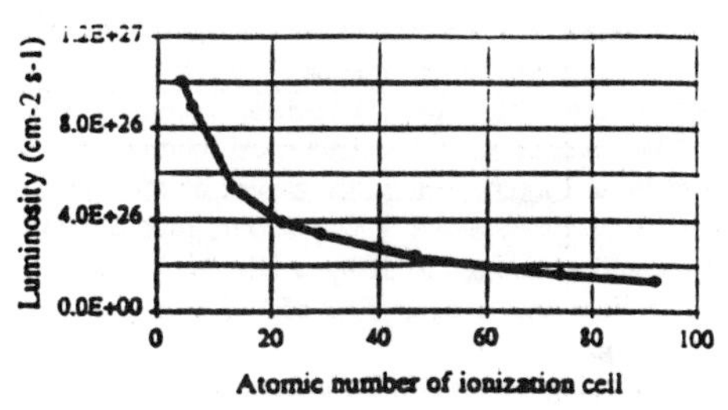

Fig. 11. Variation of luminosity with the choice of ionizing medium.

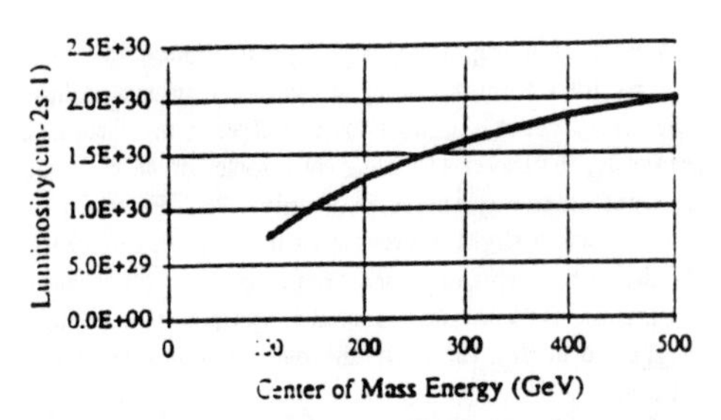

Fig. 12. Variation of luminosity with energy for a 250 GeV × 250 GeV muon collider with $\beta^* = 0.3$ cm.

Applying Eq. (7) through Eq. (17) to calculate the characteristics of a cooling system, we find that the luminosity varies with choice of the ionizing medium as shown in Fig. 11. Although the product X_R dE/dx is independent of density, the luminosity is sensitive to the density of the ionizing medium as the energy lost per cell depends on the density and thickness of the medium. For each of the points in Fig. 11 the appropriate density has been used. From this examination we confirm that the preferred ionization media are beryllium disks.

If one were to design the muon collider with a broader energy reach, for example from 100 to 500 GeV center of mass energy, one would hope to realize a higher luminosity at the higher energies as the geometrical emittance is reduced by adiabatic damping. The scaling of the luminosity, as shown in Fig. 12, is slower than linear. The calculation of Fig. 12 is based on the "Needs invention" scenario of Table 1. with β^* reduced to 0.3 cm.

In this scenario the momentum spread of the beams is largest at the lowest energy. Unfortunately, the width of a Standard Model Higgs is expected to be a rapidly increasing function of the Higgs mass with a value of 1 GeV/c^2 for $m_H = 100$ GeV/c^2. If the momentum spread were reduced at the lower energies to allow for a fine scan of the range from 100 to 200 GeV, the luminosity would fall off much more precipitously.

6. Prospects and conclusions

To obtain a muon collider with a luminosity of 10^{30} cm^{-2}s^{-1}, as desired for studies of the Higgs, one must adopt an extremely optimistic scenario (column 5 of Table 1) that includes several technological innovations (indicated by a dagger). Perhaps the easiest of these advances may be the development of very strong, precision dipoles that would enable one to design a relatively small storage ring collider with a dipole field of 6 T averaged over the entire ring.

Continuing advances in the technology of electron guns with photocathodes suggest that one may be able to obtain 50 nC bunches of electrons for injection in a S-band structure. Accelerating multiple bunches of such high charge in a S-band structure also presents difficulties. For a 50 nC, 15 ps bunch, the single-bunch beam loading in a SLAC structure operating at 20 MeV/m would be ≈ 20%. As the bunches in the macropulse are separated by hundreds of meters, multi-bunch beam breakup is not a problem. However, a head-to-tail momentum variation of 4% will be required for BNS damping of the single bunch, transverse, head-to-tail instability. Once this systematic variation is removed at the end of the electron linac, one would be left with a ±1.5% spread that must be handled by the focusing optics at the production target.

Extending the conceptual design of the CERN production target to a reliable, 5 MW design is likely to be very difficult. Of particular difficulty will be finding suitable accelerator components that can withstand the extremely high radiation environment near the target. Note that the highest power, production target in operation is the 33 kW positron production target at SLAC.

It is likely that the greatest challenge to the designer will be to find an efficient scheme for cooling the muon beam at a high initial energy. In scenarios that include beam cooling in a storage ring the momentum bite must be chosen to be consistent with the acceptance of the cooling lattice. As it should be possible to design a lattice with an acceptance of ±2%, cooling the muons at very high energy allows accepting a large momentum bite at the production target.

An idea of the scope of the project can be had by observing that in the realistic case the collider ring has a circumference of ≈ 690 m while the cooler rings (one each for the μ^+ and μ^-) have circumferences of ≈ 190 m. Operating with a gradient of 17 MeV/m, the electron linac would be 3 km long while the 20 GeV muon linac would have a length of 1.32 km. A clever design may be possible in which these same linacs could be used to accelerate the muons from the cooler ring up to the full 100 GeV per beam of the collider. In this case the major cost of the project would be the 70 GeV of S-band linac. The major complexity and technological risk is in the lattices of the cooling rings which use very high field, superconducting quadrupoles and dipoles.

In conclusion, one sees that even with optimistic assumptions, it is difficult to envision a high energy $\mu^+\mu^-$ collider which employs electro-production of muons functioning with a luminosity $> 10^{27}$ cm^{-2}s^{-1}. While the possibility of an electron-beam-driven muon collider with a luminosity ≈ 10^{30} cm^{-2}s^{-1} cannot be ruled out, it would require major advances in several of the primary constituent technologies. The areas for innovations include superconducting dipoles and quadrupoles, multi-kilo-ampere electron beam sources, and multi-megawatt muon production targets. Most critically, efficient means of both transverse and longitudinal of cooling the muon beams at high energy must be found and demonstrated, if suitably high luminosity is to be achieved.

References

[1] D.B. Cline, Opening presentation at the Workshop on High Energy Muon Colliders, Napa, California, December, 1992.

[2] W.R. Nelson, K.R. Case and G.K. Svensson, Nucl. Instr. and Meth. 120 (1974) 413.

[3] W.R. Nelson, Nucl. Instr. and Meth. 66 (1968) 293.

[4] W.A. Barletta, and A.M. Sessler, Radiation from fine self-focussed beams at high energy, in: High Gain, High Power Free Electron Lasers; eds. L. Bonifacio, L. De Salvo-Souza and C. Pellegrini (Elsevier, 1989).

[5] A. Belkacem et al., Experimental study of pair creation and radiation in Ge crystals at ultra-relativistic energies, in: Relativistic Channeling., eds. Carrigan and Ellison, Nato ASI Series B, Vol. 165 (Plenum).

[6] D. Neuffer, Part. Accel. 14 (1983) 75.

[7] Particle Data Group, Phys. Rev. D45 Part 2 (1992).

[8] K. Brown, Private communication as reported in R. Palmer, Interdependence of Parameters for TeV Linear Colliders, Proc. Workshop on New Developments in Particle Accelerator Techniques, Orsay, (June 1987).

[9] P. Sievers, Proc. Workshop on Heavy Quark Factories, Courmayeur, Italy, December, 1987.

A muon collider scenario based on stochastic cooling ☆

Alessandro G. Ruggiero

Department of Advanced Technologies, Brookhaven National Laboratory, Upton, Long Island, New York 11973, USA

The most severe limitation to the muon production for a large-energy muon collider is the short time allowed for cooling the beam to dimensions small enough to provide reasonably high luminosity. The limitation is caused by the short lifetime of the particles that, for instance, at the energy of 100 GeV is of only 2.2 ms. Moreover, it appears to be desirable to accelerate the beam quickly, with very short bunches of about a millimeter so it can be made immediately available for the final collision.

This paper describes the requirements of single-pass, fast stochastic cooling for very short bunches. Bandwidth, amplifier gain and Schottky power do not seem to be of major concern. Problems do arise with the ultimate low emittance that can be achieved, the value of which is seriously affected by the front-end thermal noise.

Since mixing within the beam bunches is completely absent, methods are required for the regeneration of the beam signal with external and powerful magnetic lenses. The feasibility of these methods are crucial for the development of the muon collider. These methods will be studied in a subsequent report.

1. Introduction

In the quest for the Higgs bosons, a muon collider may be perceived as the experimental device more affordable and more feasible than electron–positron or very large hadron colliders [1–3]. Muons have a mass ten times lighter than protons and are therefore easier to be steered on circular trajectories. On the other hand their mass is a hundred times greater than electrons and their motion is considerably less affected by the synchrotron radiation. Muons are elementary lepton particles, with no internal structure. Like the electrons, they have obvious advantages over the hadron counterpart when they are used as the main projectiles for the production of the Higgs bosons. Moreover, because of their larger mass, they are also better suited than the electrons themselves, due to a considerably larger propagator constant. Unfortunately, muons do not exist in nature and they have to be produced with the only technique we know these days: impinging an intense beam of protons or electrons on a target. Like in the case of production of antiprotons, in order to make the beam of some use for the subsequent collisions, muons also have to be collected and cooled to a sufficiently high intensity and small dimensions before they can be accelerated and injected in the collider proper. To make the situation more complicated there is also the fact that muons are intrinsically unstable particles with a very short lifetime. Accumulation, acceleration and cooling are then to be executed extremely fast if one requires that a large fraction of the particle beam survives to the collision point.

This paper deals with the requirement of betatron stochastic cooling which is to be very effective and fast. The situation being described is altogether different from the usual encountered with coasting beams [8,9]. Now the beam is tightly bunched at a very large frequency. The bunches are very narrow, having a length which is considerably smaller than the wavelength of the bandwidth of available electronic amplifiers. Thus a different method is to be developed based on the correction of the stochastic signal for all particles at the same time in one single-step. The fundamental limitation remains of the ultimate value of the final emittance that can be achieved. The limitation is caused by the thermal noise at the front-end of the amplifier.

We begin by reviewing a possible scenario of a muon collider in Section 2. This scenario assumes that stochastic cooling is done at maximum energy. The performance of the collider luminosity is evaluated in Section 3, where special emphasis is put on the effects of the beam lifetime and on the betatron emittance reduction. The requirements for the stochastic cooling proper are exposed in Section 4, where we underline the particular situation we are facing of very short beam bunches. The analysis of the cooling device itself follows in Section 5. The goal is the determination of an equation which gives the evolution of the betatron emittance with time. This is described essentially by two parameters: the cooling rate and the diffusion rate due to the thermal noise which is by far more important

☆ Work performed under the auspices of the U.S. Department of Energy.

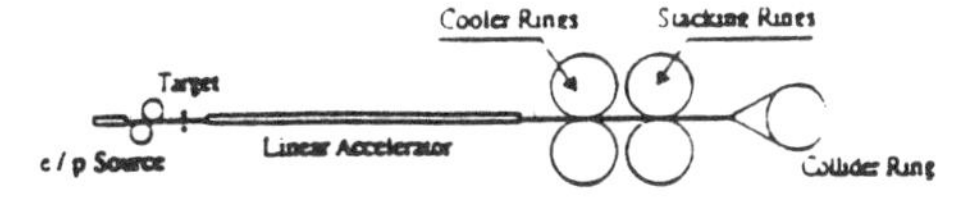

Fig. 1. A conceptual layout of the muon collider.

than any other diffusion process, for instance beam Schottky power. The derivation of the equation for the evolution of the betatron emittance is given in Section 6. A discussion leading to optimization considerations of the cooling process is presented in Section 7. Finally, an application to the muon collider is worked out in Section 8, where the dependence of the performance with a dynamical gain less than optimal and with the bunching frequency is also investigated. It is found that a luminosity of about 10^{34} cm^{-2} s^{-1} can be achieved at the very most. Conclusions are given in the last Section 9.

2. The muon collider

We expose below a possible scenario of a muon collider. A layout of the scheme is shown in Fig. 1. An intense source of either protons or electrons, at the energy of few tens of GeV with an average current of few hundreds of μA, is provided with a conventional fast cycling accelerator [4–8]. In the case of electrons, these are bunched at a large frequency, for instance 3 GHz, and are accelerated in a large-gradient linear accelerator. In the case of protons, after acceleration, the beam is debunched in a stretcher ring and then rebunched also at the frequency of 3 GHz. In either case, the primary beam is made to impinge on a sequence of targets for the production of muon pairs via decay of π mesons. The secondary beam, produced at an energy of about 1 GeV, will be collected with a large production angle yielding a normalized total emittance of about 100 π mm mrad, and with a large momentum bite of few percent. An average intensity of about 50 nA per each component of the pair production is expected, corresponding to a yield of approximately 2×10^{-4} μ-pairs per primary particle. The μ-beams are also tightly bunched at the frequency of 3 GHz so that there are only few particles per bunch (around 100). The lower number of particles per bunch is, as we shall see, a requirement for fast stochastic cooling.

Both types of beam, μ^+ and μ^-, after a preliminary bunch rotation to reduce the momentum spread, are accelerated in a large-gradient linear accelerating structure, operating also at 3 GHz, to the final energy which is in the range of 100 to 1000 GeV. At the end of the acceleration, each beam is transferred to a storage ring where fast stochastic cooling is done to reduce the betatron emittance to the final value. Each beam is then taken to a stacking ring of about the same size, where several cooled beam pulses are stacked side-wise in the momentum phase space. The stacking procedure is carefully done to avoid lengthening of the bunches and increasing of the betatron emittance. With this operation the number of particles per bunch will increase by the number of pulses being stacked. This is required to boost the magnitude of the luminosity of the collider [7]. At the end of stacking, both beams are extracted from their respective stacking ring and transferred to the collider ring proper where they are made to collide and exploited for experimentation. It is to be noticed that all storage rings involved (five, as shown in Fig. 1) operate at the same energy; thus it is expected that they have also about the same size.

Because of the relatively short lifetime of the muon particles, it is obvious that all the operations which have been described above are to be executed very fast.

3. The luminosity performance

What follows is a discussion of the luminosity performance of the muon collider. The average luminosity is given by the following expression

$$L = MN_0^2 f_{\rm bunc} F\gamma/4\pi\epsilon_n \beta^*, \tag{1}$$

where $N_0 \sim 100$ is the initial number of particles per bunch at the moment of production, $f_{\rm bunc} \sim 3$ GHz is the beam bunching frequency during acceleration and stochastic cooling, γ is the energy relativistic factor, $\epsilon_n \sim 25\ \pi$ mm mrad is the initial rms normalized emittance, at the moment of production, and β^* is the focussing amplitude parameter at the interaction point, which for a multiple pass in a collider ring can be as low as 10 cm and for the single-pass mode, where the requirements on the lattice focussing can be relaxed, it is about 1 cm. For an efficient mode of operation, it is important that the bunch length during collision is sufficiently small when compared to β^*. This is obtained with the large bunching frequency. M is the number of beam pulses which are stacked in the momentum phase space of the stacking ring. It is to be noticed that the current $I_\mu = N_0 e f_{\rm bunc} \sim 50$ nA is a constant equal to the average current of each muon beam at the moment of production. The luminosity expression above shows clearly the advantage of increasing the number of particles per bunch from N_0 to MN_0 with momentum stacking [7]. Finally F is a form factor which includes the losses of particles due to the short lifetime and the emittance reduction due to the stochastic cooling.

We can write the form factor F as the product of many other factors:

$$F = F_{\rm acc} F_{\rm sto} F_{\rm stac}^2 F_{\rm col}, \tag{2}$$

where $F_{\rm acc}$ is the square of the beam survival fraction after acceleration, $F_{\rm sto}$ reflects the effects on the luminosity of the reduction of the betatron emittance due to stochastic cooling and of the square of the fraction of beam survival after cooling, $F_{\rm stac}$ is the beam survival fraction after

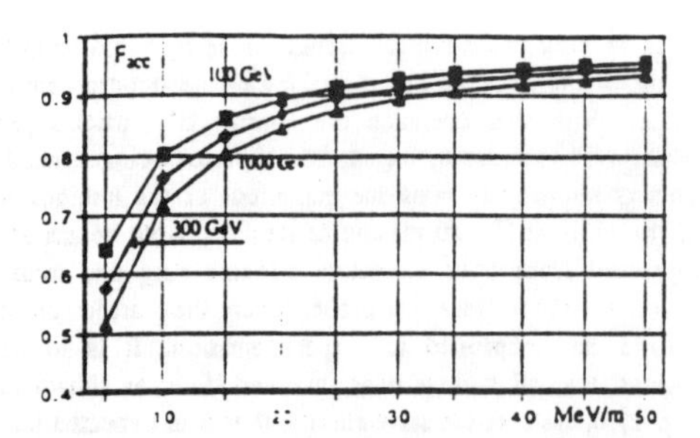

Fig. 2. The acceleration survival factor F_{acc} vs. acceleration gradient.

momentum stacking, and F_{col} represents the beam losses during collision.

The overall survival ratio after acceleration can be easily calculated by integrating the instantaneous beam survival over the acceleration cycle, combined with the fact that the lifetime increases linearly with γ. Taking into account the contribution of two beams to the luminosity, we have

$$F_{acc} = (E_{init}/E_{final})^{2E_0/cG\tau_0}, \tag{3}$$

where $E_{init} \sim 1$ GeV is the beam kinetic energy at production, E_{final} is the final energy in the collider, $E_0 = 106$ MeV is the rest energy, $\tau_0 = 2.2$ µs the lifetime of the muon at rest and G is the accelerating gradient in the linear accelerator. The behavior of F_{acc} versus the accelerating gradient is shown in Fig. 2 for various final energies. It is seen that losses are reduced with a larger accelerating gradient and a lower beam energy.

The second factor F_{sto} represents the combined effect of the beam losses during the period of time T_{sto} the beam spends to be cooled and the reduction of the betatron emittance with stochastic cooling

$$F_{sto} = (\epsilon_0/\epsilon_x)\exp(-2T_{sto}/\gamma_{final}\tau_0), \tag{4}$$

where (ϵ_0/ϵ_x) is the ratio of the initial to the final betatron emittance.

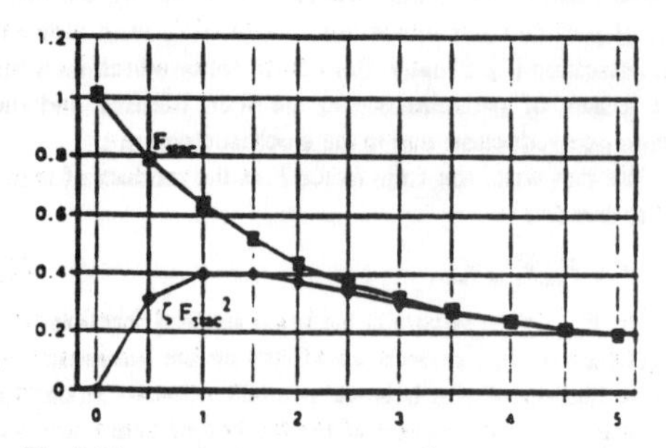

Fig. 3. The survival factors F_{stac} and F_{stac}^2 vs. the parameter.

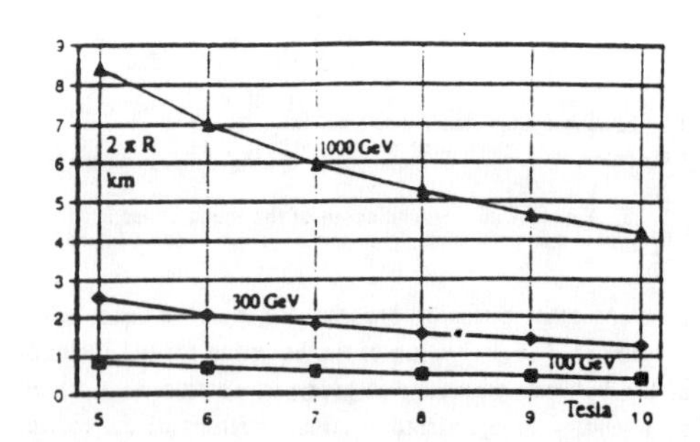

Fig. 4. Circumference of storage ring vs. bending field B for $\eta = 0.5$.

The factor F_{stac} represents the survival of the beam integrated over the M pulses being stacked in the stacking ring over a period of time T_{stac}. By denoting with f_0 the circulating frequency, we have $T_{stac} = M/f_0$ and

$$F_{stac} = [1 - \exp(-T_{stac}/\gamma_{final}\tau_0)]/(T_{stac}/\gamma_{final}\tau_0). \tag{5}$$

This quantity is plotted in Fig. 3 versus the parameter $\zeta = T_{stac}/\gamma_{final}\,\tau_0$.

The circumference $2\pi R$ of any of the storage rings can be estimated from the beam energy γ_{final}. Allowing a bending-magnet packing-factor η, denoting with B the banding field in T, and expressing the circumference in meter gives

$$2\pi R = 2.22\gamma_{final}/\eta B \tag{6}$$

from which we derive the revolution frequency

$$f_0 = (135\text{ MHz})\eta B/\gamma_{final}. \tag{7}$$

Thus is seen that

$$\zeta = 0.0034\, M/\eta B, \tag{8}$$

which does not depend on the beam energy. If we take $B = 6$ T and $\eta = 0.5$ then $\zeta = 0.0011M$. To avoid excessive luminosity losses, it is seen from Fig. 3 that at most we can allow the momentum stacking of $M \sim 900$ pulses. The quantity ζF_{stac}^2 is also plotted in Fig. 3 which shows that at most $MF_{stac}^2 \sim 350\eta B$.

For completeness we display the plots of the circumference $2\pi R$ and of the revolution frequency f_0, respectively

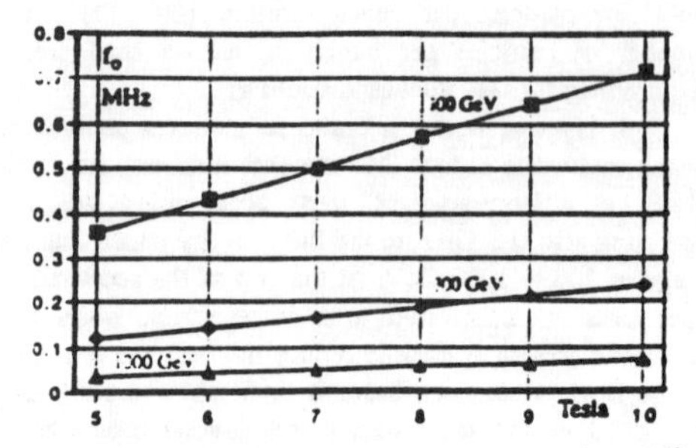

Fig. 5. The revolution frequency vs. bending field B ($\eta = 0.5$).

in Figs. 4 and 5. versus the bending field B and for various beam energy values.

Finally, F_{col} represents the loss of average luminosity due to the particle losses during collision which we assume takes a period of time T_{col}. This factor has an expression similar to the one for F_{stac} given by Eq. (5), except that T_{stac} is replaced by $2T_{col}$. It is seen that the best case is given by the single pass mode where after one interaction both beam bunches are immediately disposed. This mode is also more favorable because it does not require a complete collider ring and the final focus may correspond to a lower value of β.

4. Requirements on stochastic cooling

We can express the actual average luminosity in terms of the ideal value L_0 without stochastic cooling, without momentum stacking and for an infinitely long muon lifetime

$$L = MFL_0, \tag{9}$$

where

$$L_0 = N_0^2 f_{bunc} \gamma / 4\pi\epsilon_n \beta^*. \tag{10}$$

With the values of the parameters given in the previous two sections and for the single-pass mode, which is the most favorable, we have $L_0 = 1.0 \times 10^{15}\gamma$ cm^{-2} s^{-1}. Correspondingly, to achieve a luminosity of about $1.0 \times 10^{27}\gamma$ cm^{-2} s^{-1}, which is required for the high energy physics experimental exploitation of the collider, we need that the overall enhancement factor $MF \sim 1 \times 10^{12}$. Since at most $MF_{stac}^2 \sim 1000$, even assuming $F_{acc}F_{col} \sim 1$, we need $F_{sto} \sim 1 \times 10^9$, which is a very large requirement for the betatron stochastic cooling. The requirement is independent of the beam energy; a normalized emittance of $25 \times 10^{-9}\pi$ mm mrad is required at all energies. Assuming a cooling period T_{sto} short with respect to the muon lifetime $\gamma_{final}\ \tau_0$, one requires a reduction of the betatron emittance by nine orders of magnitude (!). Thus the fundamental question concerns the ultimate emittance that can be realistically achieved at the end of cooling.

There are major differences between stochastic cooling for the case of bunched beams we are investigating here and the usual approach for coasting beams encountered, for instance, during production and accumulation of antiproton beams [9,10]. The muon bunches have a length considerably smaller than the shortest wavelength in the frequency bandwidth of the system. It is indeed a good approximation to assume that the beam bunches have no longitudinal extension, and that all the particles are distributed on a disk with a center slightly displaced from the axis of the pickups. The beam current signal is therefore highly organized and coherent. The transverse beam position, on the other hand, has a very stochastic behavior. Because of the low number of particles, there is a random fluctuation of the beam centroid that can be related statistically to the overall transverse beam size. For the same reasons, the internal motion can be completely ignored and no mixing occurs between the detection of the beam signal at the pickups and the application of the deflection at the kickers. Moreover, the conventional analysis in the frequency domain [9,10] would be hardly applicable, since the Schottky bands remain well separated from each other. Actually from the current point of view, it is improper to refer to the beam Schottky signal. As pointed out already, the stochastic behavior appears only in the transverse displacement of the beam centroid.

The overall system, between pickups and kickers, including the power amplifier, has a bandwidth W large enough to detect and correct displacement of individual bunches. If we assume a bandwidth W extending over an octave, where the frequency at the upper end is twice than the frequency at the lower end, an optimum is given by choosing the bandwidth equal to an integer m times 2/3 of the bunching frequency. For instance, if the beam is bunched at 3 GHz, the bandwidth could be 2 GHz ($m = 1$) extending from 2 to 4 GHz. A larger bandwidth, for instance from 4 to 8 GHz ($m = 2$) or from 6 to 12 GHz ($m = 3$), is of course also possible. In this mode of operation, it is possible to process the pickup signals to allow complete rejection of a beam bunch signal on another bunch.

The lack of mixing causes a serious limitation on the effectiveness of stochastic cooling. Once the initial beam displacement has been corrected, there is no more signal from the beam that can be used. Thus everything is done in a single step with a relatively small reduction of the beam size. Between steps, the signal from the beam has to be regenerated, for instance by rearranging the particle mutual position with the aid of powerful magnetic lenses. We shall assume below that this is indeed the case. We shall investigate in a separate report the amount of particle rearrangement required and how this can be realized in practice.

Because of the low number of particles per bunch, after amplification, the beam Schottky power is not expected to be excessive. In order to obtain a very fast cooling, one requires to optimize the overall gain to correct the instantaneous beam displacement immediately in one single step. This may require a very large electronic gain. A more serious problem is associated with the thermal noise at the front-end of the amplifier, which will set a limitation of the final beam transverse dimension. The design of the cooling device is to be optimized to reduce this limitation.

5. Analysis of the cooling device

Consider a very narrow bunch made of N particles all with the same electric charge. The bunch is periodically traversing a beam position pickup made of two parallel striplines each of length l and separated by a distance d. The striplines are shorted at one end and terminated at the upstream end to their characteristic impedance R_p. During the occurrence of a traversal assumed at the instant $t = 0$, the bunch current can be represented as a pulse of zero duration, proportional to the average displacement $\bar{x}$, that is

$$I_p(t) = Ne(\bar{x}/d)\delta(t). \tag{10}$$

This current leaves at the upstream end of one pickup a voltage signal given by

$$V_p(t) = Ne(\bar{x}/2d)R_p[\delta(t) - \delta(t - 2l/c)], \tag{11}$$

that is a voltage pulse occurring simultaneously to the current pulse, followed by another at the delay of $2l/c$. Since the beam bunch duration is considerably shorter than the delay between the two voltage pulses, only the first pulse is relevant to our analysis and we shall ignore the second one which we assume can be disposed properly without disruptions to the subsequent bunches. The voltage signal is then filtered by the bandwidth of the system, mostly caused by the power amplifier, and properly amplified by the linear gain A. The resulting voltage is an oscillating and decaying signal of which only the front-end is of relevance here since it constitutes the part that is to be applied in phase with the bunch at the location of the kickers. The amount of the properly correspondent voltage is simply

$$V_A = Ne(\bar{x}/2d)R_p AW. \tag{12}$$

In the case there is only one pickup and one kicker, this is also the voltage that would appear across the kicker assuming an ideal impedance matching all along the transfer of the signal. To include the case of more pickups and kickers, we modify Eq. (12) as follows:

$$V_k = Ne(\bar{x}/2d)AW\sqrt{R_p R_k n_p / n_k}, \tag{13}$$

where $n_{p(k)}$ is the number of pickups (kickers) and R_k is the characteristic impedance of kickers. This expression gives the voltage across one single kicker. We are assuming here that both the pickups and the kickers are closely packed and that they extend over a length of the storage ring where the beam position $\bar{x}$ and the lattice functions do not vary appreciably. Moreover kickers have exactly the same geometrical configuration and size of the pickups.

Assuming small deflection angles and full correlation among the kicks, the total deflection angle each particle will be subject to at the traversal of the kickers is

$$\theta_S = eV_k l n_k / \beta^2 Ed, \tag{14}$$

where E is the particle total energy and βc is the velocity. Statistically, over many revolutions, the following relation holds between the average beam displacement $\bar{x}$ and the rms beam size σ:

$$\sigma^2 = N\bar{x}^2. \tag{15}$$

We prefer writing the expression for the deflection θ_S as follows:

$$\delta_S = g_0 \sigma / d, \tag{16}$$

where

$$g_0 = \frac{lA}{2\beta^2 Ed}\sqrt{Ne^4 W^2 n_k n_p R_k R_p}. \tag{17}$$

Another signal is induced to the kickers which is random and independent of the particle position. The finite temperature of the terminating resistors of the loop and of the preamplifiers creates at the input to the preamplifier a signal of power

$$P_T = k_B(T_A + T_R)W. \tag{18}$$

where $k_B = 8.6171 \times 10^{-5}$ eV/K is the Boltzmann constant, T_A is the equivalent temperature of the amplifier and T_R is the temperature of the resistor. Proceeding in the same way as for the beam Schottky signal, we calculate the total deflection angle due to the thermal noise:

$$\theta_T = \frac{elA}{\beta^2 Ed}\sqrt{n_k R_k P_T}. \tag{19}$$

To assess the effectiveness of stochastic cooling a useful parameter is the ratio of Schottky to the thermal power

$$S = \theta_S^2/\theta_T^2 = Ne^2 W^2 n_p R_p \sigma^2 / 4d^2 P_T. \tag{20}$$

It is seen that this ratio increases linearly with the bandwidth. This result is different from the one it was derived for coasting beams [10], in which case the ratio is independent of the bandwidth. The difference is due to the fact that now the current distribution of the beam bunch is highly organized and does not exhibit a stochastic behavior.

6. The equation for the evolution of the beam emittance

Both Schottky and thermal power kicks apply simultaneously when a particle is crossing the location of the kickers. Let us calculate the effect of both kicks to the beam emittance which we can define as follows

$$\epsilon = \sum_i \left(\gamma x_i^2 + 2\alpha x_i x_i' + \beta x_i'^2\right)/N. \tag{21}$$

Here α, β and γ are the lattice Twiss parameters.

At the kickers each particle receives the same kick, that is $x_i' \to x_i' + \theta_S + \theta_T$. The corresponding emittance change is

$$\Delta\epsilon = 2(\alpha\bar{x} + \beta\bar{x}')_k(\theta_S + \theta_T) + \beta_k(\theta_S + \theta_T)^2, \tag{22}$$

where the subscript k(p) denotes that the corresponding quantity is evaluated at location of the kickers (pickups). $\bar{x}$ and $\bar{x}'$ are the average values of the particle positions and angle. We are interested in the expectation value of $\Delta\epsilon$

over many revolutions. In this case, since there is no correlation between particle position and thermal noise, the previous equation reduces to

$$\langle \Delta\epsilon \rangle = 2(\alpha\bar{x} + \beta\bar{x}')_k \theta_S + \beta_k \theta_S^2 + \beta_k \theta_T^2. \quad (23)$$

The betatron emittance is also defined as

$$\epsilon = (\sigma^2/\beta)_{p,k}. \quad (24)$$

At the same time it can be proven that

$$\langle (\alpha\bar{x} + \beta\bar{x}')_k \bar{x}_p \rangle = -\langle \bar{x}_p^2 \rangle \sqrt{(\beta_k/\beta_p)} \sin\psi_{pk}, \quad (25)$$

where ψ_{pk} is the betatron phase advance between pickup and kicker. Manipulating some of the previous equations gives

$$\langle \Delta\epsilon \rangle = -(2g \sin\psi_{pk} - Ng^2)\epsilon + \beta_k \theta_T^2 \quad (26)$$

with the dynamical gain

$$g = g_0 \sqrt{\beta_k \beta_p / Nd^2}. \quad (27)$$

Finally, the beam emittance evolution is described by the following equation

$$d\epsilon/dt = -\lambda\epsilon + D, \quad (28)$$

where the cooling rate

$$\lambda = n_S f_0 (2g \sin\psi_{pk} - Ng^2) \quad (29)$$

and the diffusion coefficient

$$D = n_S f_0 \beta_k \theta_T^2. \quad (30)$$

with f_0 the revolution frequency. In deriving these equations we have assumed a total of n_s identical cooling systems in the storage ring. In the following we shall assume that the distance between pickups and kickers is adjusted so that $\sin\psi_{pk} = 1$.

7. Optimization of the cooling performance

An optimum cooling rate is obtained by setting $g = 1/N$ and is

$$\lambda_{opt} = n_S f_0 / N. \quad (31)$$

which corresponds to correcting the instantaneous beam bunch displacement in one single step. At the same time we can also derive the required amplifier gain

$$A = \frac{2\beta^2 Ed^2/lN}{\sqrt{e^4 W^2 n_k n_p R_k R_p \beta_k \beta_p}}, \quad (32)$$

which decreases linearly with the bandwidth and the number of particles in the bunch.

The equilibrium value of the emittance for this case is

$$\epsilon_x = D/\lambda_{opt} = N\beta_k \theta_T^2. \quad (33)$$

By combining some of the equations we obtain for the equilibrium emittance

$$\epsilon_x = \frac{4d^2 P_T}{Ne^2 W^2 n_p \beta_p R_p}, \quad (34)$$

which, similarly to the amplifier gain A, also exhibits the same dependence with bandwidth W and number of particles N. It can be seen that the equilibrium emittance corresponds to the situation where the ratio of the Schottky to thermal power $S = 1$.

It is to be noticed that an optimum bandwidth is related to the bunching frequency f_{bunc} by the relation

$$W = 2mf_{bunc}/3, \quad (35)$$

where m is a positive integer. Also the quantity

$$I_\mu = Nef_{bunc} \quad (36)$$

is the average muon current essentially equal to the one produced at the target. Thus when these relationships are taken into account we have for the amplifier gain

$$A = \frac{3\beta^2 Ed^2/I_\mu m}{\sqrt{e^2 n_k n_p R_k R_p \beta_k \beta_p}} \quad (37)$$

and for the equilibrium emittance

$$\epsilon_x = \frac{6d^2 P_T/W}{mn_p \beta_p e I_\mu R_p}. \quad (38)$$

Both of these expressions show the same dependence on the bandwidth factor m and on the average beam current. Noticing that the thermal power P_T is proportional to W, it is seen that both A and ϵ_x do not depend explicitly on how the beam is bunched. On the other hand the cooling rate λ depends very strongly with the number N of particles per bunch.

8. An application of the optimal system

As an application of these expressions we take the following values:

$d = 1$ cm,

$\beta_p = \beta_p = 200$ m,

$n_p = n_k = 1024$,

$R_p = R_k = 100\ \Omega$.

We chose $m = 3$, that is a bunching frequency $f_{bunc} = 3$ GHz, corresponding to a bandwidth $W = 6$ GHz ranging between 6 and 12 GHz. The length of the pickups is adjusted to match the bandwidth according to $l = 6$ cm/W (GHz) so that for $m = 3$ it is $l = 1$ cm. We assume also a bending field $B = 6$ T and a packing factor in the storage ring $\eta = 0.5$. Finally we set the temperature of the amplifier and resistor $T_A = T_R = 1$ K which is very likely an unrealistic value, for which moreover we cannot really

Table 1
Stochastic cooling performance

Beam energy, GeV	100	300	1000
$2\pi R$, m	700	2100	7000
n_s	8	24	80
$1/\lambda$, ms	0.030	0.030	0.030
A	1×10^9	3×10^9	1×10^{10}
$\epsilon_n = \gamma\epsilon_x$, π mm mrad	32	96	320
L_0, cm^{-2} s^{-1}	1×10^{18}	3×10^{18}	1×10^{19}
MF	1000	300	100
L, cm^{-2} s^{-1}	1×10^{21}	1×10^{21}	1×10^{21}
Ng_{max}	0.0068	0.0023	0.0007
f_{max}	40	120	400
L_{max}, cm^{-2} s^{-1}	4×10^{22}	1.2×10^{23}	4×10^{23}

foresee the behavior of the thermal noise at the front end. The summary of the results of our calculations are shown in Table 1, where we have also assumed the optimum gain $g = 1/N$.

To be observed is the increase of the circumference of the storage ring with the beam energy, and that we let the number n_s of cooling systems vary proportionally. As a consequence, the cooling time $1/\lambda$ is constant with energy, whereas the amplifier gain A and the equilibrium emittance ϵ_n increase linearly with energy. As one can see, even at the very low temperature of 1 K, thermal noise dominates over the beam signal, and the equilibrium emittance is just about comparable to the initial beam emittance at the energy of 100 GeV. For larger energies there is actually stochastic heating accompanied by an increase of the beam emittance.

Since for the optimum gain the cooling time is 0.03 ms which is considerably shorter than the beam lifetime, it is reasonable to lower the amplifier gain. The results are shown in Figs. 6–8 where the cooling rate, the amplifier gain and the equilibrium emittance are plotted versus the dynamical gain $g/g_{opt} = Ng$.

It is seen that, as the gain g is lowered, both the amplifier gain and the equilibrium emittance reduce also, but on the other hand, unfortunately, the cooling time

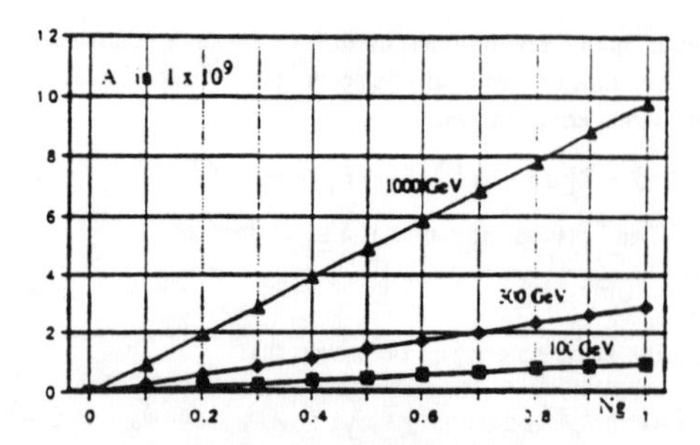

Fig. 7. Amplifier gain vs. dynamical gain Ng.

increases. If the increase is too large then the particle losses would also be too large.

The luminosity depends on the gain g only through the F_{sto} factor. We can chose an expression of the luminosity which shows the explicit dependence on g as follows:

$$L = f(Ng)L_{opt}, \tag{39}$$

where L_{opt} is the luminosity obtained with the parameters shown in Table 1, that is for the optimum gain $g = 1/N$ and

$$f(u)[(2-u)/u]\exp[-k/(2u-u^2)], \tag{40}$$

with

$$k = 2/(\gamma\tau_0\lambda_{opt}) \ll 1. \tag{41}$$

The first factor of Eq. (41) represents the reduction of the betatron emittance and the exponential factor represents the beam loss. It can be seen that the faction $f(Ng)$ has a maximum for $Ng_{max} \sim k/2$ where it takes the value $f_{max} \sim 1/k$. Thus at most the luminosity can be increased by $f_{max} \sim 0.04\gamma$, that is an increase which is proportional with the beam energy as required. The values of Ng_{max} and of f_{max} with the corresponding increased luminosity are also shown at the bottom part of Table 1. The luminosity figures are still well below the desired values.

The only other parameter that can be varied is the bunching frequency f_{bunc}. For the largest realistic bandwidth, this is equivalent to vary the frequency integer

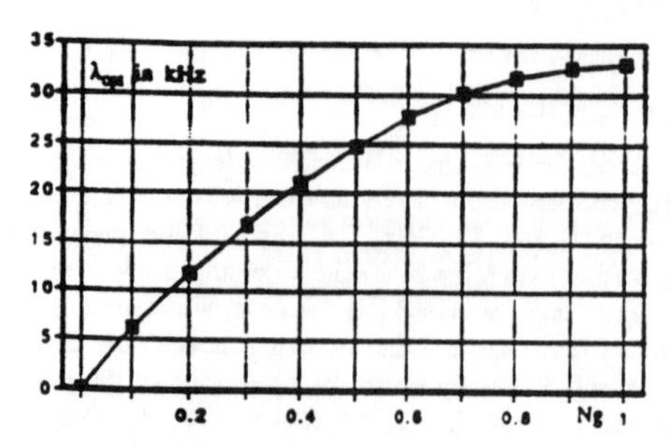

Fig. 6. Cooling rate vs. dynamical gain Ng.

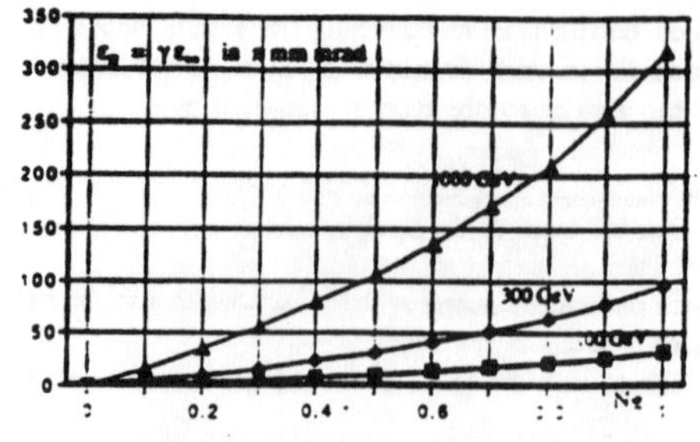

Fig. 8. Equilibrium emittance vs. dynamical gain Ng.

parameter m. If we denote with m_{opt} the value of m used previously, corresponding to $f_{bunc} = 3$ GHz, we can introduce the ratio $\mu = m/m_{opt}$. The expression for the luminosity can then be modified as follows:

$$L = f(Ng, \mu) L_{opt}, \tag{42}$$

where

$$f(u, \mu) = \mu[(2-u)/u] \exp[-\mu k/(2u - u^2)]. \tag{43}$$

The maximum of this function is the one found before and does not depend on the ratio u. That is, the optimum of the luminosity does not depend on the choice of the bunching mode number m, once optimization with respect to the dynamical gain g has been carried out.

9. Conclusions

We have determined that it is indeed feasible that the luminosity of a muon collider scales linearly with the beam energy, as it is required by physics argumentations. Unfortunately, even with the stretching of our imagination, it is seen from our results shown in Table 1 that at the very most only a luminosity of $10^{20}\gamma$ cm^{-2} s^{-1} can be obtained. This is seven orders of magnitude below what it is actually required.

The most important limitation is the effect of thermal noise to the ultimate emittance that can be achieved. This is to be coupled with the requirement on the cooling rate which is to be large compared to the inverse of the beam lifetime. To achieve very fast cooling, a large linear electronic gain is needed, which has also the effect to amplify to a larger level the front-end noise. Moreover a large cooling rate can be obtained only with a few number of particles per bunch. Even by postulating the feasibility of momentum stacking, it is rather difficult to accumulate more than 10^5 particles per bunch.

The comparison of the performance of a muon collider with respect to a proton–antiproton collider is in order. The methodology is essentially the same: both types of particle are produced from a target both need cooling to reduce their dimensions and both are to be compressed longitudinally in bunches. But the antiproton particles have an infinitely long lifetime and it is thus possible to accumulate 10^{10-11} particles per bunch after a long session of stochastic cooling. Moreover it is more convenient for stable particles to operate the collider in a storage mode with multiple passes of the same beams at the collision point.

Our estimates of the performance of stochastic cooling are based on the simple scenario of production, acceleration, cooling and collision we have proposed here. Other scenarios may be possible and we believe that an optimum configuration has still to be searched and is highly desirable. But we also believe that stochastic cooling has to be an integral part of the scheme. Then the following features are to be investigated in more details.

We have seen first that thermal noise at the front-end of the amplifier plays a crucial limiting role to the final beam emittance. We have also seen that this can be better measured with the ratio of Schottky to thermal power, given by Eq. (20). The problem is that as cooling proceeds the beam power reduces whereas the noise signal remains constant. An invention would be highly desirable where the level of noise signal can also be reduced accordingly, as for instance done in momentum cooling with notch filters [9].

The other issue relates to the complete absence of mixing of the particle motion. It is also important to demonstrate that there are ways to regenerate the beam Schottky signal. Several methods have in the meantime been proposed, like the introduction of sextupole to generate a coupling of the transverse motion with the particle momentum error, skew quadrupoles to introduce mixing between the two transverse plane of oscillations, and nonlinear magnetic lenses which cause a dependence of the betatron motion with the amplitude of the motion itself.

These two issues are of paramount importance and are to be investigated carefully if we want to keep the option of a muon collider for the search of the Higgs boson still of some interest.

References

[1] P. Chen and K.T. McDonald, Proc. Workshop on Advanced Accelerator Concepts, AIP Conf. Proc. 279 (1992) 853.

[2] Proc. Mini-Workshop on $\mu^+\mu^-$ colliders: Particle Physics and Design, sponsored by UCLA Center for Advanced Accelerators, Napa Valley, California, Dec. 9–11, 1992.

[3] Proc. Muon Collider Workshop, LA-UR-93-866, LAMPF Auditorium, Feb. 22, 1993.

[4] A.N. Skrinsky, Proc. 20th Int. Conf. on High Energy Physics, AIP Conf. Proc. 68 (1980) 1056.

[5] D. Neuffer, Particle Accel. 14 (1983) 75.

[6] R. Noble, Proc. Workshop on Advanced Accelerator Concepts, AIP, Conf. Proc. 279 (1992) 949.

[7] W.A. Barletta and A.M. Sessler, Characteristics of a high energy $\mu^+\mu^-$ collider based on electro-production of muons, LBL-33613 (1993).

[8] A.G. Ruggiero, Proc. Workshop on Advanced Accelerator Concepts, AIP Conf. Proc. 279 (1992) 958.

[9] S. van der Meer, Physics of Particle Accelerators, AIP Conf. Proc. 153, Volume 2 (1987) 1628.

[10] B. Autin, J. Marriner, A. Ruggiero and K. Takayama, IEEE Trans. Nucl. Sci. NS-30 (4) (1983) 2593.

Critical issues in low energy muon colliders – a summary

S. Chattopadhyay [a], W. Barletta [a], S. Maury [b], D. Neuffer [c], A. Ruggiero [d], A. Sessler [a,*]

[a] *Lawrence Berkeley Laboratory, University of California, Berkeley, CA 94720, USA*

[b] *CERN, CH-1211 Geneva 23, Switzerland*

[c] *CEBAF, 12000 Jefferson Ave., Newport News, VA 23606, USA*

[d] *BNL, Upton, New York 11973, USA*

We present a brief summary of the current state of conception and understanding of the accelerator physics issues for low energy muon colliders envisioned as Higgs factories, associated technological challenges and future research directions on this topic.

1. Motivation and challenges

It is well known that multi-TeV $e^+–e^-$ colliders are constrained in energy, luminosity and resolution, being limited by "radiative effects" which scale inversely as the fourth power of the lepton mass ($(E/m_e)^4$). Thus collisions using heavier leptons such as muons offer a potentially easier extension to higher energies [1]. It is also believed that the muons have a much greater direct coupling into the mass-generating "Higgs-sector", which is the acknowledged next frontier to be explored in particle physics. This leads us to the consideration of TeV-scale $\mu^-–\mu^-$ colliders. However, with the experimental determination of the top quark being heavier than the Z boson, there is increasing possibility of the existence of a "light" Higgs particle with a mass value bracketed by the Z-boson mass and twice that value. This makes a 100 GeV $\mu^+ \otimes$ 100 GeV μ^- collider as a "Higgs Factory" an attractive option [2]. The required average luminosity is determined to be 10^{30} cm^{-2} s^{-1} [2]. We note that the required luminosity for the same "physics reach" scales inversely as the square of the lepton mass and implies a significantly higher luminosity required of a similar energy $e^+–e^-$ collider, in order to reach the same physics goals.

The challenges associated with developing a muon collider were discussed at the Port Jefferson workshop [1,3], subsequent mini-workshops at Napa [2], Los Alamos [4] and at the workshop [5,6] on "Beam Cooling and Related Topics", in Montreux, Switzerland in 1993. Basically, the two inter-related fundamental aspects about muons that critically determine and limit the design and development of a muon collider are that muons are secondary particles and that they have a rather short lifetime in the rest frame. The muon lifetime is about 2.2 μs at rest and is dilated to about 2.2 ms at 100 GeV in the laboratory frame by the relativistic effect. The dilated lifetime is short enough to pose significant challenges to fast beam manipulation and control. Being secondary particles with short lifetime, muons are not to be found in abundance in nature, but rather have to be created in collisions with heavy nuclear targets. Muon beams produced from such heavy targets have spot size and divergence-limited intrinsic phase-space density which is rather low. To achieve the require luminosity, one needs to cool the beams in phase-space by several orders of magnitude. And all these processes – production, cooling, other bunch manipulations, acceleration and eventual transport to collision point – will have to be completed quickly, in 1–2 ms, and therein lies the challenge. Bunch manipulation and cooling of phase space are some of the primary concerns. In the following section, we describe the two scenarios, and associated parameters being considered at present for muon colliders.

2. Scenarios, parameters and comments

Basically, there are two scenarios that have been considered to date for muon colliders. These two scenarios start with very different approaches to the production of the secondary muon beam from a primary beam hitting a heavy target. The subsequent acceleration, cooling, stacking, bunching and colliding gymnastics are all dictated and differentiated by these production schemes, which are very different. We consider them in sequence in the following.

The first approach considers production of the muons starting from a primary "proton" beam hitting a heavy target according to the following reaction:

$$p + N \rightarrow \pi + X, \quad \pi \rightarrow \mu\nu.$$

* Corresponding author.

Since proton bunches are typically long (few ns), one basically obtains long bunches of low phase-space density unless further phase-space manipulations are done to bunch and cool the beams. The situation is similar to the use of the Proton Ring as a pion source in the Los Alamos Meson Physics Facility (LAMPF-II) or conventionally considered kaon factory sources, for example. In order to reduce the length of the produced muon beam bunches, considerable gymnastics is required of the proton ring rf system. Ultimately, of course, a bunch rotation in the longitudinal phase space to reduce bunch length comes at the expense of the relative momentum spread, $\Delta p/p$, which could be as high as 5%. The produced muon bunches will need to be cooled longitudinally from $\Delta p/p$ of 5% to about 0.1% in order to have acceptable spectral purity at the collision point. In addition, the muon bunches will have to be cooled in the transverse phase space by a significant amount in order to meet the luminosity demand at the collision point. The cooled muons are subsequently accelerated and injected into a 100 GeV $\mu^+-\mu^-$ collider where the bunches collide in at most a few hundred to a thousand turns (the number of turns, $n \simeq 300\bar{B}$ [T]). Clearly the constraint of short muon lifetime puts a premium at every stage on minimizing the time for production, cooling, acceleration and bunch processing, so as to still leave a few hundred turns in the collider to produce luminosity. Thus, it is clear that high field magnets play a crucial role in the collider. Details of this scenario have been considered by Neuffer [2,5]. In Fig. 1, we depict schematically the scenario of a muon collider based on production via protons [5].

A second approach considers production of the muons starting from a primary "electron" beam hitting a heavy target according to the following reaction:

$$e + N \rightarrow e + N + \gamma$$
$$\hookrightarrow \mu^+\mu^-.$$

In this electro-production scenario, one obtains short bunches most naturally, since it is compatible with the

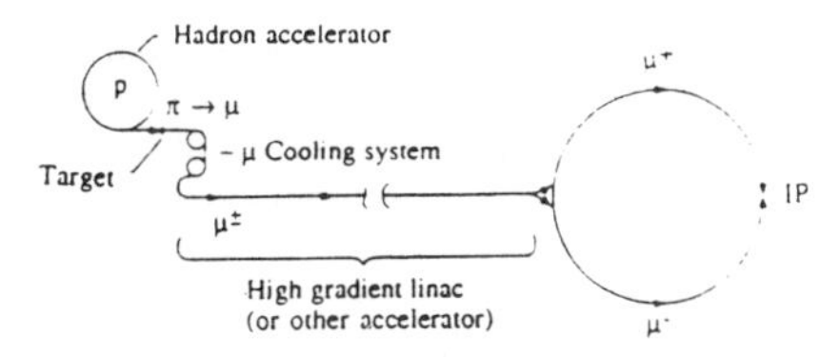

Fig. 1. Overview of a $\mu^+-\mu^-$ collider, showing a hadronic accelerator, which produces π's on a target, followed by a μ-decay channel ($\pi \rightarrow \mu\nu$) and μ-cooling system, followed by a μ-accelerating linac (or recirculating linac or rapid-cycling synchrotron), feeding into a high-energy storage ring for $\mu^+-\mu^-$ (from Ref. [5]).

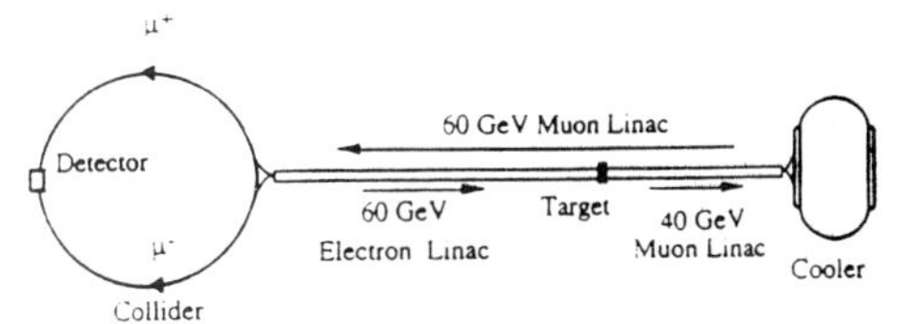

Fig. 2. A muon collider scenario with electro-production of muons (from Ref. [6]).

normal mode of operation of high energy linacs. Although one obtains the "optimum bunch format" naturally, one has to consider unprecedently high power and high repetition rate electron linacs, not explored before in order to meet the required collision luminosity. This is so because of the rather low yield of muons per electron, even at the optimum energy of incident electrons of 60 GeV, and the difficulty of packing more electrons per bunch in the linac. The low transverse phase-space density of the muons will require significant improvement via cooling, similar to the proton production scenario, and, in addition, calls for a nontrivial beam stacking scheme before collision (described in Ref. [6]). Details of this scenario have been considered by Barletta and Sessler [2,6]. In Fig. 2, we depict schematically the scenario of a muon collider based on electro-production [6].

Table 1 presents a comparison of parameters for the above two scenarios for a 100 GeV $\mu^+ \otimes$ 100 GeV μ^- collider, with an average luminosity of 10^{30} cm^{-2} s^{-1}. We assume a collider scenario with a low beta at the collision point of 1 cm, about 1000 bunches colliding in

Table 1
Parameters for a muon collider, 100 GeV × 100 GeV,

$$L = M\frac{N_-N_-f}{4\pi\epsilon_N\beta^*}\gamma \sim 10^{30}\ \text{cm}^{-2}\ \text{s}^{-1},$$

$M = 1000$; $\gamma = 1000$; $\beta^* = 1$ cm; $P = 5$ MW at the target

	Production via electrons	Production via protons
$E_{e\ or\ p}$ (GeV)	60	30
Intensity	5×10^{11}/pulse	10^{14}/pulse
Number of pulses	100 (stacked later)	1
Repetition rate	10 Hz	10 Hz
E_μ (GeV)	40	1.5
ϵ_N (π m-rad)	2×10^{-3}	2×10^{-2}
$\Delta p/p$	±3%	±3%
(μ/e) or (μ/p)	4×10^{-3}	10^{-3}
Ionization cooling	$\epsilon_n^f = 2\times10^{-5}\pi$ m-rad	$\epsilon_n^f = 2\times10^{-5}\pi$ m-rad
Bunch rotation factor	none	100

the ring and muon production limited by a 5 MW power at the target. It is clear that while powerful pion sources, bunch compression and cooling are essential for the proton-production scenario, high current electron linacs, cooling and stacking are essential for the electro-production scenario. It is fair to say from an inspection of Table 1 that, fundamentally, both scenarios are equally amenable to a muon collider configuration with comparable luminosities, given the fact that in both cases equally difficult and challenging technological problems will have to be addressed and solved.

The most difficult and challenging of these technological problems is probably that of "ultra-rapid" phase space cooling of "intense" bunches. One can consider radiation cooling via synchrotron radiation, which is independent of the bunch intensity. However, it is too slow for our purposes. The stochastic cooling rate, on the other hand, depends on the number of particles per bunch and, although too slow usually, can be made significantly faster by going to an extreme scenario of a few particles per bunch with ultra-fast phase mixing or an ultra-high bandwidth ($\sim 10^{14}$ Hz) cooling feedback loop. Both the latter cases will require significant technological inventions. A promising scheme that is both "fast" and "intensity-independent" is that of "ionization cooling", which looks feasible in principle. We have assumed ionization cooling in arriving at the parameters of Table 1. We discuss cooling considerations briefly in the next section.

3. Cooling of muons

The cooling of the transverse phase-space assumed in Table 1 is of the kind known as "ionization cooling". In this scheme the beam transverse and longitudinal energy losses in passing through a material medium are followed by coherent reacceleration, resulting in beam phase-space cooling [2,5,7]. The cooling rate achievable is much faster than, although similar conceptually to, radiation damping in a storage ring in which energy losses in synchrotron radiation followed by rf acceleration result in beam phase-space cooling in all dimensions. Ionization cooling is described in great detail in Refs. [2,5]. It seems that the time is ripe to make a serious design of an ionization cooling channel, including the associated magnetic optics and rf aspects, and put it to real test at some laboratory.

Exploration of the alternate cooling scheme of stochastic cooling takes us to a totally different regime of operation of the collider, determined by the very different nature and mechanism of cooling by an electronic feedback system. Here, the muon lifetime and the required low emittance demanded by the luminosity requirements determine the necessary stochastic cooling rate of the phase space. This rate scales directly as the bandwidth (W) of the feedback system and inversely as the number of particles (N) in the beam (stochastic cooling rate $\propto W/N$). If we limit our consideration to practically achievable conventional feedback electronics, amplifiers, etc., with bandwidth not exceeding 10 GHz, the number of particles per bunch must be less than a thousand in order to meet the desired rate. This then would imply a very different pulse format. This alone drives all the parameters back to the source and issues of "targetry" and "muon source", etc., are not critical. The critical issues for stochastic cooling are: (1) large bandwidth, (2) ultra-low noise, as the cooled emittance reaches the thermal limit of the electronics, (3) rapid mixing and (4) bunch recombination techniques.

Critical issues in the stochastic cooling scenario are discussed by Ruggiero [2,8], where he also explores a conventional cooling scheme with modest bandwidth but with a special nonlinear (magnetic) device that stirs up the phase space rapidly and provides "ultra-fast mixing". It is clear that we need new technical inventions in stochastic cooling for application in a muon collider. Another novel scheme [9] being explored currently is that of "optical cooling" where one detects the granularity of phase space down to a micron scale by carefully monitoring the incoherent radiation from the beam, which is a measure of its Schottky noise, then amplifying this radiation via a laser amplifier of high gain and bandwidth (10^7, 100 THz) and applying it back to the beam. Various issues regarding quantum noise and effective pickup and kicker mechanisms will have to be understood before it can be considered for a serious design.

4. Summary and outlook

As we have seen, both scenarios – production of muons from protons and electro-production of muons – are competitive but very ambitious and challenging. Production of muons from protons will clearly require nontrivial and sophisticated target design and configuration. In addition, in order to match the bunch length of the colliding (but secondarily produced) muon beams to the low beta function at the collision point, the primary proton beams must be bunched by a large factor (~ 100). The complicated bunch rotation and rf manipulations are cumbersome and must be done at the low energy proton end before the target, which implies an associated increase in the relative momentum spread, $\Delta p/p$. On a positive note, however, targetry with protons and rf gymnastics with proton beams are relatively familiar affairs at hadron and kaon facilities, albeit at a lower level of power and rf manipulation of the bunches. Electro-production of muons, on the other hand, requires, high peak current, high repetition rate linacs, so far unexplored, in order to meet the luminosity demand. Besides, "stacking" of many electron bunches from a linac into a single bunch poses a nontrivial problem. The significant and most attractive feature of the electro-production scenario, however, is that the "optimal pulse format" is produced directly at the target by electrons from a linac, without complex bunch compression schemes in a ring.

No matter what the optimal scenario would turn out to be, should the muon collider concept turn into reality, further consideration of such a collider at 200 GeV center-of-mass energy with an average luminosity of $\sim 10^{30}$ cm^{-2} s^{-1} would have to assume major advances in, and eventual operation of, (1) megawatt muon targets, (2) multi-kiloampere peak current electron linacs, (3) efficient transfer, compression and stacking schemes for charged particle beams, (4) high field magnets and (5), most importantly, feasible phase-space cooling technologies with low noise and large bandwidth. While "ionization cooling" looks promising, it needs experimental demonstration. A possible feasibility test of muon production and ionization cooling at existing facilities, e.g., CERN or FNAL, would be highly desirable. The "stochastic cooling" approach, however, would need fundamental invention of a new technique, as elaborated earlier. The emerging new ideas of "optical stochastic cooling", "ultra-rapid phase-mixer", etc., are ambitious, but may hold the key to the success of such high frequency stochastic cooling. Finally, the synchrotron radiation and muon decay in the collider ring vacuum chamber and detector area pose issues that cannot be overlooked.

In conclusion, surely a muon collider is exotic! But even as we contemplate the value, utility and eventual realizability of such a collider in the future, there is no doubt that the necessary conceptual and technological explorations forced upon us by these considerations are much too valuable to many fields to be simply passed up.

Acknowledgement

This work was supported by the Director, Office of Energy Research, Office of High Energy and Nuclear Physics, High Energy Physics Division, of the U.S. Department of Energy under Contract No. DE-AC03-76SF00098.

References

[1] M. Tigner, Proc. Advanced Accelerator Concepts, Port Jefferson, NY, 1992, AIP Conf. Proc. 279 (1993) 1.

[2] D.B. Cline, this issue, Nucl. Instr. and Meth. A 350 (1994) 24.

[3] P. Chen and K.T. MacDonald, AIP Conf. Proc. 279 (1993) 853.

[4] Proc. Muon Collider Workshop, February 22, 1993, Los Alamos National Laboratory Report LA-UR-93-866 (1993).

[5] D. Neuffer, Muon Cooling and Applications, Proc. of the Workshop on Beam Cooling and Related Topics, October 3–8, 1993, Montreux, Switzerland, to be published as a CERN report.

[6] W.A. Barletta and A.M. Sessler, Prospects for a High Energy Muon Collider Based On Electro-production, Proc. Workshop on Beam Cooling and Related Topics, October 3–9, 1993, Montreux, Switzerland, to be published as a CERN report.

[7] D. Neuffer, Particle Accel. 14 (1983) 75.

[8] A.G. Ruggiero, Stochastic Cooling Requirements for a Muon Collider, Proc. Workshop on Beam Cooling and Related Topics, October 3–9, 1993, Montreux, Switzerland, to be published as a CERN report.

[9] A.A. Mikhailichenko and M.S. Zolotorev, Optical Stochastic Cooling, SLAC-PUB- 6272 (1993).

III. RESULTS FROM THE SAUSALITO WORKSHOP

Physics Goals of a $\mu^+\mu^-$ Collider*

V. Barger[a], M.S. Berger[b], K. Fujii[c], J.F. Gunion[d], T. Han[d], C. Heusch[e], W. Hong[f], S.K. Oh[g], Z. Parsa[h], S. Rajpoot[i], R. Thun[j], and B. Willis[k]

[a]*Physics Department, University of Wisconsin, Madison, WI 53706, USA*
[b]*Physics Department, Indiana University, Bloomington, IN 47405, USA*
[c]*Physics Division, KEK, 1-1 Oho, Tsukuba, Ibaraki, Japan*
[d]*Physics Department, University of California, Davis, CA 95616, USA*
[e]*Physics Department, University of California, Santa Cruz, CA 95064, USA*
[f]*Physics Department, University of California, Los Angeles, CA 90024, USA*
[g]*Physics Department, Kon-kuk University, Seoul 133-701, South Korea*
[h]*Physics Department, Brookhaven National Laboratory, Upton, NY 11973, USA*
[i]*Physics Department, California State University, Long Beach, CA 90840, USA*
[j]*Physics Department, University of Michigan, Ann Arbor, MI 48109, USA*
[k]*Physics Department, Columbia University, New York, NY 10027, USA*

Abstract

This working group report focuses on the physics potential of $\mu^+\mu^-$ colliders beyond what can be accomplished at linear e^+e^- colliders and the LHC. Particularly interesting possibilities include (i) s-channel resonance production to discover and study heavy Higgs bosons with ZZ and WW couplings that are suppressed or absent at tree-level (such as the H and A Higgs bosons of supersymmetric models), (ii) measurements of the masses and properties of heavy supersymmetric particles, and (iii) study of the strongly interacting electroweak sector, where higher energies give larger signals.

I. INTRODUCTION

The physics working group focused on the new physics potential of a $\mu^+\mu^-$ collider[1]. Since the time available for analysis was limited, this report is largely directed to two areas of major current interest in particle physics:

- finding Higgs bosons or detecting strong WW scattering and thereby understanding the origin of electroweak symmetry breaking (EWSB);
- finding and studying supersymmetric (SUSY) particles.

For the same energy and integrated luminosity, it should be possible to do anything at a $\mu^+\mu^-$ collider that can be done at an e^+e^- collider. Moreover, a $\mu^+\mu^-$ collider opens up two particularly interesting possibilities for improving the physics reach over that of e^+e^- colliders:

*Report presented by V. Barger. Working group leaders V. Barger and J.F. Gunion

- s-channel Higgs production;
- higher center-of-mass energy with reduced backgrounds.

Both possibilities are simply a result of the large muon mass as compared to the electron mass. Direct s-channel Higgs production is greatly enhanced at a $\mu^+\mu^-$ collider because the coupling of any Higgs boson to the incoming $\mu^+\mu^-$ is proportional to m_μ, and is therefore much larger than in e^+e^- collisions. Higher energy is possible at a $\mu^+\mu^-$ collider due to the ability to recirculate the muons through a linear acceleration array without an overwhelming radiative energy loss. Current estimates are that an e^+e^- collider with energy above 1.5 TeV will be very difficult to construct (with adequate luminosity), whereas a 4 TeV $\mu^+\mu^-$ collider would appear to be well within the range of possibility [2]. Meanwhile, backgrounds from processes arising from photon radiation from one or both beams will be suppressed by the higher muon mass. As we shall discuss, higher energy could be crucial in at least two ways:

- improved signals for strong WW scattering — since higher energies are achievable the signal level is increased, while backgrounds are reduced due to there being less photon radiation;
- the kinematical reach for pair production of SUSY particles is extended to a possibly crucial higher mass range.

Very briefly, what new physics can an e^+e^- collider[3] do? First, neutral Higgs bosons can be discovered that are coupled to the Z-boson. The Standard Model Higgs boson or the lightest Higgs boson of the Minimal Supersymmetric Standard Model (MSSM) can be discovered (in $Z^* \to Zh$ production) if $m_h \lesssim 0.7\sqrt{s}$. Since there is a theoretical upper bound[4,5] on the lightest Higgs in the MSSM of $m_h \lesssim 130$ to 150 GeV, an e^+e^- machine with CM energy $\sqrt{s} = 300$ GeV can exclude or confirm supersymmetric theories based on grand unified theories (GUTs). However, if heavy, the other neutral SUSY Higgs bosons H, A can only be discovered via $Z^* \to HA$ production for $m_H \sim m_A < \sqrt{s}/2$; $Z^* \to ZH$ (ZA) production is not useful since the H (A) coupling to ZZ is suppressed (absent) at tree-level.

Discovery of SUSY sparticles is also possible at an e^+e^- collider for sparticle masses $m < \sqrt{s}/2$. While the energy reach of an e^+e^- collider will probably be adequate for pair producing the lightest chargino $\chi_1^\pm$, the neutralino combination $\chi_1^0\chi_2^0$ and possibly the selectron, smuon and stau $\tilde{e}, \tilde{\mu}, \tilde{\tau}$, and the lighter stop eigenstate, $\tilde{t}_1$, it could be inadequate for the heavier chargino and neutralinos and other squarks.

The figure of merit in physics searches at an e^+e^- or $\mu^+\mu^-$ collider is the QED point cross section for $e^+e^- \to \mu^+\mu^-$, which has the value

$$\sigma_{QED}(\sqrt{s}) = \frac{100 \text{ fb}}{s \text{ (TeV}^2)} \left(\frac{\alpha(s)}{\alpha(M_Z^2)}\right)^2 . \tag{1}$$

Henceforth we neglect the factor $(\alpha(s)/\alpha(M_Z^2))^2$, which varies slowly with s. As a rule of thumb, the integrated luminosity needed for the study of new physics signals is

$$\left(\int \mathcal{L}dt\right)\sigma_{QED} \gtrsim 1000 \text{ events} . \tag{2}$$

Thus $\mu^+\mu^-$ machine designs should be able to deliver an integrated luminosity of

$$\int \mathcal{L}dt \gtrsim 10 \cdot s \text{ (fb)}^{-1} . \tag{3}$$

If this is to be accumulated in one year's running, the luminosity requirement is

$$\mathcal{L} \gtrsim 10^{33} \cdot s \text{ (cm)}^{-2}(\text{sec})^{-1} . \tag{4}$$

Two possible $\mu^+\mu^-$ machines were considered at this meeting:

- Low $\sqrt{s} \simeq 400$ GeV, requiring

$$\int \mathcal{L}dt \gtrsim 1 \text{ (fb)}^{-1}, \quad \mathcal{L} \gtrsim 10^{32} \text{ (cm)}^{-2}(\text{sec})^{-1} \tag{5}$$

- High $\sqrt{s} \gtrsim 4$ TeV, requiring

$$\int \mathcal{L}dt \gtrsim 100 \text{ (fb)}^{-1}, \quad \mathcal{L} \gtrsim 10^{34} \text{ (cm)}^{-2}(\text{sec})^{-1} \tag{6}$$

Fortunately these luminosity requirements may be achievable with the designs under consideration [2]. Indeed, the luminosity of a $\sqrt{s} \sim 400$ GeV machine could be as large as $\mathcal{L} \gtrsim 10^{33}$ $(\text{cm})^{-2}(\text{sec})^{-1}$.

II. FIRST MUON COLLIDER (FMC)

Consider a $\mu^+\mu^-$ machine with CM energy $\sqrt{s}$ in the range 250 to 500 GeV. As noted above, the yearly integrated luminosity for this machine would be at least $L \geq 1$ $(\text{fb})^{-1}$ and possibly as much as $L \geq 10-20$ $(\text{fb})^{-1}$. The most interesting physics at such a $\mu^+\mu^-$ collider that goes beyond that accessible at an e^+e^- collider of similar energy is the possibility of s-channel heavy Higgs production, as illustrated in Fig. 1.

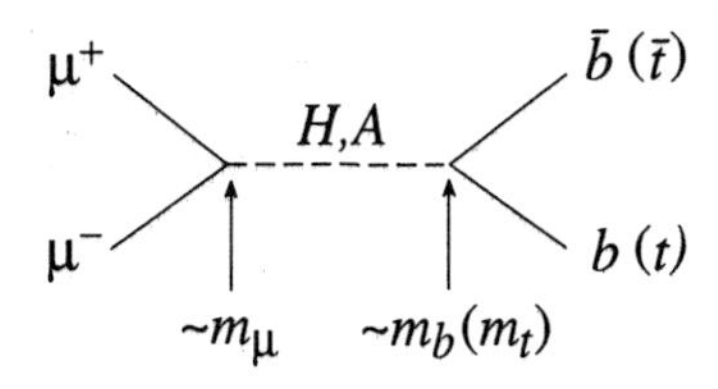

Fig. 1: s-channel diagrams for production of H, A MSSM Higgs bosons.

To discover a narrow resonance, one would ideally like to have a broad $\sqrt{s}$ spectrum. (Alternatively, a scan with limited luminosity at discretely spaced energies could be employed; our results for the case of a broad spectrum are easily altered to this latter procedure.) Once a Higgs boson is discovered, it would be best to change to as narrow as possible a spectrum and then sit on the peak to study the resonance properties.

The s-channel Higgs resonance cross section is

$$\sigma_h = \frac{4\pi\Gamma(h \to \mu\mu)\,\Gamma(h \to X)}{\left(s - m_h^2\right)^2 + m_h^2\Gamma_h^2}\,, \tag{7}$$

where X denotes a final state and Γ_h is the total width. The sharpness of the resonance peak is determined by Γ_h. For a broad energy spectrum, the relevant energy resolution is that determined by the detector. The widths of the Higgs bosons under consideration will be seen to be quite small, such that it is reasonable to consider

$$\Delta E(\text{resolution}) \geq \Gamma(\text{Higgs})\,. \tag{8}$$

Then the integrated signal over the narrow resonance is

$$S_h = \int \sigma_h d\sqrt{s} = \frac{\pi}{2}\Gamma_h \sigma_h^{\text{peak}}\,, \tag{9}$$

where the peak cross section is

$$\sigma_h^{\text{peak}} = \frac{4\pi}{m_h^2}\text{BF}(h \to \mu\mu)\text{BF}(h \to X)\,, \tag{10}$$

leading to

$$S_h = \frac{2\pi^2}{m_h^2}\Gamma(h \to \mu\mu)\,\text{BF}(h \to X)\,. \tag{11}$$

With the integrated luminosity $L = \int \mathcal{L}dt$ spread over an energy band ΔE in the search mode, the event rate for the Higgs signal is

$$N_h = S_h \frac{L}{\Delta E}\,. \tag{12}$$

In the above, we have used the general notation h for the Higgs boson.

In applying the above general formulae to the Minimal Supersymmetric Model (MSSM) Higgs bosons we note the following important facts. The couplings to fermions and vector bosons depend on the SUSY parameter $\tan\beta = v_2/v_1$ and on the mixing angle α between the neutral Higgs states (α is determined by the Higgs masses, $\tan\beta$, the top and the stop masses). SUSY GUT models predict large m_A and $\alpha \approx \beta - \pi/2$. In this case, the coupling factors of the Higgs bosons are approximately[6]

$$\begin{array}{cccc} & \mu^+\mu^-, b\bar{b} & t\bar{t} & ZZ, W^+W^- \\ h & 1 & -1 & 1 \\ H & \tan\beta & -1/\tan\beta & 0 \\ A & -i\gamma_5\tan\beta & -i\gamma_5/\tan\beta & 0 \end{array} \tag{13}$$

times the Standard-Model factor of $gm/(2m_W)$ in the case of fermions (where m is the relevant fermion mass), or $gm_W, gm_Z/\cos\theta_W$ in the case of the W, Z. The broad spectrum inclusive signal rate S_h is proportional to $\Gamma(h \to \mu^+\mu^-)$ and since the coupling of $h = H, A$ to the $\mu^+\mu^-$ channel is proportional to $\tan\beta$, larger $\tan\beta$ values give larger production.

To obtain the rate in a given final state mode X we multiply the inclusive rate by $\mathrm{BF}(h \to X)$. Here, we consider only the $b\bar{b}$ and $t\bar{t}$ decay modes for $h = H, A$, although the relatively background free $H \to hh \to b\bar{b}b\bar{b}$ and $A \to Zh \to Zb\bar{b}$ modes might also be useful for discovery. Figure 2 shows the dominant branching fractions to $b\bar{b}$ and $t\bar{t}$ of Higgs bosons of mass $m_A =$ 400 GeV $\approx m_H$ versus $\tan\beta$, taking $m_t = 170$ GeV. The $b\bar{b}$ decay mode is dominant for $\tan\beta > 5$, which is the region where observable signal rates are obtained. From the figure we see that $\mathrm{BF}(h \to b\bar{b})$ grows rapidly with $\tan\beta$ for $\tan\beta \lesssim 5$, while $\mathrm{BF}(h \to t\bar{t})$ falls slowly. For low to moderate $\tan\beta$ values, the event rates behave as

$$N(\mu^+\mu^- \to H, A \to b\bar{b}) \;\propto\; m_\mu^2 m_b^2 \left[(\tan\beta)^6 \text{ to } (\tan\beta)^4\right] \tag{14}$$

$$N(\mu^+\mu^- \to H, A \to t\bar{t}) \;\propto\; m_\mu^2 m_t^2 \left[(\tan\beta)^2 \text{ to } (\tan\beta)\right]. \tag{15}$$

It is this growth with $\tan\beta$ that makes H, A discovery possible for relatively modest values of $\tan\beta$ larger than 1. At high $\tan\beta$ the $b\bar{b}$ branching ratio asymptotes to a constant value while the $t\bar{t}$ branching ratio falls as $1/(\tan\beta)^4$, so that

$$N(\mu^+\mu^- \to H, A \to b\bar{b}) \;\propto\; m_\mu^2 m_b^2 (\tan\beta)^2 \tag{16}$$

$$N(\mu^+\mu^- \to H, A \to t\bar{t}) \;\propto\; m_\mu^2 m_t^2 (\tan\beta)^{-2}. \tag{17}$$

Consequently, the $t\bar{t}$ channel will not be useful at large $\tan\beta$, whereas the $b\bar{b}$ channel continues to provide a large event rate.

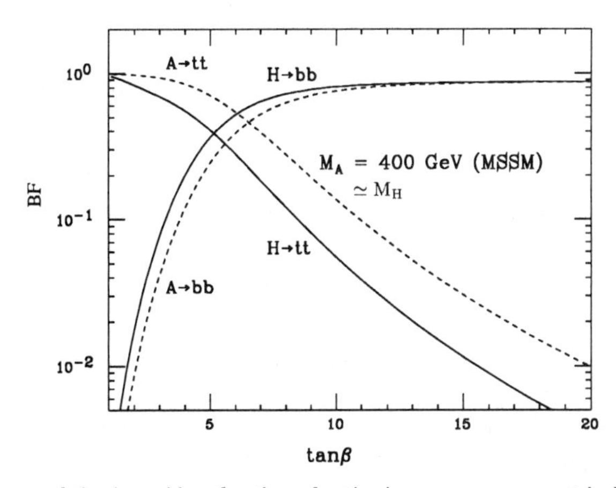

Fig. 2: Dependence of the branching fractions for the heavy supersymmetric Higgs bosons on $\tan\beta$ (from Ref. [7]).

The calculated Higgs boson widths are shown in Fig. 3 versus m_h for $\tan\beta = 5$. As promised, the H and A are typically narrow resonances ($\Gamma_{H,A} \sim 0.1$ to 2 GeV), and our approximation of $\Delta E \geq \Gamma_{H,A}$ will generally be valid; see below.

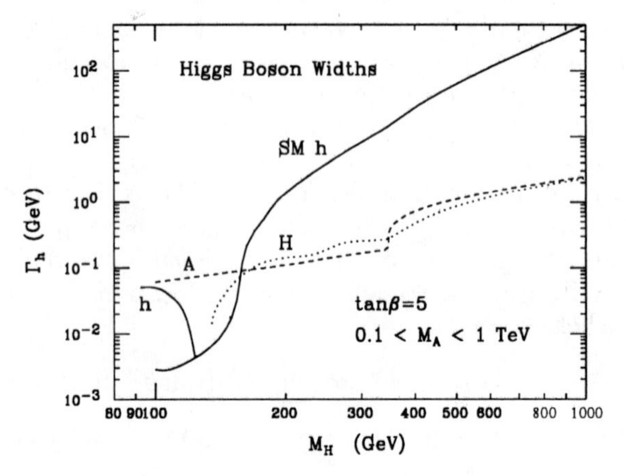

Fig. 3: The Standard Model Higgs boson and the supersymmetric Higgs boson widths (from Ref. [7]).

The irreducible backgrounds to the Higgs signals are

$$\mu^+\mu^- \to \gamma^*, Z^* \to b\bar{b}, t\bar{t}\,. \tag{18}$$

The final-state energy resolution can be estimated by taking the heavy quark energy resolution to be that of the hadron energy resolution in NLC design studies [3]

$$\frac{\Delta E_Q}{E_Q} = \frac{50\%}{\sqrt{E_Q}} + 2\%\,. \tag{19}$$

Then for example at $m_h = 400$ GeV the final state mass resolution is

$$\Delta m(b\bar{b}) \simeq 2\% m(b\bar{b})\,. \tag{20}$$

The background cross section is integrated over the bin $\sqrt{s} = m_h \pm \frac{1}{2}\Delta m(b\bar{b})$

$$B = \int \sigma_B \, d\sqrt{s}\,. \tag{21}$$

The light-quark backgrounds can be rejected using b-tagging. As a single b-tag efficiency, we assume $\epsilon_b \simeq 0.5$ and neglect mistags.

The H signal (for $\tan\beta = 5$) and the backgrounds integrated over the resolution are shown in Fig. 4 versus m_H. As a concrete example let us assume the optimistic integrated luminosity of $L = 20$ (fb)$^{-1}$ spread over an energy band $\Delta E = 50$ GeV, giving $L/\Delta E = 0.4$ (fb)$^{-1}$/GeV. The signal in the $b\bar{b}$ channel for $m_H = 400$ GeV with $\tan\beta = 5$ is

$$N(\text{signal}) = 250 \text{ events} \tag{22}$$

while the $b\bar{b}$ background is

$$B(b\bar{b} \text{ background}) = 2000 \text{ events} , \tag{23}$$

where both numbers include the single b-tag efficiency ϵ_b. The significance of the signal is

$$n_{SD} \simeq 5.6 . \tag{24}$$

Thus, discovery is possible here!

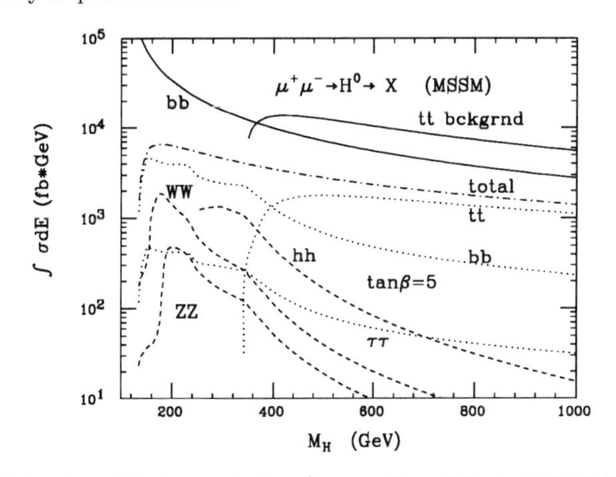

Fig. 4: Typical signals and backgrounds for $\mu^+\mu^- \to H \to X$ in the MSSM (from Ref. [7]).

Figure 5 shows the significance for H detection versus $\tan\beta$ for a variety of $L/\Delta E$ possibilities: 1 (fb)$^{-1}$/GeV, 0.2 (fb)$^{-1}$/GeV, and 0.02 (fb)$^{-1}$/GeV. Single b-tagging with $\epsilon_b = 0.5$ is assumed. Due to the sharp rise of n_{SD} with $\tan\beta$ a decrease of $L/\Delta E$ by a factor of 50 only changes the lowest $\tan\beta$ for which discovery is possible from $\tan\beta \sim 4$ to $\tan\beta \sim 7$.

Figure 6 provides an overall view of the possibilities for MSSM Higgs boson detection in terms of regions in the $\tan\beta, m_A$ plane for which $n_{SD} \geq 4$ in $h = h, H, A$ searches. This figure assumes a moderately conservative value of $L/\Delta E = 0.08$ (fb)$^{-1}$/GeV. Further, *double* b-tagging (with $\epsilon_b = 0.6$) is required in the $b\bar{b}$ mode and an additional general efficiency factor of 0.5 is included for both the $b\bar{b}$ and $t\bar{t}$ final states. Regions in which h, H, or A detection is individually possible are shown, as well as the additional region that is covered by combining the H and A signals when they are degenerate within the final state resolution. The figure shows that in the $b\bar{b}$ mode either H and A detection (for $m_A \gtrsim 140$ GeV — as preferred in GUT scenarios), or h and A detection (for $m_A \lesssim 140$ GeV), will be possible provided $\tan\beta > 3$–5. Detection of $H + A$ in the $t\bar{t}$ mode is limited to $m_H \sim m_A > 2m_t$ and $\tan\beta$ values lying between 3 and about 12.

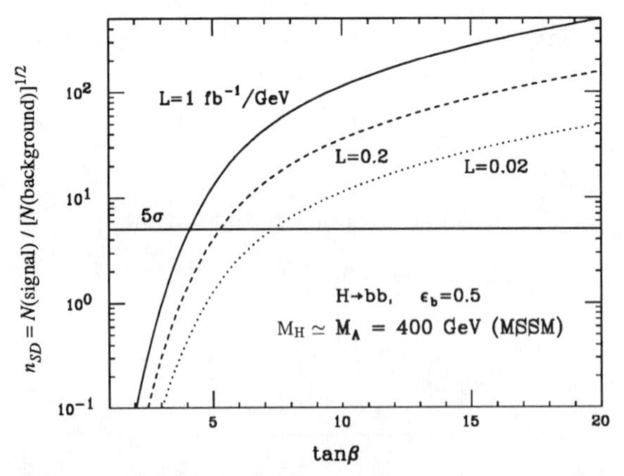

Fig. 5: The statistical significance of the Higgs H signal versus $\tan\beta$. Since in SUSY GUT models one finds H and A to be approximately mass degenerate, the combined signal will be larger (from Ref. [7]).

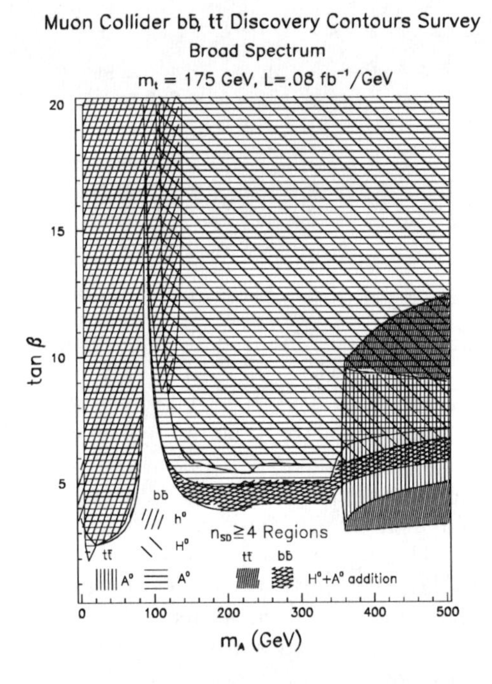

Fig. 6: Higgs boson H, A discovery regions at a muon collider in the modes $H, A \to b\bar{b}$ and $t\bar{t}$ (from Ref. [8]).

Thus, direct s-channel production allows H, A discovery up to the machine kinematical limit, so long as $\tan\beta$ is not small. This is a very important extension as compared to an e^+e^- collider. The above results assume the absence of SUSY decay modes for the H and A. Once the H, A mass becomes large it could happen that SUSY decay modes become kinematically allowed. However, at large $\tan\beta$ the enhanced $b\bar{b}$ coupling guarantees that the $b\bar{b}$ mode branching ratio will still be large. In practice, SUSY decay modes would only shift the discovery regions to slightly higher $\tan\beta$ values.

Once a signal is identified in the search mode, the luminosity can be concentrated over the Higgs peak, $\Delta E = 2\Gamma_H \simeq 1$ GeV, to study the resonance properties. Then the signal and background each increase by a factor of the luminosity energy spread ΔE_S in the broad-spectrum/scanning mode, giving an increase in the significance of a factor of $\sqrt{\Delta E_S}$ [a factor of 7 in the example of Eqs. (22–24)].

High polarization P of both beams would be useful to suppress the background to s-channel Higgs production if the luminosity reduction is less than a factor of $\left(1+P^2\right)^2/\left(1-P^2\right)$, which would leave the significance of the signal unchanged [10]. For example, $P = 0.84$ would compensate a factor of 10 reduction in luminosity.

It could also be possible to detect the lightest SUSY Higgs boson via s-channel production if m_h becomes known within a few GeV from electroweak radiative corrections or from studies of $\mu^+\mu^- \to Z^* \to Zh$ production, provided that the machine could be operated at $\sqrt{s} = m_h$. However, the $\mu^+\mu^- \to h \to b\bar{b}$ channel suffers a formidable background from $\mu^+\mu^- \to \gamma^*, Z^* \to b\bar{b}$, so the direct channel h search might well require polarized muon beams. Nonetheless, detection of the light h in direct s-channel production would be very interesting in that it would allow a determination of the $\mu^+\mu^-$ coupling of the h.

If a 0.1% energy resolution could be achieved, another interesting application of the FMC could be a precision determination of the top mass by measuring the $t\bar{t}$ threshold cross section[9]. The muon collider could present an improvement over the electron collider from the reduced initial state radiation.

III. NEXT MUON COLLIDER (NMC)

The reduced synchrotron radiation for muons, as compared to electrons, allows the possibility of a recirculating muon colliding beam accelerator with higher energy than is realizable at future linear e^+e^- colliders. A design goal under consideration is a $\sqrt{s} = 4$ TeV $\mu^+\mu^-$ machine with an integrated luminosity $L \geq 100$ (fb)$^{-1}$. The construction of the higher energy $\mu^+\mu^-$ collider could occur at the same time as the $\sqrt{s} = 0.4$ TeV machine [2].

A. Sparticle Studies

An exciting possibility is that the NMC could be a SUSY factory, producing squark pairs, slepton pairs, chargino pairs, associated neutralinos, associated $H + A$ Higgs, and gluinos from squark decay if kinematically allowed. If the

SUSY mass scale is $M_{\rm SUSY} \sim 1$ TeV, many sparticles could be beyond the reach of the NLC. The LHC can produce them, but disentangling the SUSY spectrum and measuring the sparticle masses will be a real challenge at a hadron collider, due to the complex nature of the sparticle cascade decays and the presence of QCD backgrounds. The measurement of the sparticle masses is important since they are a window to GUT scale physics.

The p-wave suppression of squark pair production in e^+e^- or $\mu^+\mu^-$ collisions, illustrated in Fig. 7, means that energies well above the threshold are needed. The threshold dependence of the cross section may be useful in sparticle mass measurements.

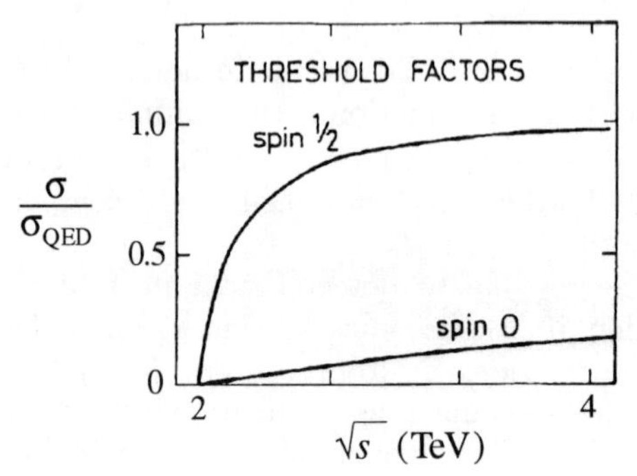

Fig. 7: Comparison of kinematic suppression for fermion pairs and squark pair production at e^+e^- or $\mu^+\mu^-$ colliders.

The cross sections for squarks (of one flavor in the approximation of L, R degeneracy), charginos, top and three generations of singlet quarks (from an E_6 GUT model, for example) are respectively

$$\begin{aligned}
\sigma_{\tilde{u}_{L,R}} &= 4\beta^3\,\text{fb} \rightarrow 250 \text{ events} \\
\sigma_{\tilde{d}_{L,R}} &= 1\beta^3\,\text{fb} \rightarrow \;\; 60 \text{ events} \\
\sigma_{\chi^\pm} &= 6\beta\;\,\text{fb} \rightarrow 500 \text{ events} \\
\sigma_t &= \;\; 8\;\,\text{fb} \rightarrow 800 \text{ events} \\
\sigma_{Q_{E_6}} &= 6\beta\;\,\text{fb} \rightarrow 600 \text{ events}
\end{aligned} \tag{25}$$

where the event rates given are for sparticle masses of 1 TeV with an integrated luminosity of 100 fb^{-1}. The production of heavy SUSY particles will give spherical events near threshold characterized by

- multijets
- missing energy (associated with the LSP)
- leptons

There should be no problem with backgrounds from SM processes.

A supergravity model with $\tan\beta = 5$, universal scalar mass $m_0 = 1000$ GeV and gaugino mass $m_{1/2} = 150$ GeV provides an illustration of a heavy sparticle spectrum, as follows:

$$
\begin{array}{ccc}
\underline{\text{sparticle}} & \underline{\text{mass}} & \\
\tilde{u} & 1000\ \text{GeV} & \\
\tilde{g} & 500 & \\
\tilde{\ell} & 1000 & \\
\chi_4^0,\ \chi_3^0,\ \chi_2^+ & 350 & \\
\chi_2^0,\ \chi_1^+ & 130 & \\
\chi_1^0 & 60 & (\text{LSP})
\end{array}
\tag{26}
$$

Consider $\tilde{u}\bar{\tilde{u}}$ production at the NMC. The dominant cascade chain for the decays is

$$
\begin{aligned}
\tilde{u}\bar{\tilde{u}} &\to (\tilde{g}u)(\tilde{g}\bar{u}) \\
\tilde{g} &\to \chi_1^{\pm} q\bar{q} \\
\chi_1^{\pm} &\to \chi_1^0 \ell\nu,\ \chi_1^0 q\bar{q}
\end{aligned}
\tag{27}
$$

The dominant branching fractions of the $\tilde{u}\bar{\tilde{u}}$ final state are

$$
\begin{array}{lr}
10\ \text{jets} + \not{p}_T & 10\% \\
8\ \text{jets} + 1\ell + \not{p}_T & 10\% \\
6\ \text{jets} + 2\ell + \not{p}_T & 2\%
\end{array}
\tag{28}
$$

Of the two lepton events, one half will be like-sign dileptons ($\ell^+\ell^+$, $\ell^-\ell^-$). The environment of a $\mu^+\mu^-$ collider may be better suited than the LHC to the study of the many topologies of sparticle events.

Turning to the SUSY Higgs sector, we simply emphasize the fact that $Z^* \to HA, H^+H^-$ will allow $H, A, H^{\pm}$ discovery up to $m_H \sim m_A \sim m_{H^{\pm}}$ values somewhat below $\sqrt{s}/2 \sim 2$ TeV. While GUT scenarios prefer $H, A, H^{\pm}$ masses above 200 to 250 GeV, such that HA and H^+H^- pair production is beyond the kinematical reach of a 400 to 500 GeV collider, even the most extreme GUT scenarios do not yield Higgs masses beyond 2 TeV. Thus, a 4 TeV $\mu^+\mu^-$ collider is guaranteed to find all the SUSY Higgs bosons.

B. Strong *WW* Scattering

If Higgs bosons with $m_H \leq \mathcal{O}(800\ \text{GeV})$ do not exist, then the interactions of longitudinally polarized weak bosons W_L, Z_L became strong. This means that new physics must be present at the TeV energy scale[11]. The high reach in energy of the NMC is of particular interest for study of a strongly interacting electroweak sector (SEWS) at a $\mu^+\mu^-$ collider via the WW fusion graphs in Fig. 8.

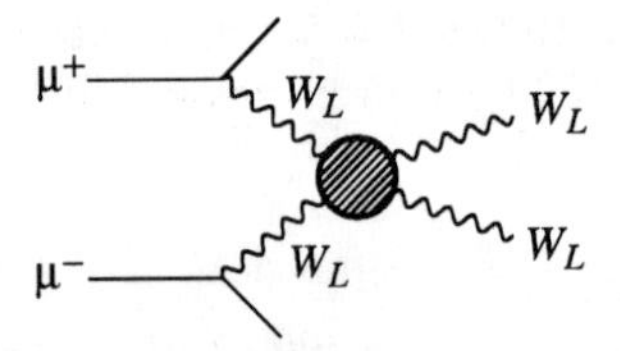

Fig. 8: Strong $W_L^+W_L^-$ scattering in $\mu^+\mu^-$ collisions.

The SEWS signals depend on the model for $W_L^+W_L^-$ scattering. An estimate of the size of these signals can be obtained by taking the difference of the cross section due to a heavy Higgs boson ($m_H = 1$ TeV) and that with a massless Higgs particle

$$\Delta\sigma_{\text{SEWS}} = \sigma(m_H = 1 \text{ TeV}) - \sigma(m_H = 0) . \tag{29}$$

The subtraction of the $m_H = 0$ result removes the contributions due to scattering of transversely polarized W-bosons. Figure 9 shows the growth of $\Delta\sigma_{\text{SEWS}}$ with energy. The Table below gives results at $\sqrt{s} = 1.5$ TeV (for the NLC e^+e^- collider) and at $\sqrt{s} = 4$ TeV (for the NMC).

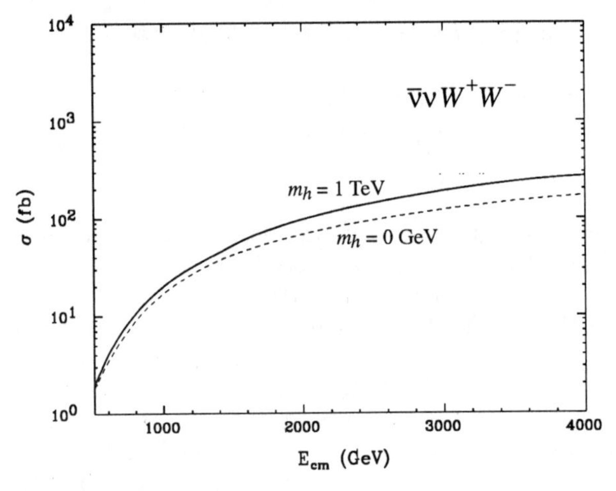

Fig. 9: The growth of the SEWS signal with E_{cm} (from Ref. [12]).

Table I: SEWS signals at future colliders

$\sqrt{s}$	$\Delta\sigma(W^+W^-)$	$\Delta\sigma(ZZ)$
1.5 TeV	8 fb	6 fb
4 TeV	80 fb	50 fb

The energy reach is thereby a critical consideration in the study of SEWS. Additionally, the backgrounds from the photon exchange process of Fig. 10 will be a factor of 3 less at a $\mu^+\mu^-$ machine compared to an e^+e^- machine[12]. The probability for γ radiation from a charged lepton

$$P_{\gamma/\ell}(x) = \frac{\alpha}{\pi}\frac{1+(1-x)^2}{x}\ln(E_\ell/m_\ell) \tag{30}$$

is logarithmically dependent on the charged lepton mass. Figure 11 shows SM cross sections with $m_H = 0$ versus CM energy. The background to SEWS from $e^+e^- \to e^+e^-W^+W^-$ can be rejected by requiring[13]

$$\text{no } e^\pm \text{ with } E_\ell > 50 \text{ GeV and } |\cos\theta_\ell| < |\cos(0.15 \text{ rad})| \,. \tag{31}$$

Some similar rejection would be necessary to suppress the $\mu^+\mu^- \to \mu^+\mu^-W^+W^-$ background at the NMC.

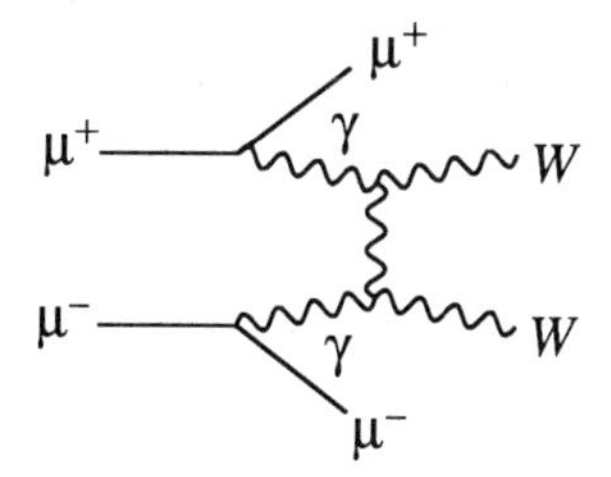

Fig. 10: The background to the SEWS signal from the photon exchange process.

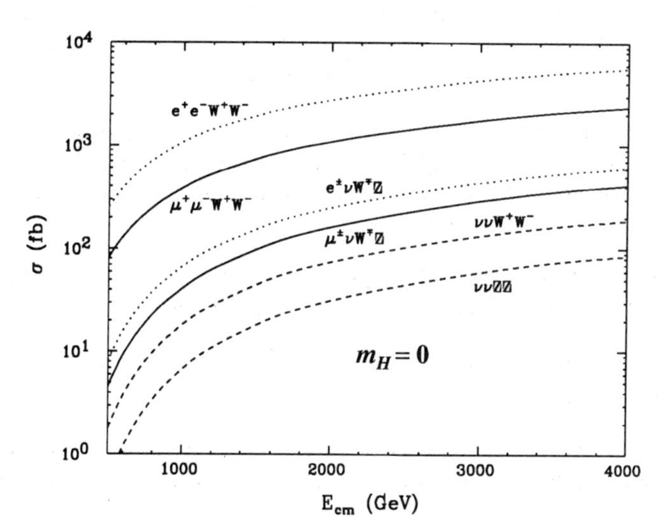

Fig. 11: The cross sections for Standard Model processes at a muon collider (from Ref. [12]).

The various SM cross sections grow with energy, for example,

$$\sigma(\mu^+\mu^- \to \mu^+\mu^- W^+W^-) = 2000 \text{ fb} \tag{32}$$

at $\sqrt{s} = 4$ TeV. These cross sections would be higher if the W is composite. Thus precision studies of WW self couplings should be possible at the NMC.

There are many other new possibilities of interest for the NMC that we have not yet addressed, including

- extra neutral gauge bosons (the NMC could be a Z' factory, with its decays giving Higgses and W^+W^- along with particle and sparticle pairs)
- right-handed weak bosons (the present limit on the right-handed weak boson of left-right symmetric models is $M_{W_R} \gtrsim 1.5$ TeV)
- vector-like quarks and leptons (present in E_6 models)
- horizontal gauge bosons X (whose presence may be detected as an interference between t-channel X exchange and s-channel γ, Z exchanges; present limits are $M_X \gtrsim 1$ TeV)
- leptoquarks

The list goes on with other exotica.

IV. CONCLUSION

In conclusion, $\mu^+\mu^-$ colliders seem to offer unparalleled new opportunities at both the low ($\sqrt{s} \simeq 400$ GeV) and high ($\sqrt{s} \simeq 4$ TeV) energy frontiers. The primary advantages of such a collider are:

- to discover and study Higgs bosons that are not coupled to ZZ or WW (e.g. the heavy SUSY Higgs bosons H, A) by employing either direct s-channel resonance production (at a low energy machine) or $Z^* \to HA$ (at a high energy machine);
- to measure the masses and properties of heavy SUSY particles in the improved background environment of a lepton collider;
- to study a strongly interacting electroweak sector with higher signal rates at higher energies.

Along with accelerator development, much work remains to be done on the physics for such machines and a continuing investigation is underway[14].

ACKNOWLEDGMENTS

This work was supported in part by the U.S. Department of Energy under Grant numbers DE-FG02-95ER40896, DE-FG03-91ER40674 and DE-AC02-76CH00016. Further support was provided by the University of Wisconsin Research Committee, with funds granted by the Wisconsin Alumni Research Foundation, and by the Davis Institute for High Energy Physics.

REFERENCES

1. *Proceedings of the First Workshop on the Physics Potential and Development of $\mu^+\mu^-$ Colliders*, Napa, California, 1992, Nucl. Instr. and Meth. **A350**, 24-5-6 (1994).
2. See summary talks in these proceedings, by R. Fernow, D. Miller, F. Mills and D. Neuffer.
3. For references to in-depth studies of physics at future e^+e^- colliders, see e.g. *Proceedings of the Workshop on Physics and Experiments with Linear Colliders*, ed. F.A. Harris, *et al.*, World Scientific (1993); *Proceedings of the Workshop on Physics and Experiments with Linear e^+e^- Colliders*, Waikoloa, Hawaii (April 1993), ed. F. Harris *et al.* (World Scientific, 1993); JLC Group, KEK Report 92-16 (1992); *Proceedings of the Workshop on Physics and Experiments with Linear Colliders*, Saariselkä, Finland (Sept. 1991), ed. R. Orava *et al.*, World Scientific (1992).
4. M. Drees, Int. J. Mod. Phys. **A4**, 3635 (1989); J. Ellis *et al.*, Phys. Rev. **D39**, 844 (1989); L. Durand and J.L. Lopez, Phys. Lett. **B217**, 463 (1989); J.R. Espinosa and M. Quirós, Phys. Lett. **B279**, 92 (1992); P. Binétruy and C.A. Savoy, Phys. Lett. **B277**, 453 (1992); T. Morori and Y. Okada, Phys. Lett. **B295**, 73 (1992); G. Kane, *et al.*, Phys. Rev. Lett. **70**, 2686 (1993); J.R. Espinosa and M. Quirós, Phys. Lett. **B302**, 271 (1993); U. Ellwanger, Phys. Lett. **B303**, 271 (1993); J. Kamoshita, Y. Okada, M. Tanaka*et al.*, Phys. Lett. **B328**, 67 (1994).
5. V. Barger, *et al.*, Phys. Lett. **B314**, 351 (1993); P. Langacker and N. Polonshy, University of Pennsylvania preprint UPR-0594-T, hep-ph 9403306.
6. J.F. Gunion and H.E. Haber, Nucl. Phys. **B272**, 1 (1986).
7. V. Barger and T. Han, results presented in the introductory talk at this workshop.
8. J.F. Gunion, results prepared in advance for this workshop.
9. For discussions of the $t\bar{t}$ threshold behavior at e^+e^- colliders, see V.S. Fadin and V.A. Khoze, JETP Lett. **46** 525 (1987); Sov. J. Nucl. Phys. **48** 309 (1988); M. Peskin and M. Strassler, Phys. Rev. **D43**, 1500 (1991); G. Bagliesi, *et al.*, CERN Orange Book Report CERN-PPE/92-05; Y. Sumino ,*et al.*, KEK-TH-284, Phys. Rev. **D47**, 56 (1992); M. Jezabek, J.H. Kühn, T. Teubner, Z. Phys. **C56**, 653 (1992).
10. Z. Parsa, *$\mu^+\mu^-$ Collider and Physics Possibilities*, to be published.
11. M.S. Chanowitz and M.K. Gaillard, Nucl. Phys. **B261**, 379 (1985); J. Bagger et al., Phys. Rev. **D49**, 1246 (1994).
12. V. Barger, K. Cheung, T. Han, R.J.N. Phillips, University of Wisconsin preprint MADPH-95-865.
13. K. Hagiwara, J. Kanzaki and H. Murayama, DTP/91/18.
14. V. Barger, M.S. Berger, J.F. Gunion and T. Han, in preparation.

Physics Potential and Development of $\mu^+\mu^-$ Colliders

David B. Cline

Department of Physics, University of California Los Angeles
405 Hilgard Ave., Los Angeles, CA 90095-1547

Abstract. We review the physics motivation for two different energy $\mu^+\mu^-$ colliders: 200×200 GeV and 2×2 TeV. We also address the major components of such a collider and the uncertainty in the luminosity. Study of $\mu^\pm$ cooling in materials as well as crystals are discussed. Some discussion is given concerning the $\mu^\pm/\pi^\pm$ hadron source.

INTRODUCTION

At the Port Jefferson Advanced Accelerator Workshop in the Summer of 1992, a group investigating new concepts of colliders studied anew the possibility of a $\mu^+\mu^-$ collider since e^+e^- colliders will be very difficult, in the several TeV range[1]. A small group also discussed the possibility of a $\mu^+\mu^-$ Collider. A special workshop was then held in Napa, California in the fall of 1992 for this study. There are new accelerator possibilities for the development of such a machine, possibly at an existing or soon to exist storage ring[2, 3]. For the purpose of the discussion here, a $\mu^+\mu^-$ collider is schematically shown in Fig. 1. In this brief note we study one of the most interesting goals of a $\mu^+\mu^-$ collider: the discovery of a Higgs Boson in the mass range beyond that to be covered by LEP I & II ($\sim 80 - 90$ GeV) and

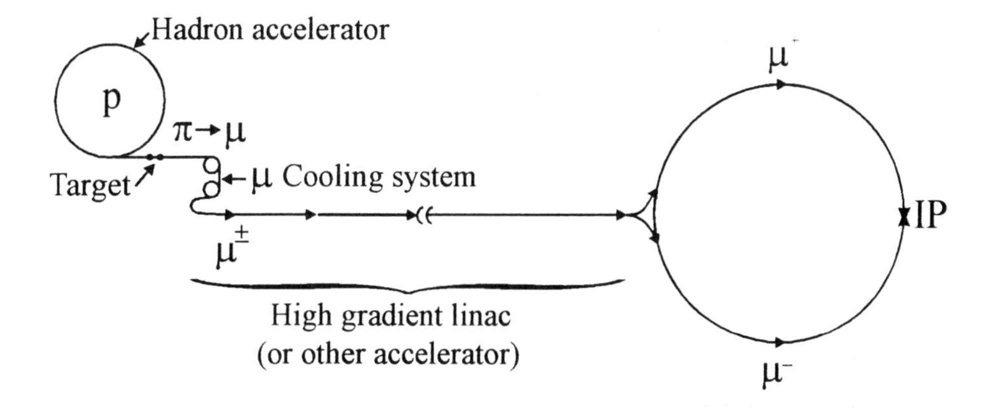

Figure 1: Schematic of a Possible $\mu^+\mu^-$ Collider Scheme – Few Hundred GeV to few TeV.

the natural range of the Super Colliders $\gtrsim 2\,\mathrm{M}_Z$[4]. In this mass range, as far as we know, the dominant decay mode of the h^0 will be

$$h^0 \quad \rightarrow \quad b\bar{b} \tag{1}$$

whereas the Higgs will be produced by the direct channel

$$\mu^+\mu^- \quad \rightarrow \quad h^0 \tag{2}$$

which has a cross section enhanced by

$$\left(\frac{M_\mu}{M_e}\right)^2 \sim (200)^2 \quad = \quad 4 \times 10^4 \tag{3}$$

larger than the corresponding direct product at an e^+e^- collider. However, we will see that the narrow width of the Higgs partially reduces this enhancement.

There is growing evidence that the Higgs should exist in this low mass range from

1. The original paper of Cabibbo, et al., shows that, when $M_t > M_Z$ and assuming a Grand Unification of Forces, $M_h < 2M_Z$ [4] [Fig. 2a].

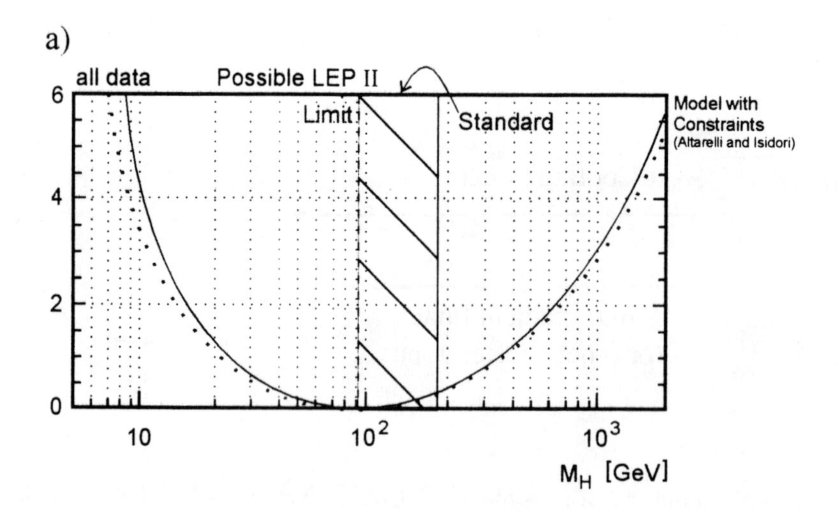

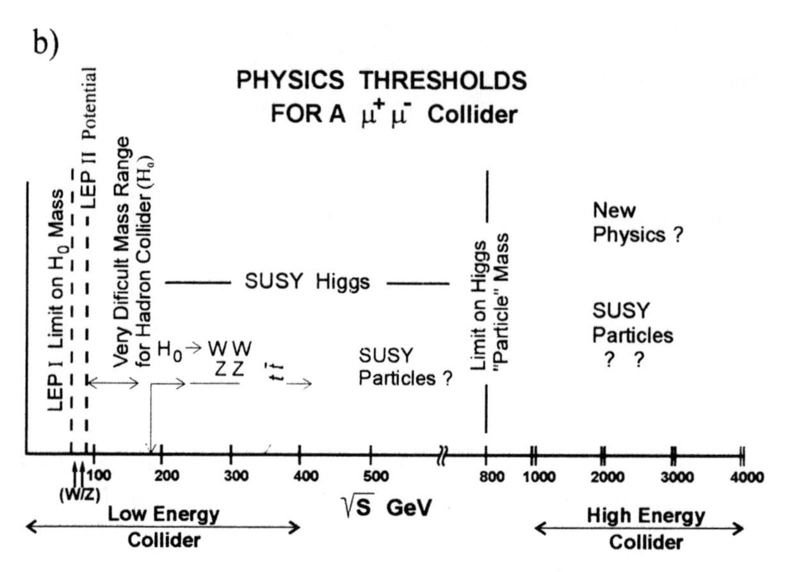

Figure 2: a) Upper and lower bounds on m_{ϕ^0} as a function of m_t, coming from the requirement of a perturbative theory[4]. b) Physics threshold for a $\mu\mu$ collider.

2. Fits to LEP data imply a low mass h^0 could be consistent with $M_t >$ 150 GeV[5].

3. The extrapolation to the GUT Scale that is consistent with SUSY also implies that one of the Higgs should have a low mass, perhaps below 130 - 150 GeV[4].

This evidence implies the exciting possibility that the Higgs mass is just beyond the reach of LEP II and in a range that is very difficult for the Super Colliders to extract the signal from background, ie. sing either $h^0 \rightarrow \gamma\gamma$ or the very rare $h^0 \rightarrow \mu\mu\mu\mu$ in this mass range, since $h \rightarrow b\bar{b}$ is swamped by hadronic background. However, detectors for the LHC are designed to extract this signal[6]. Figure 2b gives a picture of the various physics thresholds that may be of interest for a $\mu^+\mu^-$ collider.

In this low mass region the Higgs is also expected to be a fairly marrow resonance and thus the signal should stand out clearly from the background from

$$\begin{aligned} \mu^+\mu^- &\rightarrow \gamma \rightarrow b\bar{b} \\ &\rightarrow Z_{\text{tail}} \rightarrow b\bar{b} \end{aligned} \tag{4}$$

In Fig. 3, we plot the Higgs resonance signals and various types of background for this mass range. While the cross sections are fairly large, the requirements on the beam energy resolution of the $\mu^+\mu^-$ collider are very constraining. In Fig. 4 we show the relationship between the Higgs width and the required machine energy resolution. If the resolution requirements can be met, the machine luminosity of $\sim 10^{32}\,\text{cm}^{-2}\,\text{sec}^{-1}$ could be adequate to facilitate the discovery of the Higgs in the mass range of 100 - 180 GeV.

For masses above 180 GeV. the dominant Higgs decay is

$$H^0 \rightarrow W^+W^- \text{ or } Z^0Z^0 \tag{5}$$

giving a clear signal and a larger width; the machine energy resolution requirements could be relaxed somewhat!

Another possibility for the intermediate Higgs mass range is to search for

$$\mu^+\mu^- \rightarrow Z^0 + H^0 \tag{6}$$

using a broad energy sweep. The corresponding cross section is small (see Fig. 3). Once an approximate mass is determined, a strategy for the energy sweep

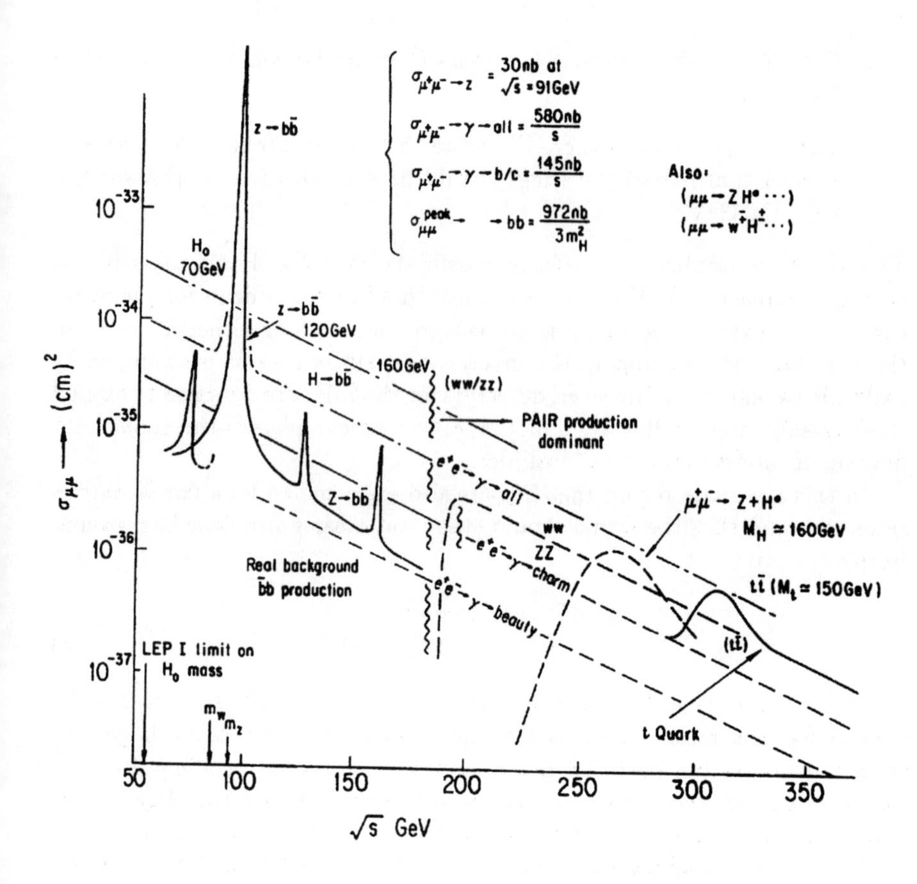

Figure 3: Estimated cross sections and backgrounds for Higgs production in the direct channel or by associated production for various Higgs mass.

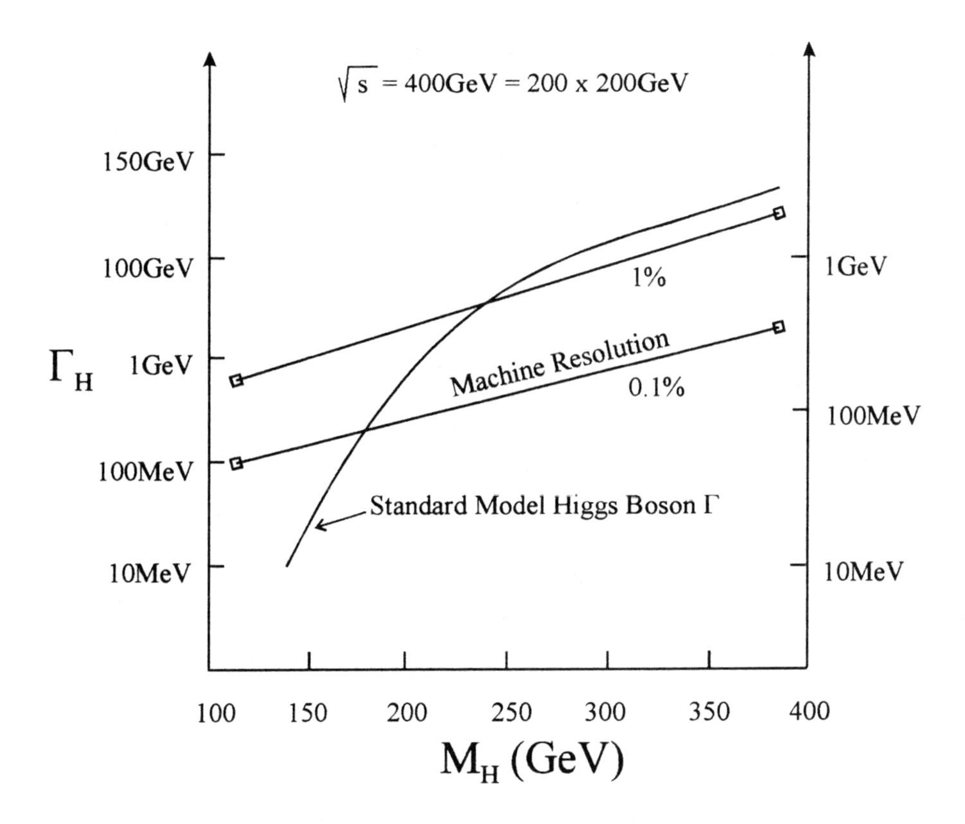

Figure 4: Higgs Search at a $\mu^+\mu^-$ collider. Required machine resolution and the expected Higgs width.

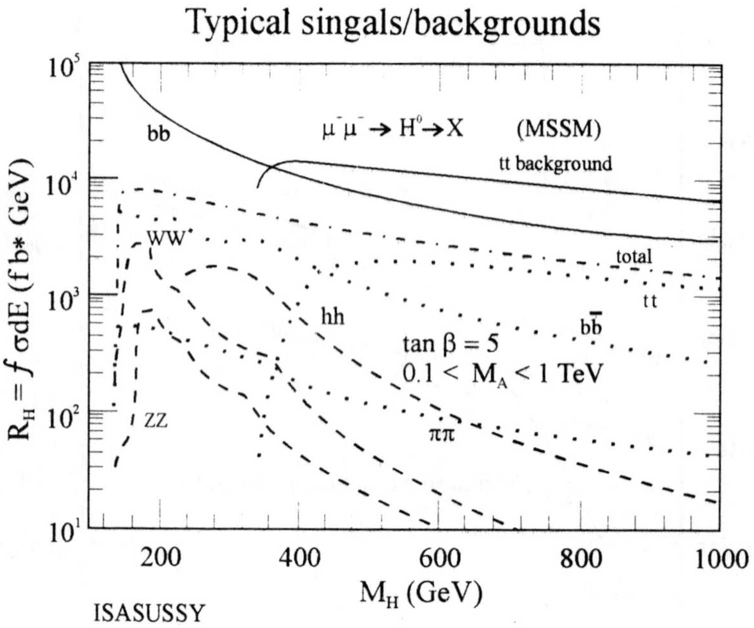

Typical singals/backgrounds

Figure 5: Total cross section for $\mu^+\mu^- \rightarrow H^0$ in the MSS M model as a function of the square root of s for the fixed m_{ϕ^0} values indicated by the numbers (in GeV) beside each line (from the talk of V. Barger at the Sausalito workshop).

through the resonance can be devised. The study of the t quark through $t\bar{t}$ production, would also be interesting.

Finally, another possibility is to use the polarization of the $\mu^+\mu^-$ particles orientated so that only scalar interactions are possible (thus eliminating the background from single photon intermediate states as shown in Fig. 3). However, there would be a trade-off with luminosity and thus a strategy would have to be devised to maximize the possibility of success in the energy sweep through the resonance.

At the Napa workshop the possibility of developing a $\mu^+\mu^-$ collider in the range of luminosity of $10^{31}\ \mathrm{cm}^{-2}\,\mathrm{sec}^{-1}$ was considered and appears feasible. However, higher luminosity will be required for the Higgs sweep. It is less

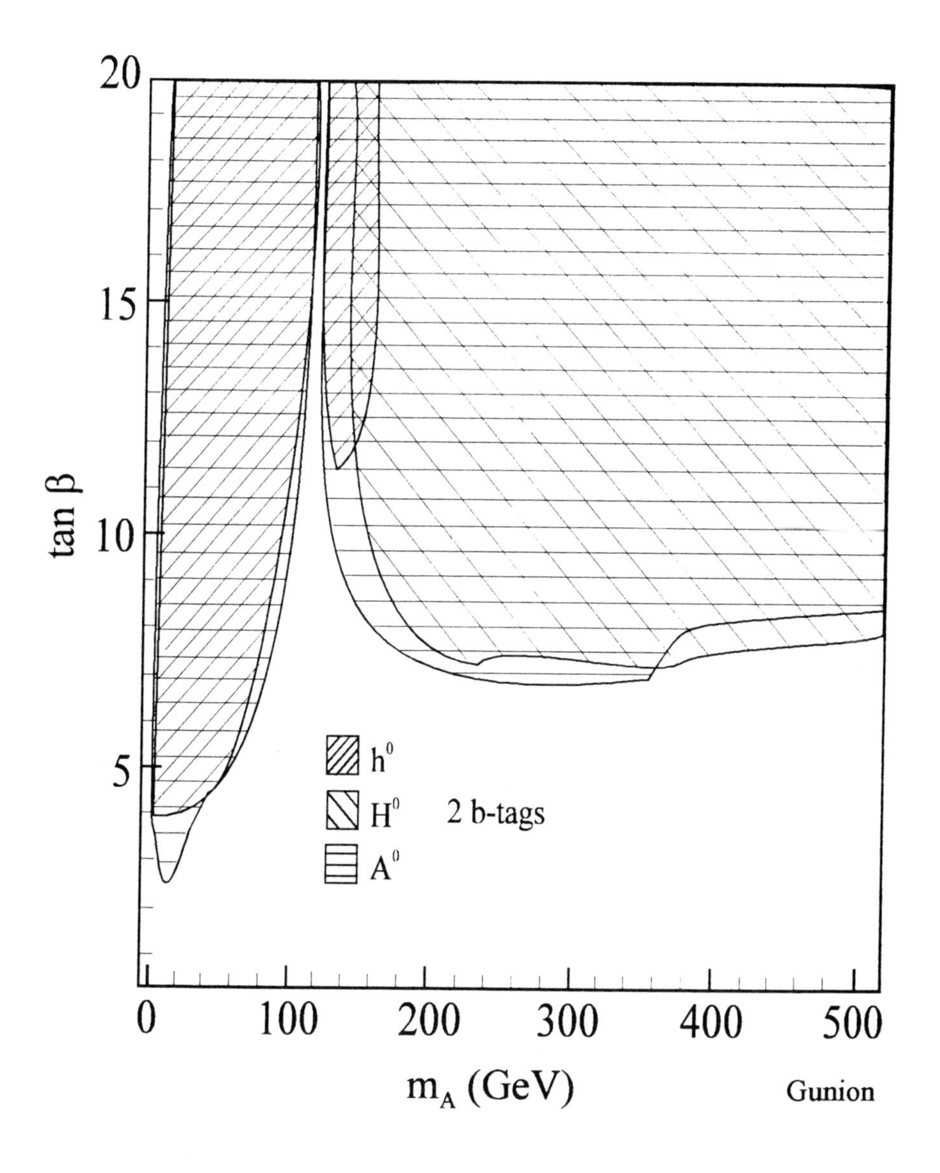

Figure 6: Muon Collider $b\bar{b}$ discovery contours survey broad spectrum (L=10Fb^{-1}, ΔE=500GeV) m_t = 175GeV, ϵ_b = 0.6 n_{SD} = 4, L=0.2fb^{-1}/GeV. From the Sausalito workshop talk of J. Gunion.

Table 1: Key issues in the Higgs search.

---

1) There is growing evidence that one Higgs particle is below $2M_Z$

2) SUSY Higgs - 3 Higgs - one near M_Z
   (possibly up to $\sim 130\,\text{GeV}$)
   Extremely hard to detect

3) Hadron machines can search for these Higgs provided:
   (i) $\int \mathcal{L}dt \geq 10^5\,\text{pb}$ (LHC)
   (ii) The background for $H \to \gamma\gamma$ is small enough

4) A $\mu^+\mu^-$ collider with $\leq \mathcal{L} \gtrsim 10^{32}\,\text{cm}^{-2}\,\text{sec}^{-1}$
   operating between 100 - 180 GeV
   could discover the Higgs provided
   sufficient energy resolution is achieved!

---

certain that the high energy resolution required for the Higgs sweep can be obtained. We summarize in Table 1 some of the key issues in the Higgs search.

$\mu^+\mu^-$ colliders could also be very important in the TeV energy range, however, since the cross sections for new particle production are much smaller, the luminosity requirements would be $\mathcal{L}_{\mu^+\mu^-} \geq 10^{33}\,\text{cm}^{-2}\,\text{sec}^{-1}$. This is the energy range where e^+e^- linear colliders are extremely difficult to develop[1]. This possibility was one focus of the 2nd workshop held in Sausalito in 1994.

Table 2: Parameters for a Muon Collider: 100GeV×100GeV[3].

$$\frac{L = M N_+ N_- f}{4\pi \epsilon_N \beta^* \gamma \sim 10^{30}\,\mathrm{cm}^{-2}\mathrm{sec}^{-1}}$$

$M = 1000$; $\gamma = 1000$; $\beta^* = 1\mathrm{cm}$; $P = 5\mathrm{MW}$ at the target

	Production via electrons	Production via protons
E_e or p (GeV)	60	30
Intensity	5×10^{11} pulse	10^{14}/pulse
Number of Pulses	100 (stacked later)	1
Repetition rate	10Hz	10Hz
E_μ (GeV)	40	1.5
ϵ_N (π m − rad)	2×10^{-3}	2×10^{-2}
$\Delta p/p$	$\pm 3\%$	$\pm 3\%$
(μ/e) or (μ/p)	4×10^{-3}	10^{-3}
Ionization cooling	$\epsilon_n^f = 2 \times 10^{-5}\,\pi$	$\epsilon_n^f = 2 \times 10^{-5}\,\pi$
Bunch rotation factor	none	100

RESULTS FROM THE NAPA SAUSALITO $\mu^+\mu^-$ COLLIDER WORKSHOP

At the Napa meeting (92) a small group of excellent accelerator physicists struggled with the major concepts of a $\mu^+\mu^-$ collider; some results are published in NIM, Oct., 1994. Table 2 lists estimated luminosity of the 200×200GeV $\mu^+\mu^-$ collider from this meeting. At the Sausalito (94) meeting a larger group of accelerator, particle and detector physicists were involved.

The proceedings will be published by AIP press in 1995.

The second workshop on the *Physics Potential & Development of $\mu^+\mu^-$ Colliders* was held in Sausalito, California, on November 16-19, 1994. There were some 60 attendees which began with a Reception and Registration. The morning of 17 November was devoted to an overview of the particle physics goals, detector constraints, the $\mu^+\mu^-$ collider and μ cooling and source issues. The major issue confronting this collider development is the possible luminosity that is achievable. Two collider energies were considered: 200×200GeV and 2×2TeV. The major particle physics goals are the detection of the Higgs Boson(s) in the s channel for the low energy collider and WW scattering as well as supersymmetric particle discovery.

At the Sausalito workshop the goal was to see if a luminosity of 10^{32} to 10^{34}cm^{-2}sec^{-1} for the two colliders might be achievable and useable by a detector. There were five working groups on the topics of (1) Physics, (2) 200×200GeV Collider, (3) 2×2TeV Collider, (4) Detector Design and Backgrounds, and (5) μ Cooling and production methods. Considerable progress was made in all these areas at the workshop.

Table 3 gives a more optimistic luminosity for the low energy collider. Table 4 gives the rather optimistic parameter list for a 2×2TeV collider from the work of Neuffer and Palmer!

The $\mu^+\mu^-$ collider has a powerful physics reach, especially if the $\mu^\pm$ polarization can be maintained. One interesting possibility is the observation of the supersymmetric Higgs Boson(s) in the direct channel. The detector backgrounds will be considerable due to high energy μ decays upstream of the detector. In the summary of working group 4 it was concluded that these backgrounds might be manageable. One key to achieving a high luminosity collider is the collection of the $\mu^\pm$ from $\pi^\pm$ decays over nearly the full phase space over which they are produced. This is far from trivial and leads to conclusions from groups 2 and 3 that the present uncertainty in luminosity is of order $10^{2\pm1}$ (at most 4 orders of magnitude, but perhaps, realistically, 2 orders of magnitude, which, unfortunately, spans the range from being uninteresting to being very interesting). Hopefully, before the next meeting this uncertainty can be reduced. Perhaps the most interesting aspect of a $\mu^+\mu^-$ collider is the need to cool the $\mu^\pm$ beams over a very large dynamic range. Three experimental programs were discussed and are being initiated to study μ cooling: at BNL, FNAL, and a UCLA group is proposing to study cooling and acceleration in crystals at TRIUMF. One major conclusion of the meeting is that $\mu^+\mu^-$ colliders are complimentary to both pp (LHC) and e^+e^- (NLC) colliders, especially for the Higgs Sector and for the study of supersymmetric particles [see Figs. 5 and 6].

Table 3: Parameter List for 400GeV $\mu^+\mu^-$ Collider, Sausalito, Nov. 1994, 200x200GeV Working Group Report.

Parameter	Symbol	Value
Collision energy	E_cm	400GeV
Energy per beam	E_μ	200GeV
Luminosity	$L = f_n\, n_s\, n_b\, N_\mu^2\, /4\pi\sigma^2$	$1 \times 10^{31}\,\mathrm{cm}^{-2}\mathrm{sec}^{-1}$
	Source Parameters	
Proton energy	E_p	30GeV
Protons/pulse	$N - p$	$2 \times 3 \times 10^{31}$
Pulse rate	$f - o$	10Hz
μ (prod./accept.)	μ/p	0.03
μ survival	$N_\mu\, /\, N_{\mathrm{source}}$	0.33
	Collider Parameters	
Number of μ/bunch	$N_{\mu\pm}$	3×10^{11}
Number of bunches	n_B	1
Storage turns	$2\; n_s$	1500 (B = 5T)
Normalized emittance	ϵ_N	10^{-4} m-rad
μ-beam emittance	$\epsilon_t = \epsilon_N\, /\, \gamma$	5×10^{-8} m-rad
Interaction focus	β_o	1 cm
Beam size at interaction	$\sigma = (\epsilon_t\, \beta_0)^{1/2}$	2.1 μm

Table 4: Parameter List for 4 TeV $\mu^+\mu^-$ Colliders (Neuffer and Palmer).

Parameter	Symbol	Value
Collision energy	E_cm	2TeV
Energy per beam	E_μ	200GeV
Luminosity	$L = f_n\, n_s\, n_b\, N_\mu^2 \,/4\pi\sigma^2$	$10^{34}\ \mathrm{cm}^{-2}\mathrm{sec}^{-1}$
	Source Parameters	
Proton energy	E_p	30GeV
Protons/pulse	N_p	$2 \times 3 \times 10^{31}$
Pulse rate	f_0	10Hz
μ (prod./accept.)	μ/p	15
μ survival	$N_\mu \,/\, N_{\mathbf{source}}$	25
	Collider Parameters	
Number of μ/bunch	$N_{\mu\pm}$	10^{12}
Number of bunches	n_B	1
Storage turns	$2\ n_s$	500
Normalized emittance	ϵ_N	3×10^{-5} m-rad
μ-beam emittance	$\epsilon_t = \epsilon_N \,/\, \gamma$	1.5×10^{-9} m-rad
Interaction focus	β_o	0.3 cm
Beam size at interaction	$\sigma = (\epsilon_t\, \beta_0)^{1/2}$	2.1 μm

THE POSSIBLE LUMINOSITY OF A $\mu^+\mu^-$ COLLIDERS

The luminosity is given by

$$\mathcal{L} = M \frac{N_{\mu^+} N_{\mu^-} F}{4\pi \varepsilon_N \beta^*} \gamma \tag{7}$$

The $N_{\mu\pm}$ depend directly on the $\mu^{\pm}$ production and capture rate (μ/p), f is related to the magnetic field of the collider, ε_N the final μ invariant emittance from the final stage of cooling, and β^* will depend on the bunch length (and the longitudinal cooling of the $\mu^{\pm}$ beams), as well as the collider lattice.

We can rewrite the luminosity as

$$\mathcal{L} \propto \frac{(\mu/p^2)\, B_{\text{collider}}\, \gamma}{\varepsilon_N\, (\text{final } \beta^*\ \mu \text{ cooling})} \tag{8}$$

In order to increase $\mathcal{L}$ we must increase (μ/p) and B and decrease ε_N (μ final) and β^*. At the Napa meeting the best judgement of the group was that $(\mu/p) \sim 10^{-3}$ and $\epsilon_N \sim 2 \times 1^{-5}$ πm-rad. $\beta^* \sim 1$cm. For the case of $\gamma = 200$ (see the paper of Neuffer) $\mathcal{L} = 2 \times 10^{30}\,\text{cm}^{-2}\text{sec}^{-1}$. If on the other had we use the optimistic values of Neuffer and Palmer of $(\mu/p) = 0.2$, $\beta^* = 1/3$cm, and $\varepsilon_N = 3 \times 10^{-5}$m-rad, we find

$$\begin{aligned} \mathcal{L} &= \left[2 \times \left(4 \times 10^4\right) \times (3)\right] \mathcal{L}_0 \\ &= 4.8 \times 10^{35}\,\text{cm}^{-2}\text{sec}^{-1} \end{aligned} \tag{9}$$

a very large luminosity, never before achieved by any collider! Clearly this must be far too optimistic. Tables 3 and 4 give some parameters of low energy and high energy colliders.

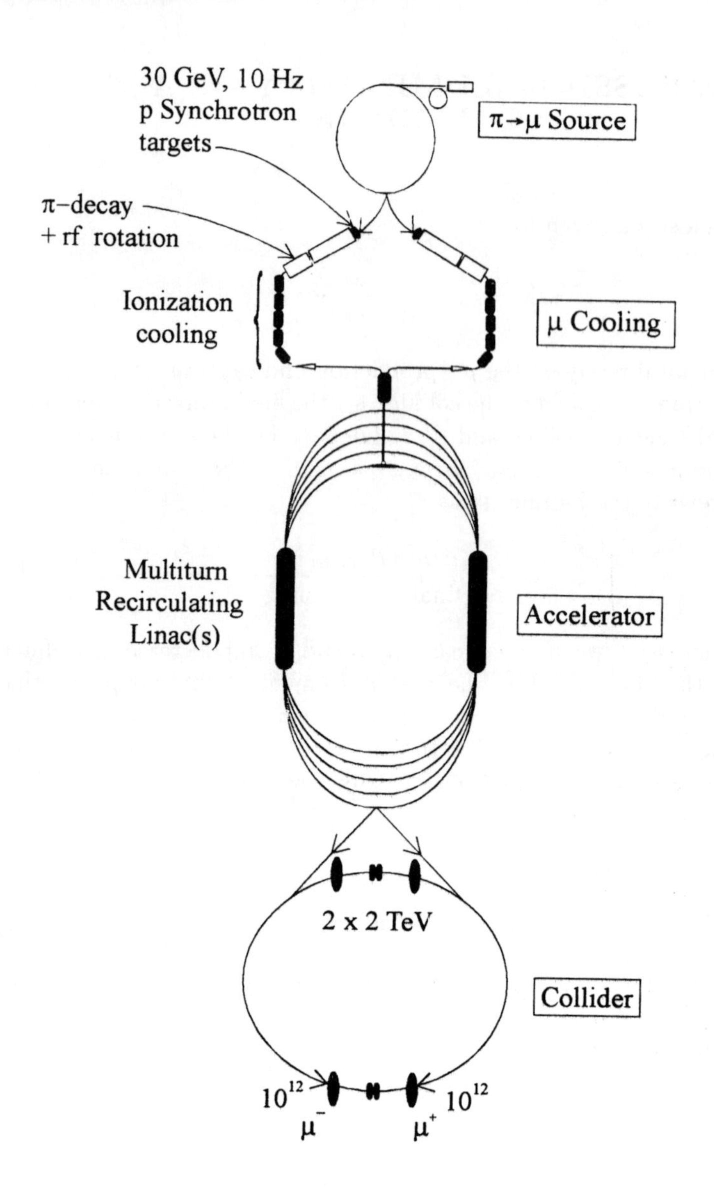

Figure 7: Schematic of high energy $\mu\mu$ collider (Neuffer and Palmer)[7].

COOLING EXPERIMENT WITH CRYSTALS

(Adapted from a paper by P. Sandler, A. Bogacz and D. Cline)

Recent results on the radiation reaction of charged particles in a continuous focusing channel indicate and efficient method to damp the transverse emittance of a muon beam[8]. This could be done without diluting the longitudinal phase-space significantly. There is an excitation-free transverse ground state to which a channeling particle will always decay, by emission of an x-ray photon. In addition, the continuous focusing environment in a crystal channel eliminates any quantum excitations from random photon emission, by constraining the photon recoil selection rules. A relativistic muon entering the crystal with a pitch angle of θ_p that is within the critical channeling angle around θ_c = 7mrad[9], will satisfy the undulator regime requirements given by the following inequality,

$$\gamma\theta_p \ll 1. \tag{10}$$

In this case, the particle will lose a negligible amount of the total energy while damping to the transverse ground state. Muons of the same energy but different θ_p will all end up in the same transverse ground state, limited by the uncertainty principle. Theoretically predicted ground state emittance is given by the following expression

$$\gamma\epsilon_{\min} = \frac{\lambda\!\!\!^{-}_{\mu}}{2}, \tag{11}$$

where $\lambda\!\!\!^{-}_{\mu} = \frac{\hbar}{m_{\mu}c}$ is the Compton wavelength of a muon.

Following the solution of Klein-Gordon equation, photons emitted in a "dipole regime", given by Eq.(10), obey the following selection rule

$$\Delta n = n_i - n_f = 1, \tag{12}$$

linking energies of the initial, E_i, and the final, E_f, state of a radiating particle according to the following formula

$$E_f = \frac{E_i}{\left[1 + \frac{1}{4}(\gamma\theta_p)^2\right]^2} \approx E_i\left[1 - \frac{1}{2}(\gamma\theta_p)^2\right], \tag{13}$$

Therefore, the longitudinal energy spread will be very small, about $\frac{1}{2}(\gamma\theta_p)^2$.

This combination of both the transverse and the longitudinal phase-space features makes a radiation damping mechanism a very interesting candidate for transverse muon cooling in an ultra strong focusing environment inside a crystal.

For positive muons channeling in a silicon crystal the characteristic transverse damping time, τ is given by the following formula

$$\frac{1}{\tau} = 2r_\mu \frac{e\phi_l}{3m_\mu c}, \tag{14}$$

where r_μ is the classical radius of a muon and $e\phi_l = 3.6 \times 10^3 \mathrm{GeV/m^2}$ is the focusing strength for silicon crystal[10]. Although the characteristic damping time, $\tau \approx 10^{-6}$sec, is long one can enhance the lattice reaction by externally modulating the inter-atomic spacing of the crystal lattice. For example, we can use an acoustic wave of wavelength l, that resonates with the Doppler shifted betatron oscillations of the beam, $\gamma\lambda_\beta$, according to the following matching condition

$$\gamma\lambda_\beta = \sqrt{2}\,l \quad , \quad \lambda_\beta = 2\pi\sqrt{\frac{m_\mu c^2}{e\phi_l}}. \tag{15}$$

If a standing acoustic wave of sizable amplitude could be generated in a crystal, then the relaxation time would shorten by the factor of one plus the ratio of the acoustic to the transverse beam power.

This phase of the experiment TRIUMF will test the cooling mechanisms summarized in the theory section above. The beam momentum will be about 250MeV/c as provided by forward decay muons[11] in M11. In this case, both channeled and unchanneled muons will penetrate the 4cm crystal and the cooling process can be compared for the two. In addition, a higher energy beam tests cooling at the energies considered for realistic collider schemes. The first step of Phase II is to measure initial and final emittances of an unmodified crystal. A schematic of our proposed experimental setup is given in Figure 8. We will track each muon individually using five sets of drift chambers. This way we can identify channeled and unchanneled particles on an event-by-event basis. We can also obtain the exact initial and final emittances for channeled and unchanneled muons separately. It is also possible to separate channeled and unchanneled muons by plotting their

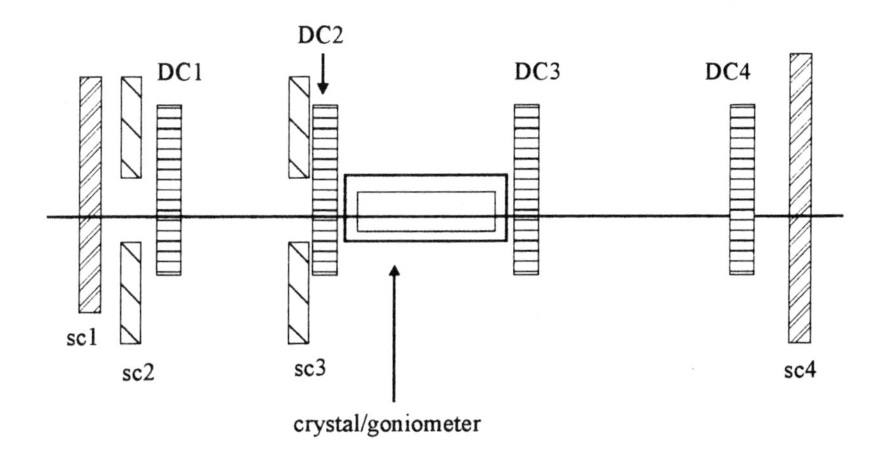

Figure 8: A schematic of Phase II: Cooling and Acceleration. Sc1-sc4 are scintillation counters, dc1-dc4 are drift chambers, and the crystal is 4cm × 1cm × 1cm silicon mounted in a goniometer.

energy loss. The trigger will be provided by scintillation counters upstream, combined with a veto counter that rejects muons which do not intersect the crystal. A time-of-flight counter will be placed in a downstream position, to be used in concert with the 1 pico second timing pulse provided by the M11 beam line. This will aid in particle identification since some positrons and pions will likely contaminate the beam.

To enhance the cooling, we will generate a strain modulation of the planar channels. An acoustic wave of 1GHz is excited via a piezoelectric transducer. We will also detect predicted channeling radiation by surrounding the crystal with CsI scintillation detectors, which are sensitive to X-rays.

The M11 beamline is presently a source of high energy pions[11]. Straightforward modification of the beamline will provide a collimated beam of forward-decay muons at high intensity – about 10^6 per second at 250MeV/c. The longitudinal momentum spread is about 2% FWHM. Assuming optimum tuning of the final focus quadrupole doublet in M11, we can achieve a spot size of 3cm×2cm with horizontal and verticle divergences of 10mrad and 16mrad respectively. The critical angle for planar channeling of μ^+ at 250MeV/c in silicon is about 7mrad, extrapolating from proton channeling data. A sizable fraction of the muons should channel through a few centimeters of the crystal.

HOW TO GET A $\mu^+\mu^-$ COLLIDER STARTED

There are many problems and also possibilities to start a $\mu^+\mu^-$ collider in the USA. For example, if crystal cooling could be used a collider of the type shown in Figure 9 might be constructed[12]. We list the major issues in the development of a collider in Table 5. In the poast the only example of such an innovative machine is the $\bar{p}p$ collider initiated by Cline, McIntyre and Rubbia[13]. In Table 6 we attempt to make a comparison between these two projects!

I wish to thank W. Barletta, D. Neuffer, A. Sessler, R. Palmer, F. Mills, and the other participants of the Napa and Sausalito workshops for many interesting discussions. Thanks also to A. Bogacz and P. Sandler for help with this paper.

Table 5:

	$\mu^+\mu^-$ Collider Development	Example in Past $\bar{p}p$ Collider (1976 – 1987) (Cline, McIntyre, Rubbia)
1)	Strong Physics Motivation HIGGS, SUSY, etc., etc. (Higgs Mass unknown - but it may be at low mass)	W/Z Discovery (M_W, Z known)
2)	Parameters Study – Are they realistic? How can we make a convincing argument	FNAL/CERN Studies (1976 - 1981) $S\bar{p}pS$/Tevatron I (LBL meeting 1978)
3)	Beam Manipulation and Cooling Rapid Acceleration Possibility? (μ Lifetime Constraint)	AA Ring and Beams ($\bar{p}$ Production yield)
4)	Demonstration of $\mu^{\pm}$ Cooling (Experiments) [Fast Stochastic Cooling?]	$p/\bar{p}$ Cooling ICE Ring Novosibirsk, FNAL (1976 - 1981)
5)	Detector Concepts And Feasibility Study	UA1/UA2 CDF/D0 Designs 1977 - 1987

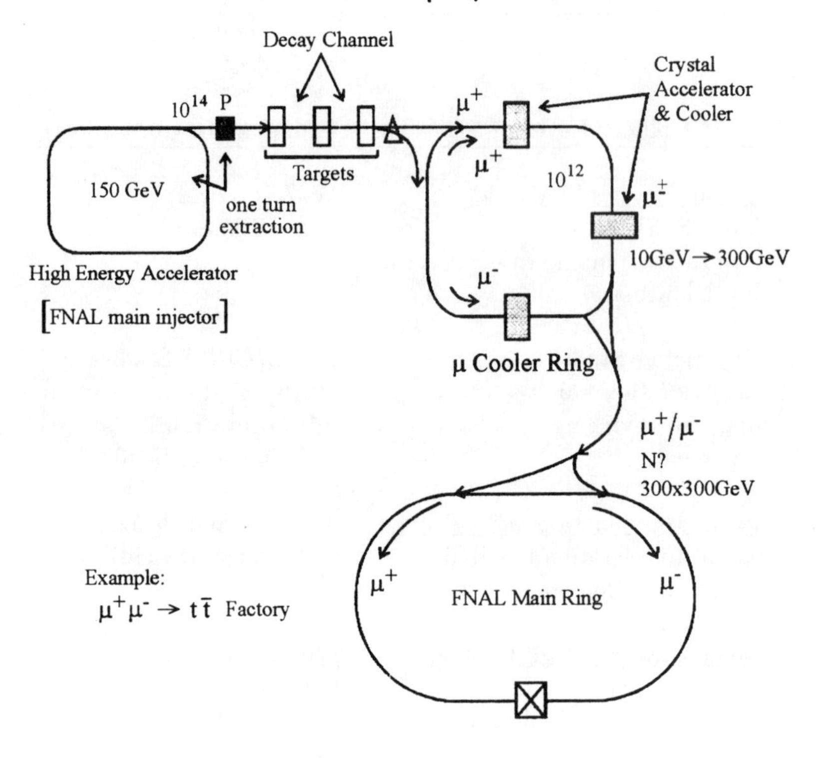

Figure 9: A scheme for a $\mu^+\mu^-$ collider using crystal cooling.

REFERENCES

[1] M. Tigner, presented at the 3rd Int. Workshop on Adv. Accel. Concepts, Port Jefferson, NY (1992); other paper to be published in AIP, (1992), ed. J. Wurtle.

[2] Early references for $\mu\mu$ colliders are: E.A. Perevedentsev and A.N. Skrinsky, Proc. 12th Int. Conf. on High Energy Accel., eds., R.T. Cole adn R. Donaldson, (1983), p. 481; D. Neuffer, Part. Accel. 14 (1984) 75; D. Neuffer, in: Adv. Accel. Concepts, AIP Conf. Proc. 156 (1987) 201.

[3] D. Cline, NIM, A350, 24 (1994) and 4 following papers constitute a mini-conf. proceeding of the Napa meeting.

[4] See for primary references to the theoretical estimates here S. Dawson, J.F. Gunion, H.E. Haber and G.L. Kane, *The Physics of the Higgs Bosons: Higgs Hunter's Guide* (Addison Wesley, Menlo Park, 1989).

[5] Adapted from a talk by D. Schaile at the WW meeting at UCLA Feb. 1995.

[6] For example the Compact Muon Solenoid detector at the LHC, CERN reports.

[7] Presentation of D. Neuffer and R. Palmer at the Sausalito workshop (to be published).

[8] Z. Hong, P. Chen and R.D. Ruth, *Radiation Reaction in a Continuous Focusing Channel*, SLAC preprint SLAC-PUB-6574, July 1994.

[9] S. Basu and R.L. Byer, Optic Letters, **13**, 458 (1988).

[10] Z.A. Bazylev and N.K. Zhevago, Sov. Phys. Usp. **33**, 1021 (1990).

[11] C.J. Oram, J.B. Warren, G.M. Marshall, and J. Dornbos, J. Nucl. Instr. & Meth. in Phys. Res., **A179**, 95 (1981).

[12] D.B. Cline, *A Crystal μ Cooler for a $\mu^+\mu^-$ Collider*, in Porc. of the 1994 Adv. Accel. Workshop, Lake Geneva, 1994.

[13] C. Rubbia, P. McIntyre and D. Cline, Proc. of 1976 Aachen Neutrino Conf., p 683 (Editors, H. Faissner, H. Reithler and P. Zerwas)(Viewag Braunschweig, 1977).

Progress Toward A High-Energy, High-Luminosity μ^+-μ^- Collider

David V. Neuffer, *CEBAF, 12000 Jefferson Avenue, Newport News VA 23606*

Robert B. Palmer, *SLAC, Stanford University, Stanford CA 94305 and Brookhaven National Laboratory, Box 5000, Upton NY 11973*

Abstract. In the past two years, considerable progress has been made in development of the μ^+-μ^- collider concept. This collider concept could permit exploration of elementary particle physics at energy frontiers beyond the reach of currently existing and proposed electron and hadron colliders. As a bench-mark prototype, we present a candidate design for a high-energy high-luminosity μ^+-μ^- collider, with $E_{cm} = 4$ TeV, $L = 3\times10^{34}cm^{-2}s^{-1}$, based on existing technical capabilities. The design uses a rapid-cycling medium-energy proton synchrotron, producing proton beam pulses which are focused onto two π-producing targets, with two π-decay transport lines producing μ^+'s and μ^-'s. The μ's are collected, rf-rotated, cooled and compressed into a recirculating linac for acceleration, and then transferred into a storage ring collider. The keys to high luminosity are maximal μ collection and cooling, and innovations with these goals are presented. Possible variations and improvements are discussed. Recent progress in collider concept development is summarized, and future plans for collider development are discussed.

I. INTRODUCTION

Recent discussions, beginning with the 1992 Port Jefferson Advanced Accelerator Concepts Workshop,[1] have indicated that the presently emphasized approaches in high-energy accelerators are reaching critical size and performance constraints. It appears that hadronic colliders (p-p) are reaching critical cost and size limitations with the CERN Large Hadron Collider (LHC, with up to 14-TeV collisions), and are performance-constrained in that they produce complicated many-particle collisions, with a rapidly-diminishing fraction (in numbers and energy) of the interactions in point-like new-particle-state production. Lepton(e^+-e^-) colliders produce simple interactions, and this magnifies the effective energy of collisions by more than an order of magnitude over hadron colliders. Extension of e^+-e^- colliders to multi-TeV energies (effectively beyond the LHC) is constrained by "beamstrahlung" and synchrotron radiation effects, which increase as $(E_e/m_e)^4$, limiting performance as well as forcing the use of two costly full-energy linacs[1], with impractically large size requirements and power demands (GW or more).

However, muons (heavy electrons, with $m_\mu = 200m_e$) have negligible radiation and beamstrahlung, and can be accelerated and stored in recirculating devices or rings. The liabilities of muons are that they decay, with a lifetime of 2.2×10^{-6}

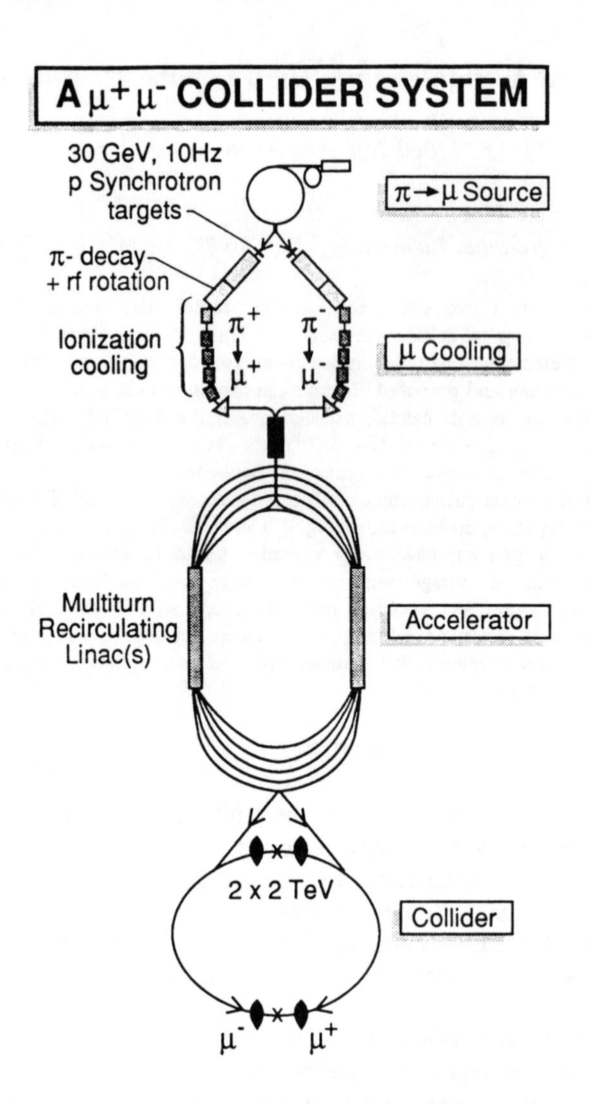

Figure 1: Overview of the μ^+-μ^- collider system, showing a muon (μ) source based on a high-intensity rapid-cycling proton synchrotron, with the protons producing pions (π's) in a target, and the μ's are collected from subsequent π decay. The source is followed by a μ-cooling system, and an accelerating system of recirculating linac(s) and/or rapid-cycling synchrotron(s), feeding μ^+ and μ^- bunches into a superconducting storage-ring collider for multiturn high-energy collisions. The entire process cycles at 10 Hz.

E_μ/m_μ s, and that they are created through decay into a diffuse phase space. But the phase space can be reduced by ionization cooling, and the lifetime is sufficient for storage-ring collisions. (At 2 TeV, τ_μ = 0.044 s.) The μ^+-μ^- collider concept had been previously introduced by Skrinsky et al.[2] and Neuffer[3]. More substantial investigations of the possibility of muon (μ^+-μ^-) colliders were initiated at the Port Jefferson meeting,[4] resulting in the first μ^+-μ^- Collider Workshop in Napa, Ca. (December, 1992)[5] This workshop and several following mini-workshops have greatly increased the level of discussion,[6, 7] stimulating the present developments and improvements which have transformed the concept into a plausible possibility, and leading to the increased degree of interest and many contributions represented in the Second Workshop on μ^+-μ^- Colliders (Sausalito, Ca., November, 1994).[8]

As a prototype, we present a design for a high-energy high-luminosity $\mu^+\mu^-$ collider, with an energy of E_{cm}=$2E_\mu$= 4TeV, and a luminosity of L = $3\times10^{34}cm^{-2}s^{-1}$, which uses only existing technical capabilities.[9, 10] In this design, we have introduced improvements, particularly in muon collection and cooling, and develop a complete scenario for a high-luminosity high-energy collider, which demonstrates that the μ^+-μ^- collider is a practical possibility. We use this example as a basis for discussion of the features and requirements of a collider system. We discuss variations and potential improvements in the various components and requirements. We then summarize the recent accomplishments in μ^+-μ^- collider development, and discuss future directions for investigation and improvement.

II. A μ^+-μ^- COLLIDER DESIGN

A. Design Overview

In this section we present details of a prototype design for a E_{cm}=$2E_\mu$= 4TeV, L = $3\times10^{34}cm^{-2}s^{-1}$, μ^+-μ^- collider. The effective energy reach of this collider is beyond that of the LHC or proposed e^+-e^- linear colliders, and the luminosity is sufficient for general-purpose high-energy physics. Table 1 shows parameters for the candidate design, which is displayed graphically in fig. 1. The design consists of a muon source, a muon collection, cooling and compression system, a recirculating linac system for acceleration, and a full-energy collider with detectors for multiturn high-luminosity collisions.

B. Muon Production

The μ-source driver is a high-intensity rapid-cycling synchrotron at KAON[11] proposal parameters (30 GeV, 10 Hz), which produces beam which is formed into two short bunches of 3×10^{13} protons. Combination of the accelerated beam bunches into two short bunches at full energy (possibly in a separate extraction

Table 1: Parameter list for a 4 TeV μ^+-μ^- Collider

Parameter	Symbol	Value
Energy per beam	E_μ	2 TeV
Luminosity	$L=f_0 n_s n_b N_\mu^2/4\pi\sigma^2$	3×10^{34} $cm^{-2}s^{-1}$
Source Parameters		
Proton energy	E_p	30 GeV
Protons/pulse	N_p	$2\times3\times10^{13}$
Pulse rate	f_0	10 Hz
μ-production acceptance	μ/p	.15
μ-survival allowance	N_μ/N_{source}	.33
Collider Parameters		
Number of μ /bunch	$N_{\mu\pm}$	1.5×10^{12}
Number of bunches	n_B	1
Storage turns	n_s	900
Normalized emittance	ε_N	3×10^{-5} m-rad
μ-beam emittance	$\varepsilon_t=\varepsilon_N/\gamma$	1.5×10^{-9} m-rad
Interaction focus	β_0	0.3 cm
Beam size at interaction	$\sigma=(\varepsilon_t\beta_0)^{1/2}$	2.1 μm

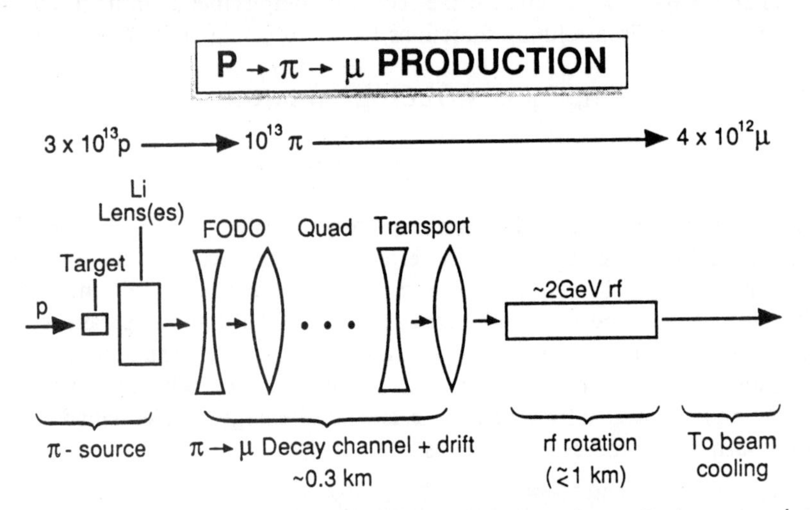

Figure 2: Overview of the $\pi \rightarrow \mu$ production transport line. Each proton bunch collides into a target, producing large numbers of π's (~1 π/interacting p) over a broad energy and angular range (E_π =0–4 GeV, $p_\perp$ < 0.5 GeV/c). The target is followed by a Li lens, which collects the π's into a transport line, which accepts a large energy width (2±1 GeV) and has a large transverse acceptance($p_\perp$< 0.4 GeV/c). This line is sufficiently long to insure $\pi \rightarrow \mu$ decay, plus rf rotation, which reduces the energy spread. The μ-beam is then matched into a beam-cooling system.

ring) simplifies the subsequent longitudinal phase space manipulations. The two proton bunches are extracted into separate lines for μ^+ and μ^- production. (Separate lines permit use of higher-acceptance, zero-dispersion $\pi \rightarrow \mu$ capture lines.) Each bunch collides into a target, producing π's (~1 π/interacting p) over a broad energy and angular range (E_π = 0.2—4 GeV, $p_\perp$ < 0.5 GeV/c). The target is followed by Li lenses, which collect the π's into a large-aperture high-acceptance transport line (an r = 0.15m, B= 4T FODO transport with a 0.8m period in a current scenario), designed to accept a large energy width (2±1 GeV) and have a large transverse acceptance($p_\perp$ < 0.4GeV). This array is sufficiently long (~300 m) to insure $\pi \rightarrow \mu$ decay, plus debunching, in which the energy-dependent particle speeds spread the beam longitudinally to a full width of ~6 m, while reducing the local momentum spread.

This is followed by an rf debuncher, a nonlinear rf system (3 harmonics are sufficient) which flattens the momentum spread.[12] We have conservatively assumed that rf gradients at these frequencies (~10—30 MHz) are limited to less than ~2 MeV/m; this implies a total rf debuncher length of ~1.4 km. The resulting μ-beam is then matched into a beam cooling system. Figure 2 shows a schematic overview of the production and collection system.

A Monte Carlo program (MCM - Monte Carlo Muon) has simulated the muon production and cooling.[13] The generation of π's in the target is calculated using a thermodynamic model or the Wang distribution,[14] validated by comparisons with production data. (The Wang distribution can be approximated by the following empirical formula, useful for initial estimates:

$$\frac{d^2N}{dP\,d\Omega} = AP_pX(1-X)e^{-BX^C - DP_t} \quad \frac{\text{pions}}{\text{sr-GeV/c}} / \text{interacting proton} \tag{1}$$

where P_p is the incident proton momentum, X = P_π/P_p is the pion/proton momentum ratio, P_t is the pion transverse momentum and A = 2.385 (1.572), B = 3.558 (5.732), C = 1.333 (1.333), and D = 4.727 (4.247) for positive (negative) pions. In this formula, pions are produced with a mean transverse momentum of ~D^{-1} or ~0.2 GeV.) In the MCM simulations, the π's are tracked through decay to μ's, and through phase-energy rotation, into the cooling system. We obtain ~0.15 captured μ's per inital proton, with ε_N =0.01 m-rad, an rms bunch length of 3m, and energy width of 0.15 GeV with an average energy of 1 GeV.

The μ capture efficiency (0.15μ/p) is larger than estimated in previous scenarios[2,3,4], and this is a result of the use of a high acceptance decay transport with a larger momentum acceptance. The transport is followed by a linac-based rf rotation, which reduces the momentum spread to a level acceptable by the subsequent transport and cooling system, while lengthening the beam bunch.

Previously, Noble[4] has noted that $\mu^+\mu^-$ collider luminosity increases with the energy spread acceptance ΔE_A by as much as ΔE_A^4, with factors accumulating from the production and the decay acceptances of both beams. The high-acceptance transport plus rf rotation has increased that acceptance by almost an order of magnitude above previous estimates. Also, for the first time, the production has been directly calculated in a realistic simulation, and not simply estimated.[13]

C. Beam Cooling

For collider intensities, the phase-space volume must be reduced by beam-cooling and the beam size compressed, within the μ lifetime. Much of the needed compression is obtained through adiabatic damping in acceleration from GeV-scale μ collection to TeV-scale collisions. Beam cooling is obtained by "ionization cooling" of muons ("μ-cooling"), in which beam transverse and longitudinal energy losses in passing through a material medium are followed by coherent reacceleration, resulting in beam phase-space cooling.[2,3,15] Figure 3 shows a schematic view of the cooling process. (Ionization cooling is not practical for protons and electrons because of nuclear scattering (p's) and bremstrahlung (e's) effects, but is for μ's and the necessary energy losses are easily obtained within the μ lifetime.) In this section we present the equations for μ-cooling, use these to deduce optimal cooling conditions, and generate a practical cooling scenario.

The equation for transverse cooling is:

$$\frac{d\epsilon_N}{ds} = -\frac{dE}{ds}\frac{\epsilon_N}{E_\mu} + \frac{\beta_\perp}{2}\frac{(0.014)^2}{E_\mu m_\mu}\frac{1}{L_R} \qquad (2)$$

(with energies in GeV), where ε_N is the normalized emittance, $\beta_\perp$ is the betatron function at the absorber, dE/ds is the energy loss, $E_\mu = \gamma m_\mu c^2$ is the muon energy and L_R is the material radiation length. The first term in this equation is the coherent cooling term and the second term is heating due to multiple scattering. This heating term is minimized if $\beta_\perp$ is small (strong-focusing) and L_R is large (a low-Z absorber). (We have anticipated the eventual optimization at relativistic energies by using relativistic approximations (v/c=1). In a more complete derivation, factors of v/c would appear. Momentum p_μ rather than E_μ would be a preferred variable.)

The equation for energy cooling is:

$$\frac{d\langle(\Delta E)^2\rangle}{ds} \approx -2\frac{\partial\frac{dE_\mu}{ds}}{\partial E_\mu}\langle(\Delta E)^2\rangle + \frac{d\langle\Delta E_\mu^2\rangle}{ds} \tag{3}$$

Energy-cooling requires that $\partial(dE_\mu/ds)/\partial E > 0$. The energy loss function, dE_μ/ds, is rapidly decreasing with energy for $E_\mu < 0.2$ GeV (and therefore heating), but is slightly increasing (cooling) for $E_\mu > 0.3$ GeV. This small natural cooling is ineffective, because of the relatively large rms energy straggling; but $\partial(dE_\mu/ds)/\partial E$ can be increased by placing a transverse variation in absorber density or a wedge absorber where position is energy-dependent. (This variation is used in two modes: a weak variation to balance cooling rates, or a thick wedge to transfer phase space.) The sum of cooling rates is invariant:

$$\frac{1}{E_{cool,x}} + \frac{1}{E_{cool,y}} + \frac{1}{E_{cool,\Delta E}} = \text{Constant} \approx \frac{2}{E_\mu}\ , \tag{4}$$

where E_{cool} is the total energy loss needed to obtain an e-folding of cooling and E_μ is the μ energy.

In the long-pathlength Gaussian-distribution limit, the heating term or energy straggling term is given by:[16]

$$\frac{d(\Delta E_\mu)^2}{ds} \approx 4\pi(r_e m_e c^2)^2 N_0 \frac{Z}{A}\rho\gamma^2(1-\beta^2/2)\ , \tag{5}$$

where N_0 is Avogadro's number and ρ is the density. Since this increases as γ^2, and the cooling system size scales as γ, cooling at low energies is desired.

To obtain energy cooling and to minimize energy straggling, we require cooling at low relativistic energies ($E_\mu \sim 300$ MeV). For optimum transverse cooling, the ideal absorber is itself a strong focussing lens which maintains small beam size over extended lengths, and a low-Z material. In this design, we use Be ($Z=4$) or Li ($Z=3$) current-carrying rods, where the high current provides strong radial focussing. For Be, $Z=4$, $A=9$, $dE_\mu/dx = 3$ MeV/cm, and $\rho = 1.85$ gm/cm^3.

The beam cooling system reduces transverse emittances by more than two

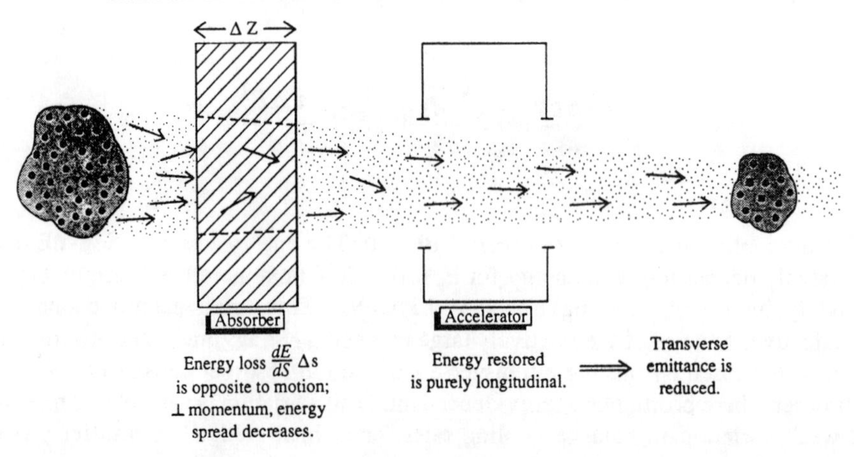

Figure 3: Schematic view of transverse "ionization cooling". Energy loss in an absorber occurs parallel to the motion; therefore transverse momentum is lost with the longitudinal energy loss. Energy gain is longitudinal only; the net result is a decrease in transverse phase-space area. Statistical beam heating by random multiple scattering opposes the coherent cooling.

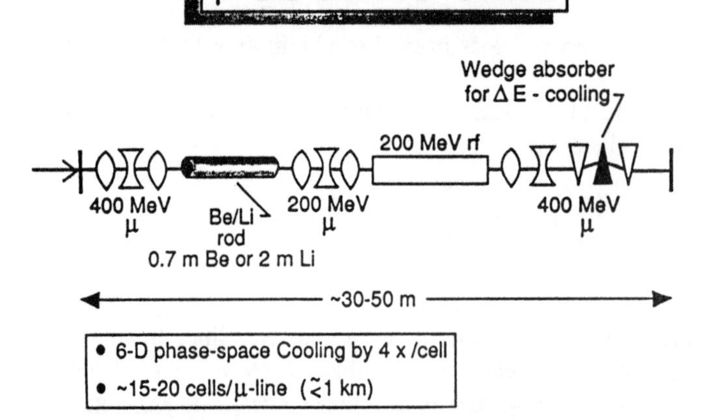

Figure 4: Overview of a μ-cooling (ionization cooling) cell. A typical cooling cell consists of a focusing cooling rod which reduces the central energy from 400 to 200 MeV, followed by a ~200 MeV linac, with appropriate optical matching. Cooling by a 3-D factor of ~4 is obtained in each cell, and 15-20 such cells (~1 km) are needed in the complete machine. For energy cooling, we introduce a dispersion (position-dependence on momentum), and use an absorber with a density gradient or wedge shape. This arrangement permits enhancement of energy-cooling[2, 3, 15], and the parameters are adjusted to obtain optimal transverse and longitudinal cooling.

orders of magnitude (from 0.01 to 3×10^{-5} m-rad), and reduces longitudinal emittance by more than an order of magnitude. This cooling is obtained in a series of cooling cells, with the initial cells reducing the energy toward the cooling optimum of 300 MeV. A typical cooling cell consists of a focusing cooling rod (~0.7 m long for Be, ~2.1 m for Li) which reduces the central energy by ~200 MeV, followed by a ~200 MeV linac (20—40 m at 10—5 MeV/m), with optical matching sections (~40–60 m total cell length) (see fig. 4). Small angle bends introduce a dispersion (position-dependence on energy), and wedge absorbers (or density gradients) introduce an energy loss dependence on beam energy. Bends are also used to provide path length dependence on momentum, in order to compress the bunch lengths. The cell parameters are adjusted to obtain optimal transverse and longitudinal cooling rates, and cooling by a 6-D factor of ~4 is obtained in each cell. ~15–20 such cells (~800 m) are needed in the complete machine.

From equation 1, we find a limit to transverse cooling when the multiple scattering balances the cooling, at $\varepsilon_N \approx 10^{-2}\ \beta_\perp$ for Be. The value of $\beta_\perp$ in the Be rod is limited by the peak focusing field to $\beta_\perp \sim 0.01$ m, obtaining $\varepsilon_N \sim 10^{-4}$ m-rad. This is a factor of ~3 above the emittance goal of Table 1. The additional factor can be obtained by cooling more than necessary longitudinally and exchanging phase-space with transverse dimensions in a thick wedge absorber. MCM simulations have demonstrated that the desired cooling and phase-space exchange can be obtained.

D. Acceleration and Collisions

Following cooling and initial bunch compression to ~.1–.3m bunch lengths, the beams are accelerated to full energy (2 TeV). A full-energy linac would work, but it would be costly and does not use our ability to recirculate μ's. A recirculating linac (RLA) like CEBAF[17] can accelerate beam to full energy in 10—20 recirculations, using only 200—100 GeV of linac, but requiring 20—40 return arcs. The μ-bunches would be compressed on each of the return arcs, to a length of 0.003m at full energy. A cascade of RLAs (i. e., 1–10, 10–100 and 100–2000 GeV), with rf frequency increasing as bunch length decreases, may be used. The cooling and acceleration cycle is timed so that less than ~half the initial μ's decay. (In Table 1, we allow a factor of 3 in total losses.)

After acceleration, the μ^+ and μ^- bunches are injected into the 2-TeV superconducting storage ring (~1-km radius), with collisions in one or two low-β^* interaction areas. The beam size at collision is $r = (\varepsilon_N\ \beta_0/\gamma)^{1/2} \sim 2\mu$m, similar to hadron collider values. The bunches circulate for ~300B turns before decay, where B is the mean bending field in T. (This is 150B luminosity turns, a factor of two smaller since both beams decay.) The design is restricted by μ decay within the rings ($\mu \rightarrow e\ \nu\nu$), which produces 1/3-energy electrons which radiate and travel to the inside of the ring dipoles. This energy could be intercepted by a liner inside the magnets, or specially designed C-dipoles could be used and the electrons intercepted in an external absorber. The design constraints may limit B; we have chosen B = 6T (900 turns) in the present case. μ-decays in the interaction areas will also provide some background levels in detectors. The limitations in detector design are being studied.

III. VARIATIONS AND IMPROVEMENTS

A. Design Overview

We have presented a candidate scenario for a high-energy high-luminosity μ^+-μ^- collider. In this section we discuss variations in the scenario, both in outline and details, which should be considered and studied. This discussion should not be considered complete. The most important innovations have probably not been invented yet, and will result from the thoughts and investigations of our future research, including the μ^+-μ^- workshops.

The luminosity estimate of the scenario should not be considered as an absolute limit. Luminosity could be readily increased by reducing losses, or by increasing the source repetition rate (increasing the source to above KAON-class intensity). In the present workshop,[8] potential difficulties and possible improvements were identified. Within present uncertainties in implementation (μ-production, cooling, possible improvements or regressions, etc.), the achieved luminosity could be as much as two orders of magnitude larger or smaller. More precision is needed.

The 4 TeV energy was set as a benchmark; however it is at a size and scope similar to existing facilities. Higher or lower energy colliders are possible, and we are not near absolute luminosity limitations. The μ^+-μ^- collider concept naturally increases in luminosity with energy. One factor of E_μ increase results from transverse emittance adiabatic damping. Since the beam also decreases in size longitudinally by adiabatic damping, smaller interaction-region β^* should also be possible, permitting further enhancement. If the injector intensity/cost is also allowed to increase (beyond KAON-class) in proportion to E_μ, then beam intensities increase as E_μ, and luminosity increases by at least another factor of E_μ. These increases of luminosity with energy, by at least factors of E_μ^2, should be followed up to the 100 TeV scale, beyond which μ synchrotron radiation becomes large. (Radiation damping would then initially permit further improvements.)

Lower energy machines (100–400 GeV Higgs, top factories, etc.) are also possible, particularly if there are strong physics motivations.[18] Lower-energy machines could also be a useful step toward a high-energy machine. These would have lower luminosity, because of the adiabatic damping factors. However, if a

KAON-class muon source is also used, a relatively high luminosity ($L \sim 10^{33} cm^{-2}s^{-1}$) is possible. (The source intensity need not decrease with decreasing E_μ.)

B. Muon Production

The scenario for production and collection of muons needs much further study and development. Production calculations should be more precisely calibrated with experimental observations. Many options for targetting, collection and, transport, as well as variations on the rf debunching and compression scenario, should be explored.

In the present scenario the entire beam of the proton accelerator ring has been compressed into two short bunches before delivery onto target. While this appears possible (extrapolating from AGS experience and using a separate bunching ring), it is difficult, and it has not been demonstrated that bunches that are short enough for the subsequent rf rotation are readily obtained. Scenarios with a larger number of initial bunches (possibly shorter) may be preferable. As discussed in the previous workshop, bunch combination in π-decay or in the cooling system will be possible.

In the present scenario, the debunching process obtains relatively long bunches (up to 10m long), and therefore relatively low frequency rf (10–30 MHz) is required in the debunching rf. (An induction linac rather than resonant-cavity rf could also be used.) Acceleration gradients at these frequencies are currently relatively limited; research enabling gradient increase would permit a more compact facility and better performance. The μ-bunch length also implies that low-frequency-rf must be used in the initial cooling, and complicates subsequent bunch compression. Scenario variations with shorter bunches may be more readily implemented, and should be explored.

Variations on initial proton beam and π collection energies should be considered. Sources based on existing accelerators (AGS, Fermilab booster and main injector) could be developed, and used in initial colliders. A high-energy primary beam could obtain amplified production by using multiple interactions through a cascade of production targets. A lower energy proton beam may be more efficient in single π production. A GeV p-beam (π-factory, similar to LAMPF) could enable use of low-energy stopped π sources, where the μ-beam would be naturally cooler at production.

C. Beam Cooling

In the present scenario, we cool only with ionization cooling in conducting Be (or Li) rods, along with phase space exchange, and that is sufficient for high luminsosity. While portions of the process have been Monte Carlo simulated, the complete scenario has not been specified and studied. Longitudinal damping with wedge absorbers, and bunch compression, has not been fully integrated with transverse cooling.

The ionization cooling scenario needs to be optimized in full detail, and verified by more complete simulations. The detailed design will be dependent on bunch lengths, rf gradients, and optics optimization.

Previous active conductor lenses have been Li lenses, and that experience can be extended into the present case. Be is much denser, has higher electrical and thermal conductivity (permitting higher fields), and has similar multiple scattering, but has had safety problems in other applications. Research leading to validation of Be for rod cooling would be desirable.

Other techniques (such as ionization cooling in focussing transports or rings, or using plasma lenses, or high-frequency "optical" stochastic cooling[19]) may permit improvements, and are being studied. Also, with a low-energy source, μ-cooling at low (thermal) energies, which has somewhat different behavior, could be implemented.[20] Study of low-energy sources, with low-energy cooling, is needed.

D. Acceleration and Collisions

The bunch-compression and acceleration scenario also needs to be optimized and simulated. A detailed scenario including all stages of bunch compression has not yet been specified. Variations such as rapid-cycling synchrotrons and fixed-field alternating gradient (FFAG) machines should be considered. Hybrid recirculating-linac rapid-cycling devices are possible, in which each arc of the recirculating linac contains rapid cycling magnets, permitting multiturn passes within each recirculating arc. (A low-cost scenario requiring only 20 GeV of rf, using an injector and three RLA stages with rapid-cycling in the last stage, has been generated.)

Complete lattices are needed for the accelerator(s) and for the full-energy collider, where the low-β^* interaction region must be integrated with the high-field arcs, and modifications must be included to accomodate μ-decay in the arcs and near the IR's. Tracking of complete lattices for a muon lifetime should be investigated, and any instability limits should be identified.

It would be desirable to collide polarized beams. Muons are naturally polarized when produced in π-decay; however, polarization selection may reduce intensity, and we do not yet know if that polarization can be maintained through a debunching, cooling, acceleration and collision cycle. Further study is needed.

IV. SUMMARY AND CONCLUSIONS

A. Recent Accomplishments

In this section we recount some highlights of the progress in the past two years in development of the μ^+-μ^- collider, most of which are described in greater detail in the μ^+-μ^- collider workshop proceedings:[8]

The concept of a high-energy collider has been greatly improved, and a reasonable first scenario for a high-luminosity collider has been developed, based only on existing technical capabilities.

Since this scenario is confined to existing technology, first approximate estimates of the size and scope of a facility can be made. While, remembering the SSC, we would not like to state a crude cost number, the size and scope of the concept are within that of existing and planned facilities, and within plausibly estimated high-energy physics resources.

rf rotation immediately following $\pi \rightarrow \mu$ production has been incorporated into that scenario.[12] This greatly enhances μ acceptances; and therefore greatly increases luminosity. For the first time, the muon source has been simulated in a Monte Carlo simulation, with particle tracking from the π-production target through the decay transport, and followed by rf rotation. Relatively large μ-acceptance is obtained ($\sim$0.2 μ/p).

Also, for the first time, Monte Carlo simulations of the ionization cooling process have been initiated, by Palmer and Gallardo, and by van Ginnekin. These simulations have included increasingly complicated, and hopefully more realistic, models of the multiple scattering and energy straggling effects. Initial results are in reasonable agreement with the the rms cooling model.

For the first time, serious plans toward initial ionization cooling experiments have been developed. Proposals for initial experiments at TRIUMF, Fermilab and BNL are being developed, and collaborations are forming for their implementation.

Detailed studies of stochastic cooling of muons have been developed, by Ruggiero, and Barletta and Sessler. While conventional stochastic cooling is relatively ineffective because of the short muon lifetime, very-high-frequency cooling, such as proposed in "optical stochastic cooling", shows some promise of eventual utility.

Studies of alternative sources (photoproduction, or low-energy stopped-π) have been initiated.

As is also discussed in ref. 8, the underlying physics motivation for a muon collider has been more clearly formulated. The simple single-particle interactions characteristic of lepton-antilepton collisions are obtained, with the important advantage that the collisions occur within a small energy width (without the large energy broadening due to beamstrahlung expected in e^+-e^- linear colliders). The possibility of s-channel Higgs production is being studied. Also, the detector complications implied by μ-decay have also been studied, and initial detector configurations have been developed.

B. Future Directions

In this section, we outline some of the effort which we hope will be accomplished in the near future, before the next workshop.

The present scenario is a first proof-of-principle design concept. Much further optimization and design and concept development will be implemented, and the enlarged μ^+-μ^- collider collaboration developed at the collider workshops will enable this. One goal of the research will be complete source to collider simulations. These detailed calculations will identify key difficulties and potential improvements in the present scenario. We also hope to have a complete design concept for the μ^+-μ^- collider detector, and that the detector is integrated into the collider ring design.

Other scenarios, with substantial differences in source, cooling, and acceleration, will be developed, to a level that reasonable comparisons and optimizations can be made.

The presently developing experimental collaborations should have made substantial progress in experimental design and construction. A first demonstration of ionization cooling should be obtained. Other experiments may be needed. More accurate measurements of π-production under the same conditions used in the collider would be useful, and could be important in developing an accurately optimized scenario. rf acceleration development would also be desirable, both in the low-frequency rf systems needed in the debuncher and cooling, and in the high-frequency rf needed in the accelerator and collider.

The most important innovations cannot, of course, be predicted, but we expect substantial inventions and dramatic changes in the future. The progress over the past two years has transformed the μ^+-μ^- collider concept into a credible possibility. A similar degree of progress over the next two years could be the basis for the next-generation collider, which would enable particle-physics exploration at energy frontiers beyond the reach of existing and proposed accelerators.

Acknowledgments

We acknowledge extremely important contributions from our colleagues, especially D. Cline, J. Gallardo, R. Fernow, F. Mills, A. Ruggiero, A. Sessler, S. Chattopadhyay, W. Barletta, R. Noble, D. Winn, S. O'Day, and I. Stumer.

References

1. M. Tigner, in *Advanced Accelerator Concepts*, AIP Conf. Proc. **279**, 1 (1993).
2. E. A. Perevedentsev and A. N. Skrinsky, Proc. 12th Int. Conf. on High Energy Accel., 485 (1983), A. N. Skrinsky and V.V. Parkhomchuk, Sov. J. Nucl. Physics **12**, 3 (1981).
3. D. Neuffer, Particle Accelerators, **14**, 75 (1983), D. Neuffer, Proc. 12th Int. Conf. on High Energy Accelerators, 481 (1983), D. Neuffer, in *Advanced Accelerator Concepts*, AIP Conf. Proc. **156**, 201 (1987).
4. R. J. Noble, in *Advanced Accelerator Concepts*, AIP Conf. Proc. **279**, 949 (1993).
5. Proceedings of the Mini-Workshop on μ^+-μ^- Colliders: Particle Physics and Design, Napa CA, D. Cline, ed., published in Nucl Inst. and Meth. A **350**, 24(1994),
6. Muon Collider Mini-Workshop, Berkeley CA, April 1994.
7. Proceedings of the Muon Collider Workshop, February 22, 1993, Los Alamos National Laboratory Report LA-UR-93-866 H. A. Theissen, ed.(1993)
8. Proceedings of the Second Workshop on Physics Potential & Development of μ^+-μ^- Colliders, (Sausalito, Ca., November 17-19,1994) , D. Cline, ed., to appear in AIP Conf. Proc.(1995).
9. D. Neuffer and R. Palmer, Proc. 1994 European Particle Accelerator Conference (London, England, June25-July1, 1994).
10. R. Palmer, D. Neuffer, J. Gallardo, Proc. Advanced Accelerator Concepts Workshop, (Lake Geneva, WI, June 1994), to appear as AIP Conf. Proc.
11. KAON Factory proposal, TRIUMF, Vancouver, Canada (1990)(unpublished).
12. The rf debuncher, developed here by R. Palmer, is similar to one proposed for antiproton acceptance by F. Mills (1982).
13. R. Palmer, J. Gallardo et al., in ref. 8.
14. C. L. Wang, Phys Rev **D9**, 2609 (1973) and Phys. Rev. **D10**, 3876 (1974).
15. D. Neuffer, Nucl. Inst. and Meth. A **350**, 27(1994).
16. U. Fano, Ann. Rev. Nucl. Sci. **13**, 1 (1963).
17. CEBAF Design Report, CEBAF, Newport News VA (1986) (unpublished).
18. D. Cline, Nucl Inst. and Meth. A **350**, 24 (1994).
19. A. A. Mikhailichenko and M. S. Zolotorev, Phys. Rev. Lett.**71**, 4146 (1993).
20. H. Daniel, Muon Catalyzed Fusion **4**, 425(1989), M. Muhlbauer, H. Daniel, and F. J. Hartmann, Hyperfine Interactions **82**, 459 (1993).

MONTE CARLO SIMULATIONS OF MUON PRODUCTION

Robert B. Palmer, Juan C. Gallardo, Richard C. Fernow, Yağmur Torun
Brookhaven National Laboratory
P. O. Box 5000, Upton, New York 11973-5000

David Neuffer
CEBAF
Newport News, VA, 23606

David Winn
Fairfield University,
Fairfield, CT, 06430-5195

Abstract

Muon production requirements for a muon collider are presented. Production of muons from pion decay is studied. Lithium lenses and solenoids are considered for focussing pions from a target, and for matching the pions into a decay channel. Pion decay channels of alternating quadrupoles and long solenoids are compared. Monte Carlo simulations are presented for production of $\pi \rightarrow \mu$ by protons over a wide energy range, and criteria for choosing the best proton energy are discussed.

1 INTRODUCTION

The luminosity $\mathcal{L}$ of a muon collider is given by

$$\mathcal{L} = \frac{\mathrm{N^2 \gamma n_t f}}{4\pi\beta^* \epsilon_\mathrm{n}} \tag{1}$$

where N is the number of muons per bunch, γ is the energy per beam divided by the muon mass, n_t is the effective number of turns made by the muons before they decay, f is the repetition frequency, β^* is the Courant-Snyder parameter at the focus, and ϵ_n is the r.m.s. normalized emittance of the beams (assumed symmetric in x and y). In order to obtain a reasonable event rate, the luminosity must be proportional to the center of mass energy squared and may be taken to be of the order of $\mathcal{L} \geq 10^{33}\ \mathrm{E_{cm}^2\,cm^{-2}s^{-1}}$ where E_{cm} is the energy in TeV. Possible parameters that would yield a luminosity close to this requirement, at 4 TeV in the center of mass, are given in Table 1.

Beam energy	TeV	2
Beam γ		19,000
Repetition rate f	Hz	30
Muons per bunch N	10^{12}	2
Bunches of each sign		1
Normalized rms emittance ϵ_n	mm mrad	50
Average ring mag. field	Tesla	6
Effective turns before decay n_t		900
β^* at intersection	mm	3
Luminosity $\mathcal{L}$	$cm^{-2}s^{-1}$	10^{35}

Table 1: Parameters of a 4 TeV center of mass $\mu^+\mu^-$ Collider

The value of the emittance ϵ_n used is limited by the technology used to cool the beams, and is chosen here to be consistent with that believed (1) obtainable with ionization cooling. The intersection point (IP) β^* will be limited by the design of the chromatic correction system at the IP, and by the achievable bunch length. The value given here is believed to be technically possible. The number of turns n_t is set by the average bending field in the collider ring. The value taken here corresponds to a mean field of 6 Tesla, which is probably as high as is technically feasible. The repetition rate f is taken to be 30 Hz, and is constrained by proton source, power consumption, and radiation considerations.

With the above parameters, the required luminosity is achieved, but the number of muons per bunch N is large ($2\ 10^{12}$). Stacking many smaller muon bunches to achieve such a population would be hard because of the life time limitations of the muons. Thus if the proton bunch population is to be kept reasonable, the production must be efficient, i.e. we require a high value of the number of captured muons per initial proton $\eta_\mu = n_\mu/n_p$.

Since pion multiplicity rises with proton energy, the value of η_μ tends also to rise with the proton energy used. But if the energy is allowed to rise, then the cost and energy consumption of the proton source will also rise. Thus the requirement is for the highest number of muons per proton η_μ, at the lowest possible proton energy E_p.

2 THEORY

2.1 Introduction

We consider muons made by the decay of pions generated by the interaction of a proton beam with a metal target. With 30 GeV protons, the pion multiplicity is of the order of one. Almost every pion made, if it does not interact in the target or otherwise get lost, will decay into a muon. The potential value of η_μ is thus of the order of one. At proton energies lower than 30 GeV, the multiplicity falls, but remains substantial until below the N^* resonance (at $E_p = 0.73\,GeV$). The challenge is to target efficiently and capture as large a fraction of the pions as possible; then to capture as large a fraction as possible of the muons from their decay.

Previous estimates (2) of the efficiency of such capture were low ($\approx 10^{-3}$), and only moderate luminosity was possible. Such estimates assumed conventional focussing technology and decay channels with restricted momentum acceptances ($\pm 5\%$). In this case, the values of η_μ are proportional to the square of the momentum acceptances; one factor from the fraction of pions accepted from the target, and another from the fraction of muons accepted from the decay of those pions. Since the luminosity per pulse is proportional to the square of the number of muons, the luminosity goes as the fourth power of the acceptances. If these acceptances can be raised, then a major improvement in η_μ may be expected.

In this paper we discuss two methods of capturing pions from the production target: a) lithium lenses followed by a decay channel consisting of alternating quadrupoles, and b) a high field solenoid matched adiabatically to a lower field solenoid decay channel.

2.2 Lithium Lenses and Quadrupole Channel

2.2.1 Capture from target

The Courant-Snyder parameter β in a long lithium lens, assuming uniform current density, is

$$\beta \;=\; \frac{p\,\theta_{max}}{B_{max}\;c} \tag{2}$$

where p is the particle momentum in eV/c, c is the velocity of light. If θ_{max} is the maximum angular amplitude accepted, and B_{max} is the field on the surface, its radius a is

$$a \;=\; \beta\,\theta_{max} \;=\; \frac{p\,\theta^2_{max}}{B_{max}\;c} \tag{3}$$

Inserting the target inside the lithium lens maximizes the yield of captured pions. The required length ℓ_1 of the lens is half the betatron wavelength, $\lambda/2$; if we express the maximum angle θ_{max} as $\hat{p}_t/p$ then,

$$\ell_1 \;=\; \frac{\pi}{2}\;\frac{\hat{p}_t}{B_{max}\;c}, \tag{4}$$

which, for a given transverse momentum acceptance, is independent of the momentum p. The radius a_1 of the lens is

$$a_1 = \frac{\hat{p}_t^2}{B_{max}\ c}\ \frac{1}{p} \tag{5}$$

which rises as the momentum falls. For $B_{max} = 10\ T$, and a nearly ideal $\hat{p}_t = 0.6\ GeV/c$, then $\ell_1 = 0.31\ m$. The radius will be 12 cm at 1 GeV/c, falling to 3 cm at 4 GeV/c.

The required current I is

$$I = \frac{2\pi\ \hat{p}_t^2}{\mu_o\ c\ p} \tag{6}$$

which falls as the momentum rises, independent of the surface field. For $\hat{p}_t = 0.6\ GeV/c$, the current falls from a value of 6 MA at 1 GeV/c to 1.5 MA at 4 GeV/c.

These currents and radii are far larger than those in currently used lenses (eg FNAL: a=1 cm, I=0.5 MA), but they may still be possible. The temperature rise and central pressures are proportional only to the surface field, and this we kept constant. However, it is clear that this type of focussing is better suited to the higher momenta.

At a fixed momentum, the normalized total emittance of the resulting beam $\epsilon_{tn}(tgt)$ is set by the target length ℓ_{tgt}

$$\epsilon_{tn}(tgt) = \frac{\ell_{tgt}}{2}\ \frac{\hat{p}_t^2}{m_\pi\ p} \tag{7}$$

where m_π is the pion mass in units of eV/c^2 and the momenta p and $\hat{p}_t$ are in units of eV/c. At 1 GeV/c, a target length of 0.18 m (for Cu), and a maximum $\hat{p}_t$ of 0.6 GeV/c, the emittance has a value of 0.23 m, but has fallen to a 0.06 m at 4 GeV/c.

For a point source, with momentum spread $\frac{dp}{p} = \delta$, the total normalized emittance $\epsilon_{tn}(mom)$ is

$$\epsilon_{tn}(mom) = \frac{\pi}{2}\ \frac{\hat{p}_t^3\ \delta}{B_{max}\ c\ m_\pi\ p} \tag{8}$$

which also falls with momentum. This equals the emittance for target length when

$$\delta = \frac{\ell_{tgt}\ B_{max}\ c}{\pi\ \hat{p}_t} \tag{9}$$

which for the above parameters is ± 0.28.

2.2.2 Quadrupole channel

The focussing strength k (defined to be the inverse of the focal length) of a thin quadrupole lens is

$$k = \frac{\ell_q \, B_{pole} \, c}{ap} \tag{10}$$

where B_{pole} is the pole field at the pole radius a, and the momentum p is in eV/c. If the phase advance per cell is ψ, and a half cell length is ℓ_h, then

$$s \;=\; \sin\frac{\psi}{2} \;=\; \frac{\ell_h \, k}{2} \tag{11}$$

The average β is $2\ell_h/\psi$ and the maximum β is

$$\beta_{max} \;=\; \frac{2}{k}\sqrt{\frac{1+s}{1-s}} \tag{12}$$

Thus, the unnormalized acceptance of a quadrupole channel varies inversely with the momentum

$$\mathcal{A} = \frac{a^2}{\beta_{max}} = \frac{a\ell_q B_{pole} c}{2p}\sqrt{\frac{1-s}{1+s}} \tag{13}$$

$\mathcal{A}$ is zero until the momentum has risen to the value p_{min} corresponding to the onset of the stop band at a phase advance per cell of π. This minimum accepted momentum is

$$p_{min} = \frac{\ell_q^2}{F}\frac{B_{pole} c}{2a}, \tag{14}$$

where $F = \ell_q/\ell_h$ is the fraction of length full of quads. For nominal capture of 1 GeV/c pions we require a p_{min} of the order of $p/4 = 0.25\ GeV/c$. Then for $B_{pole} = 6\ T$, $a = 0.15\ m$ and $F = 1/2$, then we obtain $\ell_q = 0.14\ m$. For higher momenta ℓ_q should be increased as the square root of the momentum.

The unnormalized acceptance rises to a maximum at a momentum of approximately 4 times this minimum, and then falls slowly. The normalized acceptance $\mathcal{A}_n = \gamma\beta\mathcal{A}$ rises continuously, approaching an asymptotic value of

$$\mathcal{A}_n(quad) = \frac{\ell_q a B_{pole} c}{2m_\pi} \tag{15}$$

Fig. 1 shows the normalized acceptances for lattices with quadrupole lengths ℓ_q of 14, 20 and 30 cm, corresponding to optimized designs for proton energies of 10, 30, and 100 GeV. A fixed value of $F = 1/2$ is assumed. It is seen that for longer quadrupoles, the maximum acceptance is increased, but the minimum momentum rises.

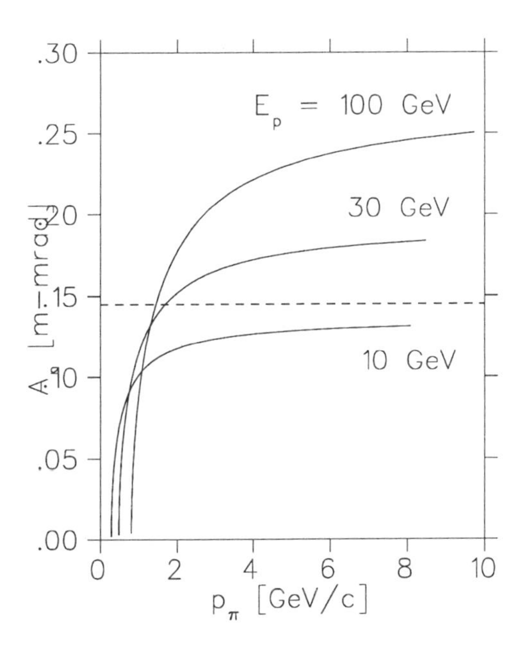

Figure 1: Decay Channel Acceptances. Continuous lines are for quadrupole channels optimized for proton energies $E_p = 10$, 30, and 100 GeV by adjusting the quadrupole lengths; dashed line is acceptance for a solenoid channel with the same field.

2.2.3 Matching

Matching from the first lens into the channel is challenging. The beam size after the lens is usually significantly smaller than the aperture of the channel. An adiabatically tapered lithium lens transition would be ideal, providing a match at all momenta, but would involve too much beam loss from interactions in the lithium. Instead we consider a two lens match that is similar in function to the multiple horns used in neutrino beams (3).

We consider first a single lens focus. The solid line in Fig. 2a shows the ratio of outgoing angles over incoming angles, for a single thin lens of focal length 20 cm, set to focus 2 GeV/c momentum particles. θ_{out}/θ_{in} is less than 1/3 over the momentum range of about 1.5 GeV/c. The dotted lines show the same ratios for particles starting at 9 cm in front or behind the nominal source, i.e. they represent particles coming from the ends of an 18 cm target.

Fig. 2b shows the same thing for a two thin lens system. The second lens has a focal length 3 times that of the first, and both are placed at distances from the source equal to their focal lengths. It is seen that there are now two momenta for which the outgoing angles are zero, and the range of momenta for which θ_{out}/θ_{in} is less than 1/3 is now 5 GeV/c, compared with only 1.5 GeV for the single lens case. The effect of displacing the source is to smear the distributions, but leave a net reduction of angles over the same wide momentum range.

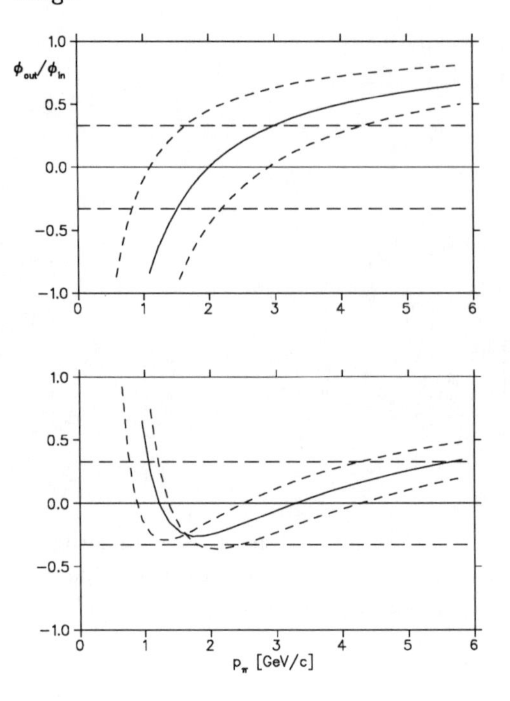

Figure 2: θ_{out}/θ_{in} vs momentum in thin lens focus systems, solid lines are for target center, dashed lines are from target ends; a) for single lens focussing; b) for two lens focussing.

Using this approach to match from the initial lithium lens of length ℓ_1, we place the second lens at a distance 2 ℓ_1. The radius a_2 is set equal to that of the decay channel, and the length is

$$\ell_2 \approx \frac{a_2\, p}{3\, \ell_1\, B_2\, c} \tag{16}$$

which for a maximum field of $B_2 = 3.3\,T$ gives $\ell_2 = 15\,cm$ at 1 GeV/c, rising linearly with momentum.

2.2.4 Parameters of Systems

Using the above criteria we have designed lithium-lens and quadrupole channel systems for the mean captured momenta, corresponding to initial proton momenta of 10, 30 and 100 GeV/c.

Proton mom.	GeV/c	10	30	100
Nom. π mom.	GeV/c	1.3	2.4	5
First Li Lens				
Surface field	Tesla	10	10	10
Total length	cm	40	40	40
Nominal length	cm	31	31	31
Radius	cm	10	5	2.5
Space	cm	60	60	60
Second Li Lens				
Surface field	Tesla	3.3	3.3	5.7
Total length	cm	20	36	40
Radius	cm	15	15	15
Decay channel				
Pole tip fields	Tesla	6	6	6
Quad. length	cm	15	20	30
Gap length	cm	20	20	20
Decay length	m	400	800	1600

Table 2: Lithium and Quadrupole Channel Parameters

2.3 Solenoid Focussing

2.3.1 Capture from Target

If the capture is done inside a solenoid, then the calculations are very simple. The required maximum radius a_{sol} is

$$a_{sol} = 2\,\frac{\hat{p}_t}{B_{sol}\;c} \tag{17}$$

where $\hat{p}_t$ is the is the maximum transverse momentum in units of eV/c. This radius is seen to be independent of momentum p. For the near ideal $\hat{p}_t = 0.6\ GeV/c$ and a conventional superconducting field $B_{sol} = 10\,\mathrm{T}$, the radius would be 40 cm which is very large and would correspond to a very large emittance. But hybrid solenoids can be made with fields as high as 45 Tesla (4). If we used a field of 33 T, then the radius with the same assumptions, would be only 7.5 cm. At lower momenta, a somewhat lower maximum $\hat{p}_t$ may be acceptable, for $a_{sol} = 7.5\ cm$, and $B_{sol} = 28\,\mathrm{T}$ the maximum transverse momentum captured is 0.31 GeV/c.

The solenoid length ℓ_{sol} to focus pions, at a fixed momentum p is

$$\ell_{sol} = \frac{\pi}{2}\,\beta = \pi\,\frac{p}{B_{sol}\;c} \tag{18}$$

which rises with momentum. For a 28 T field, the length is 37 cm for 1 *GeV*, rising to 150 cm for 4 GeV. But one notes that for a long solenoid, the beam will be captured in the solenoid at all momenta, and for all source positions, an almost ideal situation. In this case, the normalized acceptance of the resulting beam is

$$\mathcal{A}_n(\mathrm{sol}) = \frac{a_{sol}\hat{p}_t}{m_\pi} = \frac{2p_t^2}{B_{sol}cm_\pi} = \frac{a_{sol}^2 B_{sol} c}{2m_\pi} \tag{19}$$

In practice we cannot maintain such a solenoid over any significant length and must match it into a lower field decay channel.

2.3.2 Decay Channel

For a solenoid decay channel we chose a field of 7 Tesla, which is easily attainable with superconducting magnets. The value is a little higher than that taken for the quadrupole pole tip fields, since the fields seen by the conductors in a quadrupole will be somewhat higher. We take the radius of the channel to be the same as that in the quadrupole case, i.e. 15 cm.

The acceptance of this channel will be seen to be the same as that for the 7.5 cm, 28 Tesla capture solenoid. If a suitable wide band matching can be provided, then all pions captured by the target solenoid will be transferred to the channel, independent of their momentum.

2.3.3 Match

The match between the target capture solenoid and the decay channel solenoid can be made without significant loss if the field and radius are varied so as to maintain the same acceptance, and if the rate of change of β with z is small compared with one. Defining $d\beta/dz = \epsilon$, we obtain (5)

$$B = \frac{B_o}{1 + \alpha z} \tag{20}$$

$$a = a_o \sqrt{1 + \alpha z} \tag{21}$$

where

$$\alpha = \frac{cB_o\epsilon}{2p} \tag{22}$$

It is found that for values of ϵ less than 0.5 there is negligible loss of particles. Fig. 3 shows the field and dimension profiles of such a match.

2.3.4 Parameters of Solenoid Systems

In table 3, parameters are given of solenoid capture and channel systems optimized for pion momenta of 0.5, 0.9, 1.3, and 2.8 GeV/c; corresponding approximately to the peak pion production from proton momenta of 3, 10, 30 and

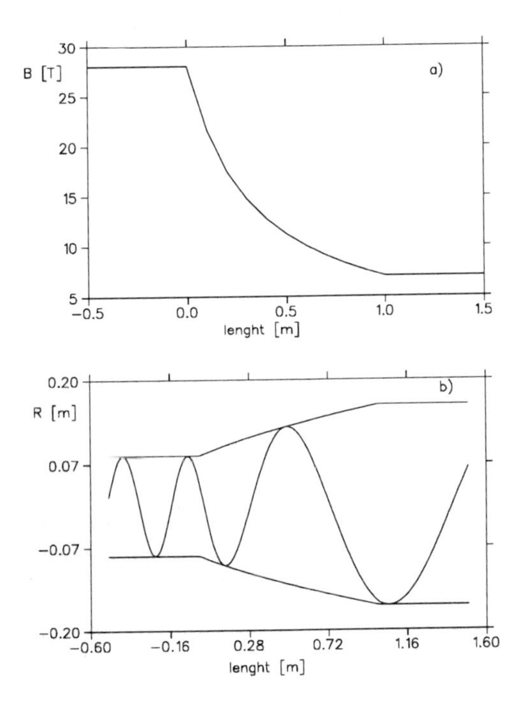

Figure 3: Solenoid Matching: a) Magnetic field vs. length; b) Radius vs. length, with a typical trajectory of a particle.

100 GeV/c. These pion momenta are lower than those given for the lithium lens-quadrupole case, see Table 2; reflecting the loss of low momenta particles from the channel cutoff in the lithium lens-quadrupole case.

Proton mom.	GeV/c	3	10	30	100
Nom. π mom.	GeV/c	0.5	0.9	1.3	2.8
Target length	cm	24	24	24	24
Capture field	Tesla	28	28	28	28
Solenoid radius	cm	7.5	7.5	7.5	7.5
Transition length	m	0.5	0.9	1.3	2.8
Channel field	Tesla	7	7	7	7
Channel radius	cm	15	15	15	15
Decay Length	m	100	200	340	500

Table 3: Solenoid Focus and Channel Parameters

3 MONTE CARLO RESULTS

3.1 Monte Carlo Program

A Monte Carlo program has been written to give a first approximation to the performance of the capture systems described above. The program, at this time, contains many approximations and the results obtained from it must be taken with some caution.

- Pion production spectra use the Wang (6) formulation. These distributions were derived by fitting proton proton interaction cross sections. The π multiplicities are slightly lower than those given by H Boggild and T. Ferbel (7) and significantly lower than those given by a nuclear calculation for Cu (8). Our estimates should thus be conservative.

- The pion momentum distributions given by the Wang formula are peaked somewhat lower than those given by the nuclear calculation for hydrogen, but higher than that given by the calculation for Cu. Clearly, the use of the Wang formula is unsatisfactory, but the qualitative results obtained are probably correct.

- Particles were followed using the paraxial approximation. This is a reasonable approximation for the production at 10 GeV and above, but is a poor approximation for pions produced by protons below 3 GeV.

- The program assumed that all particles are relativistic. This again is a poor approximation for proton production below 3 GeV.

- The initial proton beam was assumed to have an rms transverse radius of 1 mm, and a divergence of 1 mrad. The target was taken to be Cu with an interaction cross section of 0.782 barns. Pions passing through the target were reabsorbed with a cross section 2/3 of the above, and no tertiary pion production was included. Coulomb scattering, energy loss and straggling were calculated from Gaussian formulae. The pion decay lifetime was taken to be $2.603 \times 10^{-8}\ s$, the branching ratio into muons was assumed to be 100%. The kinetic energy distribution of decay muons was taken to be flat. Pions or muons which exceeded the aperture of focus components were assumed lost.

3.2 Results of Simulations

Table 4a gives the muon production for different initial proton energies, for the two capture systems described above: a) lithium lenses and quadrupole channel; b) solenoid capture and channel. The "capture % " given is the fraction of pions that decay into muons that remain always within the focus channel. These fractions, particularly in the case of solenoid focussing, are relatively insensitive to the details of the pion production as given by the Wang formula. The "μ/p" ratio is the product of the "capture % " and the average charged pion multiplicity ("π mult."), and gives the final number of muons produced per initial proton. The average μ momentum is that found in the decay channel at the end, and the "rms mom. %" is the rms muon momentum spread divided by the average.

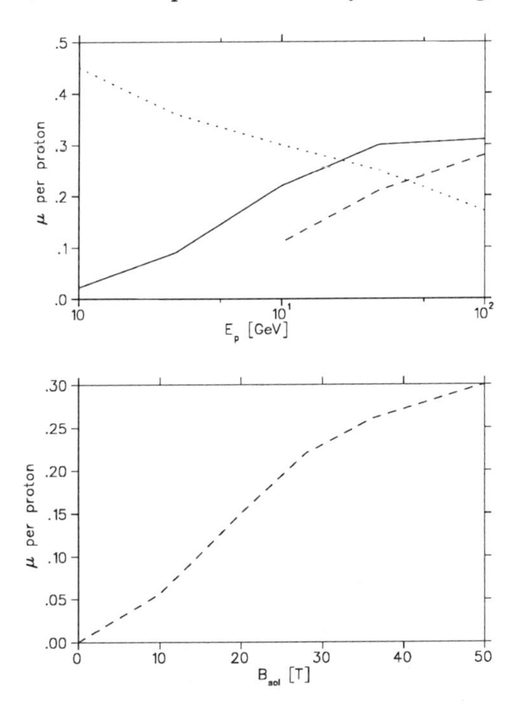

Figure 4: Monte Carlo Results: a)Muons per proton vs. proton energy; dashed line is for lithium lens system; solid line is for all solenoid systems; dotted line is capture efficiency for solenoid systems; b)Muons per proton vs. solenoid field.

Fig. 4a shows the calculated number of muons per proton for the different cases. It is seen that the lithium lens + quadrupole channel systems are universally worse than the solenoid systems; although the difference is shrinking as higher proton energies are used. In the all-solenoid cases the production increases with proton energy, but the increase levels off above 30 GeV. The dotted line in this figure shows the capture efficiency for the solenoid case. In reality this efficiency would be less at the lowest energies because of production in the backward direction that is not represented by the Wang formulation.

Table 4b and Fig. 4b show the dependence of muon production on the capture magnetic fields. All cases are for solenoid focussing from 10 GeV protons. As expected, the production rises monotonically with field, but the gain begins to saturate at fields above about 30 Tesla.

Table 4c compares capture efficiency (at 10 GeV with solenoid focussing) for a 60 cm long, 1 cm radius beryllium target with that from the shorter copper targets assumed above. It is seen that there is little difference between them. It must be noted however that this comparison does not include any differences in the production multiplicities or distributions.

Method	p Energy GeV	capture %	π mult.	μ/p	ave μ mom. GeV/c	rms mom. %
Li lens + quads	100	15	1.8	0.28	8	120
	30	17	1.2	0.21	2.3	70
	10	16	0.7	0.11	1.1	60
Solenoid	100	17	1.8	0.31	1.5	110
	30	25	1.2	0.30	1.0	140
	10	30	0.7	0.22	0.6	80
	3	35	0.3	0.09	0.34	60
	1	46	0.05	0.023	0.14	66

a) Dependence on proton Energy

Solenoid Field Tesla	Channel Field Tesla	capture %	μ/p	ave μ mom. GeV/c
20	5	22	0.15	0.46
28	7	30	0.22	0.60
36	9	37	0.26	0.61

b) Dependence on capture Magnetic Field

Tgt Material	Tgt length cm	Tgt rad. mm	capture %	μ/p	ave μ mom. GeV/c
Copper	24	3	31	0.22	0.6
Beryllium	60	10	29	0.20	0.6

c) Dependence on target materials

Table 4: Muon Production from Monte Carlo Studies

3.3 Choice of Proton Energy

Using a high field solenoid for capture, we find that the capture efficiency rises as the energy falls, at least down to a proton energy of 3 GeV. Similarly, the number of muons made per unit of beam energy also rises, at least down to an energy of the order of 3 GeV. Thus from the point of view of muon production economy and efficiency, it seems desirable to use this relatively low energy.

But as the proton energy falls, a larger number of protons are needed to obtain the required number of muons if we assume a single bunch targetting. Problems might arise from targeting larger bunches, and severe space charge problems arise in the proton ring used to bunch them prior to extraction and targeting. The tune shift in a ring whose mean bending field is B_{ave}, for a Gaussian bunch of length σ_z, is given by

$$\Delta\nu = \frac{N\ r_o\ m_p}{\sqrt{2\pi}\ 2\ \sigma_z\ \gamma_p\ \epsilon_n\ B_{ave}\ c} \tag{23}$$

where N is the number of protons in the bunch, r_o is the classical radius of the proton, m_p is the proton mass in electron volts, γ_p is the energy of the proton divided by its mass, ϵ_n is the rms normalized emittance of the protons and c is the velocity of light.

If longitudinal emittance of the muons is to be kept as low as possible then the proton bunch length must be less than a value, obtained from a Monte Carlo study of phase rotation, which is proportional to the average pion momentum,

$$\sigma_z \approx 3.0\,[m]\sqrt{\frac{1}{\gamma_p}} \tag{24}$$

For B_{ave} of 4 Tesla, an rms normalized emittance of 62 mm mrad (95 % emittance of 375 π mm mrad), and values of N such as to give 10^{13} muons using a 28 Tesla solenoid system, then we obtain the tune shifts given in Table 5 and Fig. 5. In this table and figure the proton beam power is also given.

Proton energy	GeV	3	10	30
Protons required	10^{13}	2 x 11	2 x 4.5	2 x 3.3
Proton beam power	MW	3.2	4.3	9.5
Bunch len. req.	m	1.7	1	0.58
Tune Shift		0.8	0.17	0.07

Table 5: Proton Source parameters

Assuming a largest acceptable tune shift of about 0.2 would indicate a preferred proton energy of 10 GeV. At this energy, the total required number of

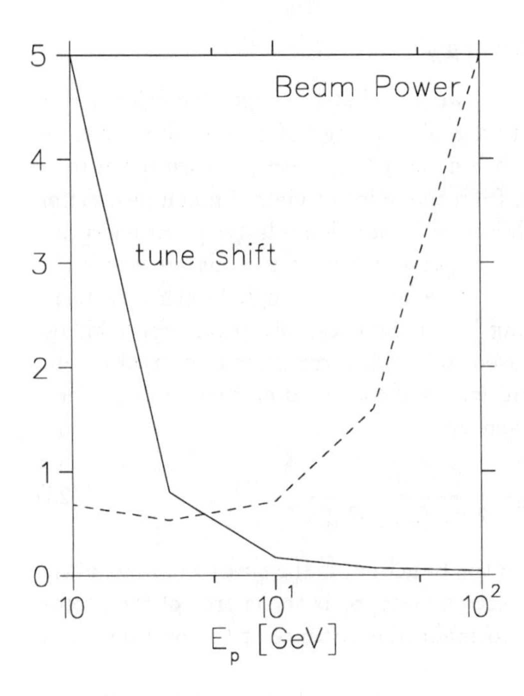

Figure 5: Choice of Proton Energy: the continuous line shows the tune shift, and the dashed line the relative proton beam power, each plotted vs. the proton beam energy.

protons (0.9 x 10^{14}), and the required repetition rate of 30 Hz are also close to the specification of the spallation source studied at ANL (9) and may thus be taken as reasonable values. It is noted however, that in such rapid cycling machines, the average bending fields are far less than the value of 4 T assumed above. Thus, in order to bunch without excessive space charge tune shift, we would require an additional fixed field superconducting bunching ring. This tune-shift limit is valid for a single bunch in a circular machine. The limit can be evaded in a linear machine or in schemes where multiple bunches are arranged to be targetted simultaneously, for example, by using dog-legs as "delay-lines", multiple rings or multiple synchronized kikers.

4 CONCLUSION

4.1 Caveats

It must be emphasized that the above calculations contain many approximations. In particular, pion production spectra used are those given by an approximate formula that was obtained by fitting proton-hydrogen data, rather than proton-Cu as assumed in most cases here. The tracking program used the paraxial approximation and assumed that all particles were relativistic. These are poor approximations for proton production much below 3 GeV.

4.2 Solenoids vs. Lithium Lenses and Quadrupole Channels

Despite the approximations used in this study, it seems clear that the use of high field solenoids for both capture and decay channels is to be preferred over the use of lithium lenses and quadrupole channels.

- Given a capture solenoid approaching 30 Tesla, the absolute capture of muons per initial proton appears higher than with any plausible lithium lens system, at all energies. It is clearly superior for proton energies of 10 GeV and below.
- The use of a solenoid, instead of lithium lenses, allows the use of the same target, capture and decay channel for both signs of pions. This would be a significant saving. It may not, however, allow the use of muons of both signs from a single proton bunch. If rf is used to bunch rotate the muons, the polarity of this rf has to be different for the two signs.
- The technology of high field solenoids is more mature than that of large lithium lenses. Life time is less likely to be a problem, although questions of radiation damage must be studied.

4.3 Choice of Proton Energy

The proton beam energy required per muon, falls with proton energy down to a value around 3 GeV, but the number of protons required per bunch rises and the tune shift problems in the proton accelerator also rise as the energy falls. A reasonable compromise appears to be around 10 GeV.

4.4 Proposed parameters

On the basis of the above considerations, we propose the use of

- Solenoid capture, using a 28 Tesla 7.5 cm radius magnet.
- 10 GeV protons, using two bunches of $3 - 5 \times 10^{14}$ protons.
- We expect at least 0.2 muons per proton, thus generating $6 - 10 \times 10^{12}$ muons of each sign. This appears adequate to assure final bunches of 2×10^{12} in the collider, yielding, with the other parameters given in table 1, a luminosity of 10^{35} cm^{-2} s^{-1}.

5 Acknowlegement

This research was supported by the U.S. Department of Energy under Contract No. DE-ACO2-76-CH00016 and DE-AC03-76SF00515.

6 REFERENCES

1. D. V. Neuffer, R. B. Palmer, Proc. European Particle Acc. Conf., London (1994)
2. R. J. Noble, *Advanced Accelerator Concepts*, AIP Conf. Proc. **279** (1993) 949
3. R. B. Palmer, *Magnetic fingers*, Proc. of the Informal Conference on Experimental Neutrino Physics, Cern 65-32 (1965), C. Franzinetti editor.
4. Physics Today, Dec (1994), p21-22
5. R. Chehab, J. Math. Phys. **5** (1978) 19.
6. C.L. Wang, Phys. Rev. **D10** (1974) 3876.
7. H. Boggild and T. Ferbel, Annual Rev. of Nuclear Science, 24 (1990) 74
8. D. Kahana, private communication.
9. Yang Cho, et. al., ANL-PUB-081622, *Proc. Int. Collab. on Advanced Neutron Sources*, Abbingdon, UK., May 24-28 (1993).

Comparison of the Wang and Wachsmuth models for π production with measurements at 12 GeV/c

R.C. Fernow

Brookhaven National Laboratory, Upton, NY 11973

We converted the invariant cross section measurements of Blobel *et al* at 12 GeV/c into the form $d^2\sigma / d\Omega\ dp$ as a function of the LAB total momentum p and p_T. We adjusted the parameters of the pion production models of Wang and of Wachsmuth-Hagedorn-Ranft to obtain the best fit to the data. Neither model gave a statistically accurate fit to the data.

In an earlier study[1] of inclusive pion production at 28 GeV, we found that the Wang model[2] agreed with measured π^- cross sections to ±25% for lab production angles less than 10°. The model seriously overestimated π^- production at large angles. The model underestimated the limited amount of π^+ data by about 25%. We have subsequently decided that a lower incident proton momentum may be preferable for the driver for the μμ collider. The lower proton momentum brings the use of the Wang formula in the μμ collider monte carlo into even more question. For that reason we examine the comparison of the model with the measurements of Blobel et al[3] for p p -> π X at 12 GeV/c.

According to the Wang formula, the number of produced pions per interacting proton per interval of pion solid angle and momentum bite is

$$\frac{d^2\sigma}{dp\,d\Omega} = A\,p_{max}\,x\,(1-x)\,\exp(-B\,x^C - D\,p_\perp) \tag{1}$$

where p_{max} is the maximum possible longitudinal momentum for the pion in the LAB frame, and

$$x = \frac{p_l}{p_{max}} \tag{2}$$

$p_\perp$ is the transverse momentum of the pion, and {A,B,C,D} are fit parameters. p_{max} should be approximately equal to p_b.

An alternative model of Wachsmuth[4], based on the formalism of Hagedorn and Ranft, predicts that

$$\frac{d^2\sigma}{dp\,d\Omega} = \sigma_{inel}\frac{p^2}{E}a_1\,(1-x^2)\exp(-a_2x^2)\,[\exp(-a_3p_\perp^2) + a_4\exp(-a_5p_\perp^2)] \qquad (3)$$

where σ_{inel} is the pp inelastic cross section at 12 GeV/c and $\{a_1...a_5\}$ are the fit parameters.

1. Data reduction

We examine here data from Blobel et al on inclusive pion production from p - p interactions at 12 GeV/c in a hydrogen bubble chamber. The measured data in this paper is presented as two tables showing the π^+ and π^- invariant cross section (I) for intervals of $p_\perp$ and the rapidity y^* , where

$$I = E_\pi\frac{d^3\sigma}{dp_\pi^3} \qquad (4)$$

and y^* is defined from

$$\tanh(y^*) = \frac{p_\parallel^*}{E_\pi^*} \qquad (5)$$

The superscript * refers to quantities measured in the center of momentum (CM) frame.

The quantity analogous to Eq. 1 in the LAB frame was determined from[1]

$$\frac{d^2\sigma}{dp\,d\Omega} = \frac{p^2}{E}I \qquad (6)$$

It is useful to change from the variables $\{y^*, p_\perp\}$ used by Blobel et al to the variables $\{p, \theta\}$ in the LAB frame, where p is the magnitude of the pion momentum and θ is the pion production angle with respect to the incident proton direction. Using Eq. 5 and

$$E_\pi = \sqrt{p_\parallel^{*2} + p_\perp^2 + m_\pi^2} \qquad (7)$$

we find that

$$p_{\parallel}^{*} = \tanh(y^{*}) \sqrt{\frac{p_{\perp}^{2} + m_{\pi}^{2}}{1 - \tanh^{2}(y^{*})}} \qquad \textbf{(8)}$$

We calculated the transformation between the CM and LAB frames (β_{CM} = 0.925) and the maximum pion momentum in the LAB (p_{max} = 10.40 GeV/c) using a procedure outlined in a previous note[1]. A simple Lorentz transformation gives $p_{\parallel}$ in the LAB for each bin.

Positive pions were separated from protons in the data using ionization information. Since this is only reliable for tracks with LAB momenta less than 1 GeV/c, Blobel et al only measured events with $y^{*} \leq 0$ (backwards hemisphere in the CM). However, because of the symmetry of the p p initial state in the CM, the invariant cross section distribution must be symmetric around $y^{*} = 0$. Therefore, in generating our plots in the LAB frame, we plotted each cross section twice, once at the momentum corresponding to y^{*} and once at the momentum corresponding to $-y^{*}$.

2. Inclusive π cross sections

The inclusive π cross sections are shown as a function of the total π momentum in Fig. 1. Separate curves are shown for each p_T bin. The peak cross section for the π^{+} data is 44 mb/sr/GeV. The peak cross section for the π^{-} data is 19 mb/sr/GeV, roughly half the value for π^{+} . The measured total inelastic cross section and pion multiplicities are given in Table 1.

Table 1 Global results at 12 GeV/c	
σ_{inel} [mb]	29.75 ± 0.25
$< N\pi^{+} >$	1.44 ± 0.03
$< N\pi^{-} >$	0.71 ± 0.02

The pion multiplicities are defined as the ratio of the inclusive cross section to the total inelastic cross section.

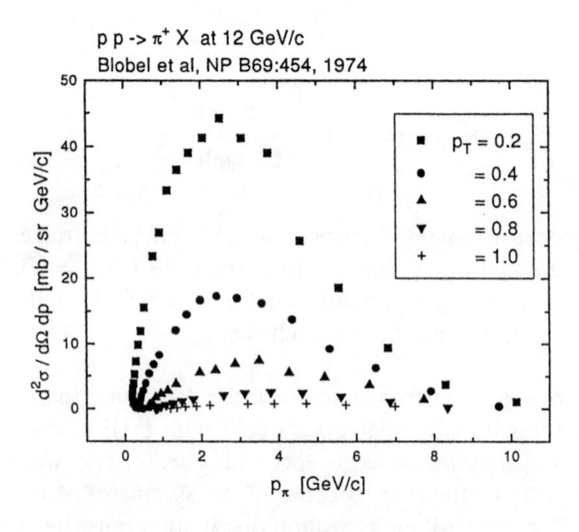

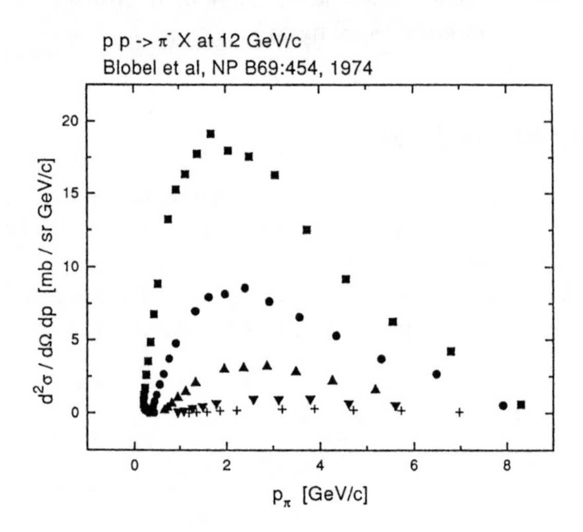

Fig. 1 Measured inclusive cross sections for π production at 12 GeV/c.

The Wang model underestimated the peak π^+ cross section by about 25%. It fit the peak π^- data to better than 10%, but underestimated the high momentum tail of the distributions. This level of agreement is similar to that in our earlier analysis at 28 GeV [1]. However, the disagreement was far larger than the size of the error bars on the points, so we next attempted to improve the fits by readjusting the parameters.

3. Optimization of the model parameters for 12 GeV/c

We attempted to improve the agreement between the Wang model and the data by refitting the four parameters in Eq. 1 to a subset of the 12 GeV/c data. The optimization used all data with $p_T \leq 0.6$ GeV/c and $y^* > 0$. The old and new parameter values are given in Table 2.

Table 2 Wang model parameters

		A	B	C	D
π^+	old	77.8	3.56	1.33	4.73
	new	90.1 ± 8.4	3.35 ± 0.12	1.22 ± 0.04	4.66 ± 0.03
π^-	old	51.4	5.73	1.33	4.25
	new	50.4 ± 3.1	5.08 ± 0.16	1.34 ± 0.02	4.33 ± 0.03

Fig. 2 shows a comparison of the adjusted Wang model with the π^+ and π^- data at $p_T = 0.2$ and 0.4 GeV/c. The model still underestimates the peak π^+ cross section . The fit results given in Table 3 show that in the statistical sense, the model provides a terrible description of the data (χ^2 per degree of freedom on the order of 50).

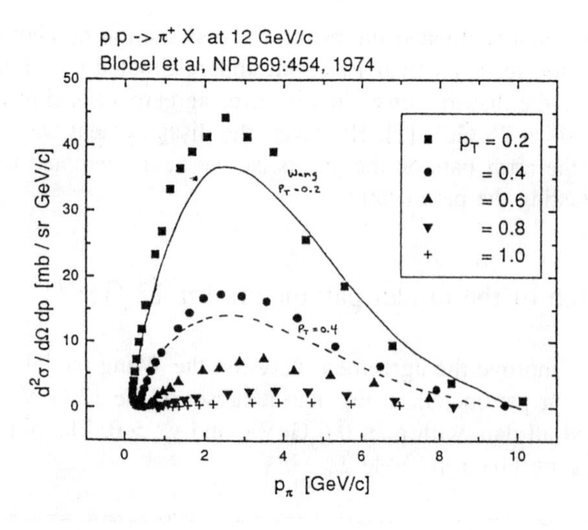

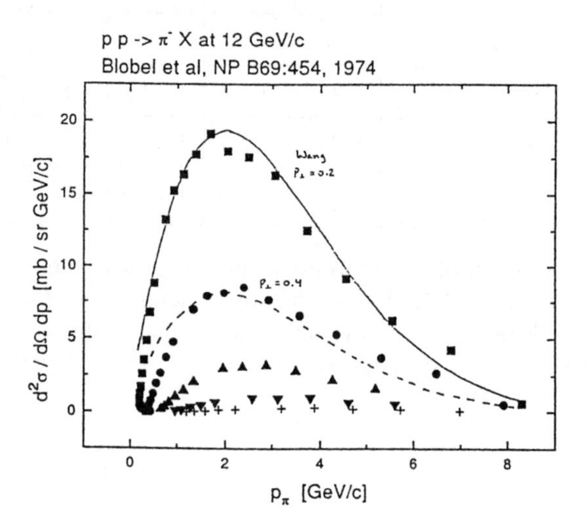

Fig. 2 Comparison of the Wang model with measured π momentum spectra.

Table 3 Wang model fit results			
π^+	p_T [GeV/c]	old χ^2 / dof	new χ^2 / dof
	0.2	173	53
	0.4	147	49
	0.6	165	78
	0.8	99	57
	1.0	74	36
π^-	0.2	11	14
	0.4	54	21
	0.6	65	35
	0.8	157	181
	1.0	106	138

We also attempted to fit the Wachsmuth expression given in Eq. 3 to the data. The best fit parameters are given in Table 4.

Table 4 Wachsmuth model parameters					
	a_1	a_2	a_3	a_4	a_5
π^+	0.988	6.06	16.91	0.525	3.84
π^-	0.547	9.20	11.69	0.331	2.65

Fig. 3 shows a comparison of the adjusted Wachsmuth model with the π^+ and π^- inclusive data at $p_T = 0.2$ and 0.4 GeV/c. There is still significant disagreement for the peaks of the curves and in the high momentum tails. Given that this model has an extra parameter to vary, Table 5 shows that it doesn't provide a much better fit than the Wang model.

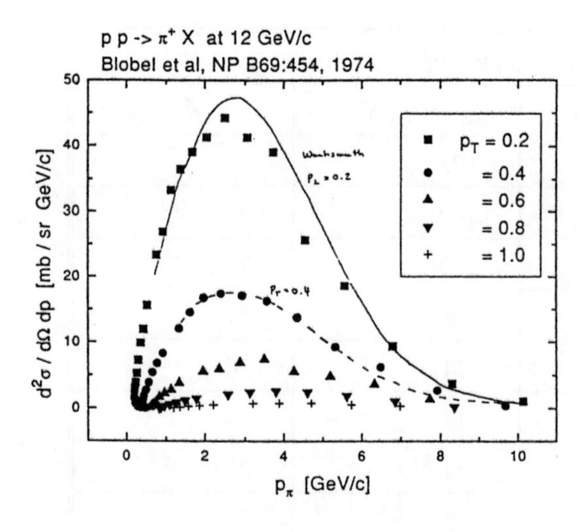

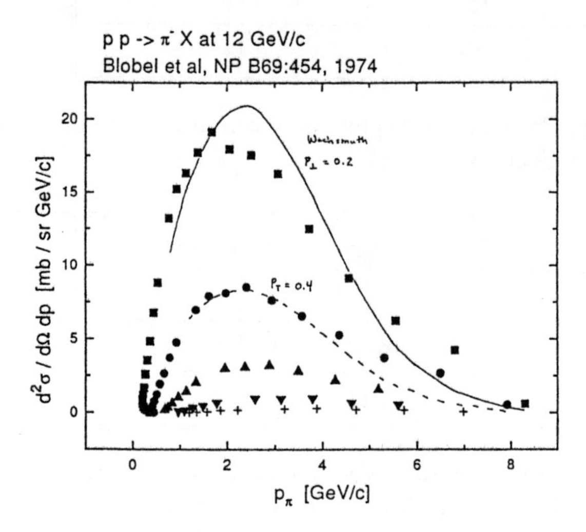

Fig. 3 Comparison of the Wachsmuth model with measured π momentum spectra .

Table 5 Wachsmuth model fit results		
π^+	p_T [GeV/c]	χ^2 / dof
	0.2	51
	0.4	19
	0.6	87
	0.8	124
π^-	0.2	116
	0.4	55
	0.6	110

Acknowledgements

This research was supported by the U.S. Department of Energy under Contract No. DE-AC02-76CH00016.

Notes and references

[1] R.C. Fernow, Comparison of the Wang formula for π production with measurements at low momenta, $\mu\mu$ Collider Tech. Note 04-94.

[2] C.L. Wang, Empirical formula for pion production in proton-proton collisions up to 1500 GeV/c, Phys. Rev. D7:2609-13, 1973; C.L. Wang, Pion production in high energy collisions, Phys. Rev. D10:3876-8, 1974.

[3] V. Blobel et al, Multiplicities, topological cross sections, and single particle inclusive distributions from pp interactions at 12 and 24 GeV/c, Nuc. Phys. B69:454-92, 1974.

[4] H. Wachsmuth, Physics of neutrino beams, in Weak Interactions, Proc. 1971 course of Enrico Fermi School, Baldo-Ceolin (ed), North-Holland , 1977, p.143-63.

Targets and Magnetic Elements for Pion Collection in Muon Collider Drivers

R.C. Fernow, J. Gallardo, Y.Y. Lee, D. Neuffer†, R.B. Palmer, Y. Torun and D.R. Winn*

Brookhaven National Laboratory
Box 5000
Upton, NY 11973

Abstract. We review quasi-achromatic magnetic focussing elements which collect pions produced in a target for transport into a π->μ decay channel, with features appropriate to the development of high energy muon colliders. We discuss how the collection and target requirements of a muon collider are different from and similar to existing secondary particle collection systems. We briefly discuss target technology issues.

Introduction

A muon collider at an energy of 2-4 TeV in the center of mass needs to collect of order a few million or more times as many muons per second as the present antiproton source at FNAL collects antiprotons, before the cooling process, in order to produce a large enough luminosity to discover the physics of interest at these energies ($L\sim10^{34}$ cm^{-2}sec^{-1}, and beyond). This large increase in secondary particle collection is feasible with: (i) an increase in the source driver intensity (targetted protons/sec to produce π->μ) and a lower proton driver energy, (ii) the intrinsically higher yield of pions per proton compared with antiproton yields, and (iii) higher efficiency collection of the secondaries. In this note, we discuss possible magnetic secondary particle collection elements which directly follow a target, which collect pions for transport into a muon decay channel (FODO and/or solenoid strong focussing). We discuss features appropriate to the development of high energy muon colliders, and how the requirements of a muon collider are different from and similar to existing situations.

The formation of high intensity, large momentum bite beams of secondary particles using magnetic elements has been widely developed at accelerator laboratories, especially for neutrino and muon beams, and for antiproton collection and cooling. The ideal collection system would have achromatic focussing at all production angles, for all momenta, azimuthal symmetry, and have a large depth of focus. To first order, the system of elements focusses point to parallel, from the target into the decay channel.

† CEBAF, Newport News, VA, 23606

* Also Department of Physics, Fairfield University, Fairfield, CT 06430-5195

The magnetic element will have an outer radius r_{max} sufficient to accept the pion beam divergence anticipated from the target. The accepted divergence is in general a target-energy dependent quantity, which may even include backward pions for a proton beam near threshold (E_p-m_p=T_p~750 MeV) and decreases with target proton energy (θ_{max}~30° at E_p~30 GeV). To increase the number of accepted pions, one would therefore like as large an r_{max} as feasible.

On the other hand, one would like to minimize the r_{max} necessary, both for cost (magnetic volume -> stored energy -> mechanical strength-> cost), and for minimizing beam cross-section. Thus one would like to make the distance of the magnetic element to the target proportional to r_{max} as small as possible.

In general, the maximum bend angle to be provided by the lens element is given by:

$$\theta_b \sim \sin\theta_b \sim 0.3Gr_{max}L/p \sim r_{max}/Z, \qquad (1)$$

where G is the effective lens gradient, L is its length, and Z is the distance from the target center to the lens center. To this approximation, θ_{max} does not depend on r_{max}, except when r_{max} is small, the effective focal length is small, and the different acceptance between the ends of the target becomes important. A large r_{max} also results in sensitivity to focussing errors. For a fixed r_{max}, the acceptance is therefore maximized by making the gradient as high as possible, and the lens as short as possible. To avoid focussing errors, it is also desirable for the target to lens distance to exceed the target length. Clearly, for many technologies, r_{max} and G_{max} are not independent. In general, the bend angle, and hence the focussing, is not independent of momentum, as in equation 1 above.

In the case of muon colliders, the targetting time (proton bunch length) is desired to be very short, typically 1 ns or less in order to reach a high luminosity in the collider. Therefore the active time of the lens may be shorter for this application than in many other secondary collectors, potentially lowering the cost of the energy store necessary to erect the focussing field, but raising the technical difficulty of the proton source and the target peak powers. Moreover, the repetition rate is desired to be high, depending on the driver, as high as 10-100 Hz in the case of a fast synchroton or other recirculating proton driver accelerator (say 3-30 GeV), or perhaps as high as CW in the 100's MHz in the case of a ~1 GeV linac, which elevates the average power of the energy-store for the lens fields .

A second feature of muon collider pion collection elements is the very large pion momentum bite needed, as much as ±100% about an average momentum in order to generate a sufficient flux of muons before ionization cooling.

We have therefore considered quadrupole triplets, magnetic horns (2 elements), Li lenses (2 elements), solenoids (tapered), and plasma channels.

Quadrupole Triplets

Quadrupole triplets are the simplest elements to construct with large R_{max}, but do not produce perfect images of the target in all transverse directions. They cannot match the field gradients of the other collection devices, when run DC. We have not considered pulsed triplets at this time.

Magnetic Horn Systems

Magnetic horns were developed to generate neutrino beams with a large density of particles per area at ~ 0.1-1 km from the target. They are excellent for collecting pions in terms of B·dl, but not over a short enough distance to create a beam of small cross-section. For example, the BNL horn system which produces toroidal magnetic fields of about 6 T with 300 kA of pulsed currents, requires nearly 8 m of space (Fig. 1). The resulting pion beam is ~ 30 cm or more in radius, for a modest momentum bite of ~15%. A typical magnetic horn system uses 2 horns, the first of which somewhat overfocusses the large angle particles, and the second of which brings them parallel.

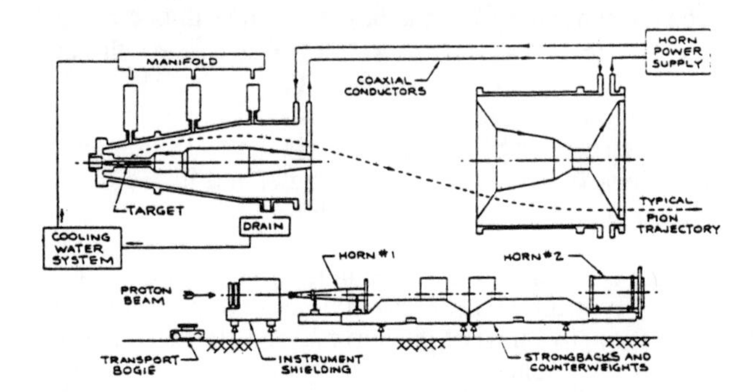

Figure 1. Magnetic horn system at BNL, from reference [2]. Note the large size.

A typical horn has a current distribution shown in Figure 2a and creates an axisymmetric parabolic magnetic field as shown in Figure 2b [1], with a dependence

$$B_\phi = \mu_0 I / 2\pi r. \tag{2}$$

A particle from the target traversing the horn, diverging at a polar angle θ, receives a deflection:

$$\theta = (1/B_0) \int B_\phi ds. \tag{3}$$

Since the path length in a parabolic horn is proportional to r^2, the deflection will be proportional to r, as in a linear focus lense. The current is chosen to so that the deflection cancels the initial divergence θ.

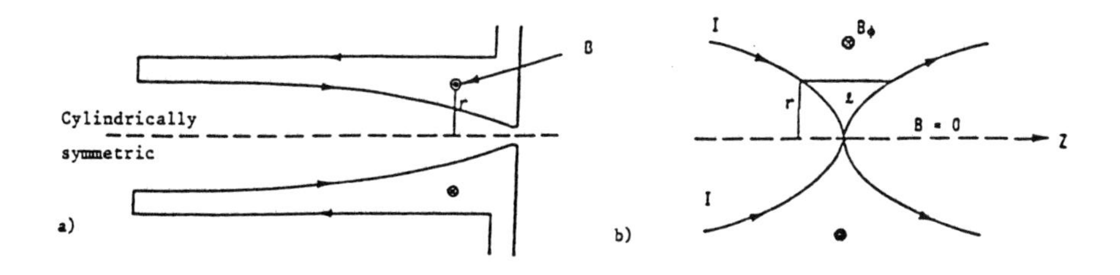

Figure 2a, b: The current profile and parabolic magnetic field shape in a typical parabolic horn, from reference [1].

Magnetic horns cannot be thought of as thin lenses because of their length. A typical horn system is ~ 2.5-3 times worse than an ideal lens [2]. Also, about 30% of the particles are lost in the Al web of the horn. Current magnetic horns have lifetimes of ~10^4-10^5 pulses before failure because of stress fatigue. Even 10^6 shots would be unacceptable for a muon collider with a 30 Hz driver repetition rate (~ 10 hours of operation). Magnetic horns are proven in general, but cannot match the gradients of a system of Li lenses.

Lithium Lenses

In the most analogous case to that of the muon collider, the antiproton accumulators at Fermilab[3] and CERN[4], the primary collection element of choice is a lithium lens. A Li lens is a cylindrical piece of Li with a large axial current. Lithium is chosen for the obvious advantage of long nuclear absorption and radiation lengths, with relatively good resistivity and skin depth. The field produced by a constant current density J is:

$$B = \mu_o J r/2 \tag{3}$$

where r is the radius at which the field is sampled by the pions. The field is thus proportional to the radius, which results in focussing, and has an important advantage over A.G. focussing in that the horizontal and vertical (x,y) planes are focussed in the same device. The lens gradient is therefore the constant:

$$G = \mu_o J/2 \tag{4}$$

Figure 3 shows the field configuration and trajectory of a particle within the lens. A particle diverging with angle $\tan\theta = dx/ds$ will be determined by an equation of motion:

$$\frac{d\sin\theta}{ds} = (e/p)B = [e\mu_o J_o/2p]\, x \tag{5}$$

Solving for x, x', for small angles, gives:

$$x = x_o + x_o' \sin\sqrt{k}s, \quad x' = x_o \cos\sqrt{k}s, \tag{6}$$

which is the equivalent to a lens of strength:

$$k = 1/2\,(e\mu_o/p)\,J \tag{7}$$

A particle produced at $x = z = 0$ with a production angle θ_x will project to $z = 0$ with an $x = -z\,\theta_x/2$. For a given θ_x, $z < \pm L\theta_x/2$, where L is the target length along z. The critical strength per unit length which removes the depth of focus problem of a normal lensed target which is given by

$$k = (\pi/L)^2, \tag{8}$$

where L is the target length. With this strength, the beta of a subsequent beam transport is matched at

$$\beta = 1/\sqrt{k} \tag{9}$$

The further inclusion of smearing from proton beam size and multiple scattering will demand a short target for efficient use of the lens. The disadvantage compared with quadrupoles is the passage of the particles through lithium and the windows of the device.

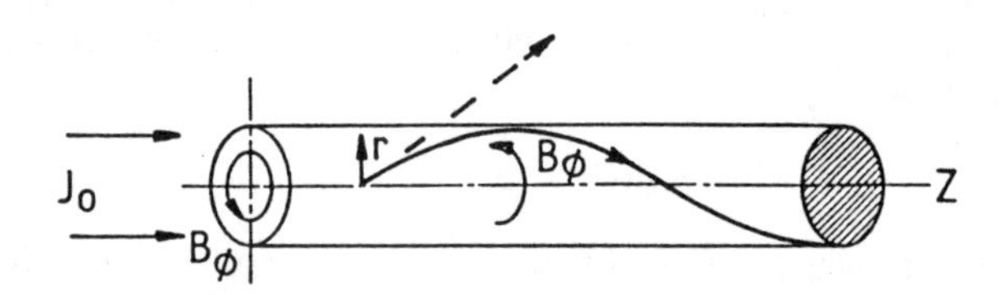

Figure 3: Field configuration and particle trajectories of a Li lense, from reference [1].

At CERN and FNAL, gradients of 800 T/m have been achieved, but in an r_{max} of only 1 cm in a 10 cm length. A prodigious current density of ~ 130 kA/cm^2 is necessary, with a total current of ~400 kA. The time constant must be long enough to allow the fields to penetrate the Li, but not so long as to deposit enough energy by Ohmic losses (~10 kJ/pulse) to exceed the physical limitations of the Li or the electrical leads. This half-sine wave time is ~0.1-0.3 ms minimum and < ~1-3 ms maximum. At FNAL, the magnetic (**JxB** body forces) and thermal stresses (~150 MPa) produce stresses in the cooling jacket of order 0.5 GPa. These will scale with the lens area ~ r^2. Figure 4 shows a schematic of the CERN Li lense [5].

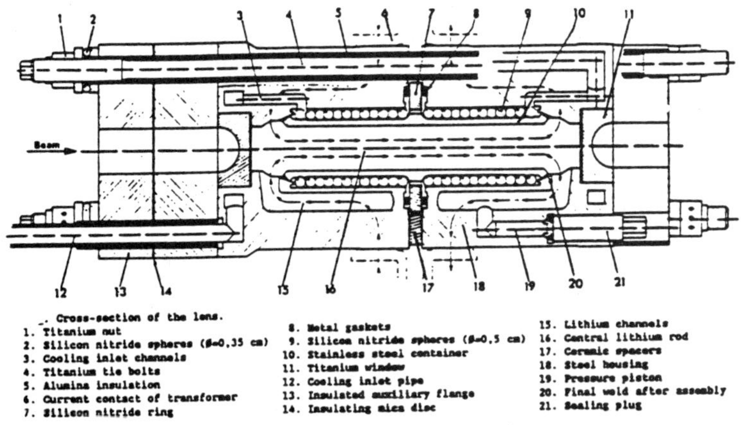

Figure 4. Lithium lense from CERN, from reference [5]. © 1985 IEEE

For the muon collider, for 30 GeV protons incident, we are considering lenses with a total current ~8-10 times those currently in use (4 MA), a length 2.5-3 times current technique (25-30 cm) , radius ~8-15 times existing lenses (8-15 cm radius), current density ~0.2-0.1 times current technique (15-4 kA/cm^2), with repetition rates ~20-25 times higher (~10 Hz). Figure 5 shows schematic of a possible target collection system with 2 lenses of 8(15) cm radius and lengths 25(30) cm long. The lenses are preceeded by an 18 cm long Cu target which is itself a lens, by passing 3 MA of axial current through it.

Using this lens system, a Monte Carlo using the Wang formula[6] for pion production for 30 GeV incident protons was run, using a 6 mm radius beam spot. At 30 GeV, about 1.2 π/proton are collected. The maximum angle of collected pions was about 500 mrad, with an average (maximum) p_T of about 0.3 (0.8) GeV, and an average pion momentum of 2.4 GeV. Reducing the energy by a factor of 3 reduces the pion capture by about a factor of 2. (The 6T pole-tip quadrupole FODO channel with a 40 cm periodicity for pion decay will be described elewhere. With 30 GeV protons incident, the number of muons collected per proton after the decay channel may be as large as $N_{\mu/p} = 0.25$. This implies that ~1 MW of 30 GeV protons is sufficient to generate ~ 5 x 10^{13} collected muons/sec.)

Substantial development work is necessary to prove the feasibility of this extrapolation of Li lens technologies. However, we note that:

(1) The resistance of the Li will be ~25-75 times lower than that of the FNAL lens, and the I^{2}R losses per pulse in the Li will be similar to or lower than the FNAL case, over a volume of Li hundreds of times larger than existing lenses, despite the larger current. With the repetition rates expected, the specific Power/Volume lost in the Li would be an order of magnitude lower than existing lenses;

(2) Much larger volumes of Li and liquid Li have been investigated experimentally for reactor cooling and fusion reaction heat exchange; and

(3) A gradient of 125 T/m would be achieved in the first lens, well-below the gradient used at FNAL of ~800 T/m. At FNAL, a surface field on the Li of ~8 T is used, similar to the 10 T is used in the muon collider design.

If lower production energies are chosen (<30 GeV E_p), then the peak current parameters of the Li lens may be relaxed.

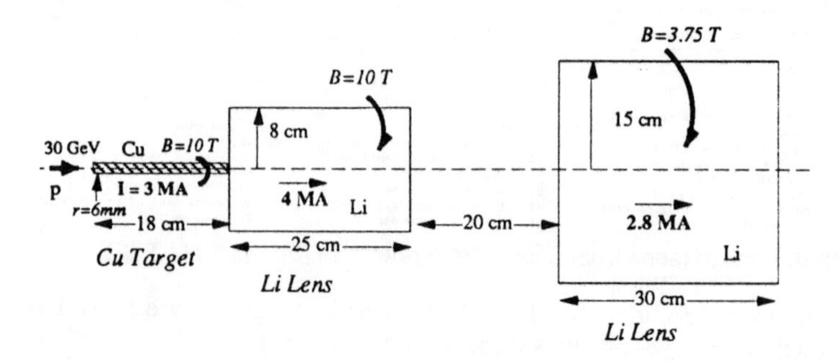

Figure 5. Example of a Li lens schematic system proposed for a muon collider pion collection element.

Solenoids

Solenoidal focussing becomes promising at low particle momenta where $eL\beta cB\sin\theta$ is significant compared with p_T or at large angles where the bending force is effective, where L is the length along z of a solenoid of field $B=B_z$. Modern solenoids may achieve very large axial fields, and for pions (as compared with heavier particles), the solenoid becomes attractive an attractive lens compared with quadrupole triplets, especially because of the symmetric focussing and long depth of focus.

Particles enter the lens first through radial magnetic fields and then an axial field. The radial field gives an azimuthal force $ev_z \times B_r$ and the resulting v_ϕ leads to a radial force when crossed with B_z inside the solenoid, leading to a lens condition of net deflection towards the axis independent of the sign of the charge or transit direction. As it leaves the lens, the conservative force cancels out this radial velocity, but leaves a fixed rotation. For constant B_z, the rotation in the lens from the conserved azimuthal forces $\Delta\phi$ is approximately [7]:

$$\Delta\phi = \phi - \phi_o = (qB_z/2p)\Delta z \qquad (10)$$

The longitudinal derivative of a track in the uniform solenoid is proportional to:

$$v_r/v_z \sim \int dz\, (qB/p)^2 r/4 \qquad (11)$$

from which a focal length f is derived as:

$$f^{-1} = \int dz\,(qB_z/p)^2/4 \qquad (12)$$

with

$$\theta = v_r L/v_z r \qquad (13)$$

and with lens strength

$$k = eB_z/2p \qquad (14)$$

The focal length is inversely proportional to the square of the gyroradius and the field must increase proportionally to the momentum. The lens gives a real focus, independently of the direction of the magnetic field. The focussing is second order, resulting from a small azimuthal velocity crossed into the axial field, with strength $\propto B_z^2$. The lens is said to be "weak" if $eB_zL/p = 2kL \ll \pi$. The effective gradient is $G \sim B\sin\theta_{max}/r_{max} \sim B(r_{max}/L)/r_{max} \sim B/L$, for small r/L. Figures 6a, 6b show the geometry of the the transverse motion of a charged particle in a uniform solenoidal field(6a), and the radial fields and forces (6b) of a end of a solenoid[8].

While solenoidal lensing is most effective on electrons because of the mass effect, the pion is light enough that solenoidal focussing for pion collection can be competitive at momenta characteristic of pion production kinematics at lower proton energies, from threshold to roughly 10 GeV. Moreover, the target can be immersed in the lens.

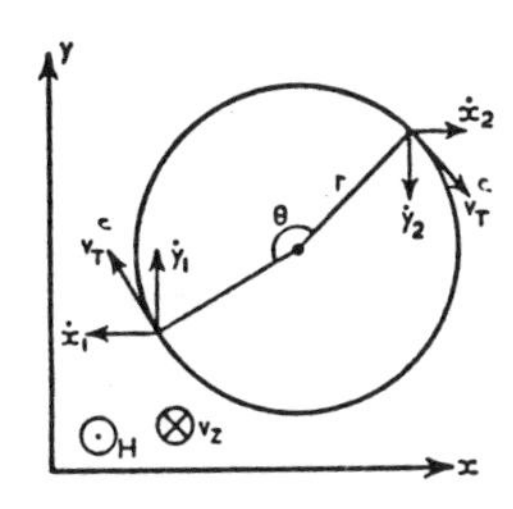

Figure 6a: Geometry of the the transverse motion of a charged particle in the uniform field of a solenoid , from reference [8] .

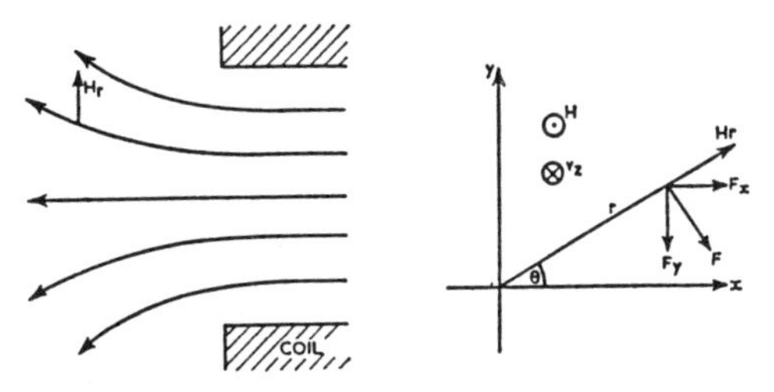

Figure 6b: Geometry of the the fields and forces on a charged particle at the end of a solenoid (axial view), from reference [8] .

High field, large volume solenoids are now possible. DC superconducting solenoids with large bores ~50 cm - 2m radius achieve ~8-4 T with lengths up to ~5-10 meters possible (as used in several varieties of plasma or fusion experiments and in high energy colliding beam detectors). With warm bores, a ~1.5 cm radius, 40 T field is possible over a 10 cm axial length [9].

We considered a solenoid system which starts at 30 T, 8 cm radius, 24 cm long, with an immersed 24 cm Al target, and tapers from 30 T to 7 T over a 1 m length, tapering up to an exit bore of 30 cm in diameter (Fig. 7). The final solenoid is 7T with a 30 cm diameter bore, of arbitrary length. A Monte Carlo of the collection efficiency at E_p=10 GeV gives the same number of pions captured per proton as in the Li lens case described above, $N_{\pi/p}$ = 0.67 at 10 GeV. (After a solenoidal decay channel, to be described elsewhere, the N_μ per proton roughly matches or slightly exceeds that of the Li Lens system described above for targetted protons between 30-100 GeV. For incident protons of 10 GeV and below, the collection efficiency $N_{\mu/p}$ ~0.23 (10 GeV) exceeds that of a Li lens by factors of 2 or more.)

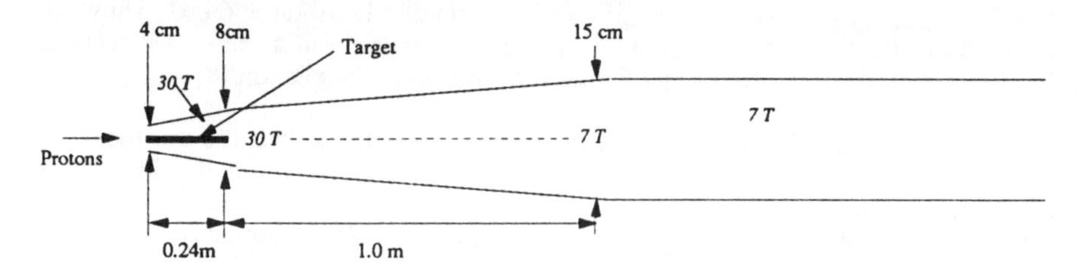

Tapered Solenoid Lens

Figure 7. Schematic of a tapered solenoid pion collection system.

The non-pulsed nature of a superconducting or superferric solenoid is extremely attractive, but at present, it is unkown whether such a system can operate in the radiation field and loss conditions downstream of a target which is accepting 1-2 MW of average beam power, and peak powers of ~10^{14} W. However, it seems feasible to make large diameter conductors which are heavily shielded, with warm bores. Alternatively, pulsed solenoids (example: Bitter-type) of non-superconducting materials may be contemplated.

Plasma Lenses

A plasma channel conducting large axial currents produces a toroidal magnetic field similar to that of a lithium lens, but with a much lower mass density than Li, and the potential for larger currents over large axial lengths [10]. An example which might be applicable is the spark- or Z- channel, as sketched in Fig. 8. Examples conducted ~200 kA over 50 cm, and 75 kA over 5 m at a gas pressure of ~1-10 Torr. It consists of annular plates serving as cathode and anode for the discharge, inside a vacuum chamber filled with a light or heavy gas (Argon + others). A laser preionizes and heats a conducting channel to strike the arc. The channel radii are typically 1 cm and larger, up to ~10 cm, well-matched to typical target conditions where a Li lens would be used.

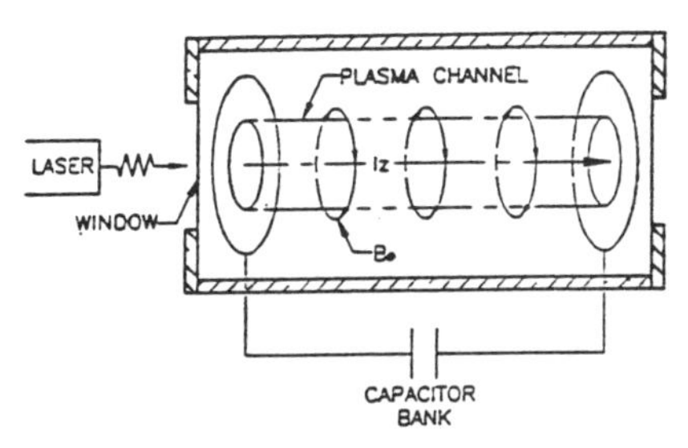

Figure 8. Schematic of a plasma channel (spark- or Z- channel) which may be adaptable as a lens for a muon collider [10].

A Z-pinch has conducted 2.8(10) MA over 20(3) cm with pulses 10's of ns long in milliTorr pressures, but in very small radii (~ well-less than 1 mm in the pinch), and could only be used very near the target, in the later stages of ionization cooling, or possibly for the muon storage ring insertion section focussing. (This has some similarities to the plasma lens final focus concept at electron linac colliders).

These speculative examples may even offer the possibility of "electron cooling" pions or muons when the velocity of the plasma conduction electrons matches that of the muons or pions[10], in the case of the Z-channel. Recent estimates require 3 km at 10 MA to bring a muon into equilibrium with the plasma.

Target Issues

The shortest possible target maximizes the phase space density of pions, and thus targets of heavy nuclei are desired. Unfortunately this also maximizes the specific energy density deposited in a target which may lead to failure. In Cu, a typical number is of order 1.5 MeV per cm of target per GeV per proton, for proton energies above ~10 GeV[11]. The energy deposition scales as ~Z^2 (mainly because of electromagnetic processes); in W, this number rises by a factor of ~9.

In one example, a 10 Hz, 30 GeV, 3 x 10^{13}/pulse, 1 ns long bunch proton beam has been proposed for the muon collider driver. In 7 cm of Cu, 16 kW would be dissipated, which is feasible but difficult (13 times that of the CERN antiproton source). However, the specific energy density may easily exceed the local ability to extract heat. Using a large beam spot of σ=1 mm, about 7 times larger than the beam spot at FNAL, about ~0.06 GeV/g/proton is deposited in Cu, ~ 300 kJ/g per pulse at 3 x 10^{13} protons per pulse. Using:

$$\Delta E \sim C_v \Delta T \sim C_v(T\text{-}300°K), \qquad (15)$$

one finds T ~ 1,100 K°, approaching the melting limit in Cu but quite tolerable.

The short deposition time (~ 1 ns) compared with the sound velocity (scales as thermal conduction) prevents cooling from affecting this conclusion. Indeed, the short deposition time leads to thermal shock, and a resulting overpressure. The target maximum pressure P from an energy deposition E in a material with density ρ is estimated from Gruneison's equation[12] and Gruneison's constant γ as:

$$P = \rho\gamma E \tag{16}$$

In the FNAL target for the antiproton collider, this pressure reaches 7 GPa, under beam conditions which are ~35x less in total deposition than might be envisaged for a muon collider at 1 mm spot radius. It may be required to use a rotating or replaceable target. A beam spot of 1 mm may be at the limits of tolerability for most efficient imaging for pion collection, although in Monte Carlo, 6 mm spot-sizes seem to be tolerable, and reduce the instantaneous energy density to tolerable limits. To reduce aberration in the collection system, a shorter W-Rh target with interspersed graphite disc shock dampers as used at CERN may be contemplated[13].

Another target technology to increase pion collection is to pulse a conducting target (Cu) with a large axial current (3 MA), creating a toroidal lens-target, as shown schematically in Fig.5, and studied in the Monte Carlo of pion collection. This type of combined lens-target has been successfully tested at CERN at 100 kA, using W, Cu, Al and related alloys, in lengths of about 10 cm[14]. Prestressing and cooling proved essential. The extra Ohmic losses and radiative enhancement of target dissipation from captured electrons have not been been included in our target design at this time.

Conclusion and Experimental Plans

A pion collection system using Li lenses is attractive for proton energies on the production target above ~20-30 GeV. Between threshold and 10-20 GeV for the proton driver energy, solenoid collectors may offer some advantages over Li lenses, provided that the radiation hardness of superconducting systems can be demonstrated. The hardware proposed is a reasonable extrapolation from existing technologies.

For the future, we will consider proposals to:

(a) Design and construct a large Li lens, at least 20 cm in diameter and 25 cm long, with at least 1 MA current.

(b) Design and construct a 20-30 T solenoid, >10 cm diameter x >1 m long (stored energy ~1 MJ), with sufficient radiation shielding, thermal protection and thermal reservoir to accept pulses on a 5 cm Cu target directly in front of the solenoid.

(c) Design and construct pulsed Cu or other targets in lengths between 7 cm and 18 cm, with at least 1 MA of axial current.

(d) Conduct beam tests with the targets, lenses and solenoids at proton energies between 0.8-30 GeV to measure the pion collection optics, stability and reliability.

A by-product of tests could be a measurement of the pion yields at energies and angles appropriate to muon collider cooling channels, at each proton driver energy, and for several targets. Surprisingly, a comprehensive set of directly comparable data on pion yields (both charges, similar target materials and lengths, similar phase space) over a kinetic energy range of 0.8-30 GeV incident protons apparently is not available in the literature.

Acknowledgements

We thank the Department of Energy for support in all phases of this work.

References

[1] E.J.N.Wilson, "Antiproton Production and Accumulation", in AIP Conf.Proceedings 153, Physics of Particle Accelerators, Vol.2, 1663 (1987) M.Month, Ed., AIP, New York

[2] A.S.Carroll, "Focussing Horn Upgrades", *BNL Neutrino Workshop*, BNL 53079, 145 (Feb.1987) M.Murtagh, ed

[3] M.Church & J.Marriner, "The Antiproton Source: Design and Operation", *Ann.Rev.Nucl.Part.Sci.* 43:253 (1993)

[4] R.Bellone et al., In *Proc.8th Int.Conf.on H.E.Accelerators*, Novosibirsk, 1986, 2:272 (1986) Novosibirsk: Nauka

[5] P.Sievers et al, Development of Lithium Lenses at CERN, IEEE Trans.Nuc.Sci.Vol.NS-32, No.5, 3066 (1985)

[6] C.L.Wang, *Phys.Rev.Lett.* 25(70)1068

[7] S. Humphries, *Principles of Charged Particle Acceleration*, 125-127, [1986], Wiley, NY, NY

[8] A.P.Banford, *The Transport of Charged Particle Beams*, 130 (1966) E.&F.N.Spon,Ltd, London

[9] National High Magnetic Field Lab, Tallahassee, FL, as reported in the Search and Discovery Section, Physics Today, pages 21-22, Dec. 1994

[10] Ady Hershkovitch, BNL, private communications, 1994-1995

[11] M.Church & J.Marriner, "The Antiproton Source: Design and Operation", *Ann.Rev.Nucl.Part.Sci.* 43:253 (1993)

[12] Z.Tang and K.Anderson, Fermilab Tech.Note TM-1730 (1991)

[13] C.Johnson et al., *Proc.IEEE Part.Accel.Conf.*, Washington 1987, 3:1749 (1987)

[14] T.Eaton et al., "Conducting Targets for pbar Production of ACOL", IEEE Trans.Nuc.Sci., Vol.NS 32, No.5, 3060 (1985)

Muon Cooling and Acceleration Experiment using Muon Sources at TRIUMF

Pamela H. Sandler, S. Alex Bogacz and David B. Cline

Center for Advanced Accelerators, Department of Physics and Astronomy, University of California Los Angeles, Los Angeles, CA 90024-1547

Abstract. We propose to demonstrate an effective method for cooling and accelerating muons by channeling them in a crystal structure. Charged particles trapped in the high-field environment of crystal channels may be damped to the ground-state transverse emittance by reaction with the lattice (4). The damping time for this is characteristically 10^{-6} seconds. However, the reaction may be enhanced by modulating the inter-atomic spacing of the crystal lattice. Acceleration gradients around GeV/m are a result of the high fields in the channel. Three crystal-acceleration schemes are explored in this paper. One involves coupling to the plasma formed by conduction electrons found in the crystal. Other schemes provide acceleration by inverse FEL coupling with a high-power optical driver. To demonstrate crystal cooling and acceleration, the experiment will be done with muon beams at TRIUMF in Vancouver Canada. The characteristics of these beams are favorable for testing muon-crystal channeling, cooling and acceleration.

INTRODUCTION

Leading schemes for future high energy $\mu^+\mu^-$ colliders(1,2) rely on fast cooling and high gradient acceleration of short-lived muons ($t_{rest} = 2.1 \times 10^{-6}$ sec.). This experiment aims to prove that both processes can be integrated and achieved in the ultra-strong focusing environment of a solid state system. Practical demonstration of transverse cooling in a modulated crystal channel and verification of theoretically predicted cooling efficiencies are the first crucial steps towards meeting the challenges of $\mu^+\mu^-$ colliders (3). Furthermore, experimental measurement of high-acceleration gradients around GeV per meter promised by the high fields in a crystal channel would make $\mu^+ \mu^-$ colliders a real possibility (3).

THEORETICAL OVERVIEW

Cooling

Recent results on the radiation reaction of charged particles in a continuous focusing channel (4) indicate an efficient method to damp the transverse emittance of a muon beam. This could be done without diluting the longitudinal phase-space significantly. There is an excitation-free transverse ground state (4) to which a channeling particle will always decay, by emission of an x-ray photon. In addition, the continuous focusing environment in a crystal channel eliminates any quantum excitations from random photon emission, by constraining the photon recoil selection rules. A relativistic muon entering the crystal with a pitch angle of θ_p that is within the critical channeling angle (5) around $\theta_c = 7$ mrad, will satisfy the undulator regime requirements given by the following inequality,

$$\gamma\theta_p << 1. \tag{1}$$

In this case, the particle will lose a negligible amount of the total energy while damping to the transverse ground state. Muons of the same energy but different θ_p will all end up in the same transverse ground state, limited by the uncertainty principle. Theoretically predicted ground state emittance is given by the following expression

$$\gamma\varepsilon_{min} = \frac{\lambda_\mu}{2}, \tag{2}$$

where $\lambda_\mu = \frac{\hbar}{m_\mu c}$ is the Compton wavelength of a muon.

Following the solution of Klein-Gordon equation (4), photons emitted in a "dipole regime", given by Eq.(1), obey the following selection rule

$$\Delta n = n_i - n_f = 1 \ , \tag{3}$$

linking energies of the initial, E_i, and the final , E_f, state of a radiating particle according to the following formula

$$E_f = \frac{E_i}{\left[1 + \frac{1}{4}(\gamma\theta_p)^2\right]^2} \approx E_i\left[1 - \frac{1}{2}(\gamma\theta_p)^2\right] \ , \tag{4}$$

Therefore, the longitudinal energy spread will be very small, about $\frac{1}{2}(\gamma\theta_p)^2$.

This combination of both the transverse and the longitudinal phase-space features makes a radiation damping mechanism a very interesting candidate for transverse muon cooling in an ultra strong focusing environment inside a crystal.

For positive muons channeling in a silicon crystal the characteristic transverse damping time, τ, is given by the following formula

$$\frac{1}{\tau} = 2r_\mu \frac{e\phi_1}{3m_\mu c}, \tag{5}$$

where r_μ is the classical radius of a muon and $e\phi_1 = 3.6 \times 10^3$ GeV/m² is the focusing strength for silicon crystal (15). Although the characteristic damping time, $\tau \approx 10^{-6}$ sec, is long one can enhance the lattice reaction (6) by externally modulating the inter-atomic spacing of the crystal lattice. For example, we can use an acoustic wave of wavelength l, (see Figure 1a) that resonates with the Doppler shifted betatron oscillations of the beam, $\gamma\lambda_\beta$, according to the following matching condition

$$\gamma\lambda_\beta = \sqrt{2}\, l \quad , \quad \lambda_\beta = 2\pi\sqrt{\frac{m_\mu c^2}{e\phi_1}}. \tag{6}$$

If a standing acoustic wave of sizable amplitude could be generated in a crystal, then the relaxation time would shorten by the factor of one plus the ratio of the acoustic to the transverse beam power.

Acceleration

According to previous calculations (7,8), one can achieve acceleration gradients of GeV per meter in the high fields found in a crystal channel. The first paper (7) explores the idea of particle acceleration by self-coupling to a plasma formed by the conduction electrons in a crystal. This can also be accentuated by resonant excitations of the plasma.

In another scheme (8), a high gradient acceleration (GeV/m) is provided by inverse free-electron laser (FEL) coupling with a high-power optical driver. A strain modulated silicon crystal acts as a microundulator for a channeled muon beam. This crystal is then placed in an optical cavity – between two axicone mirrors powered by a GWatt laser at visible frequencies.

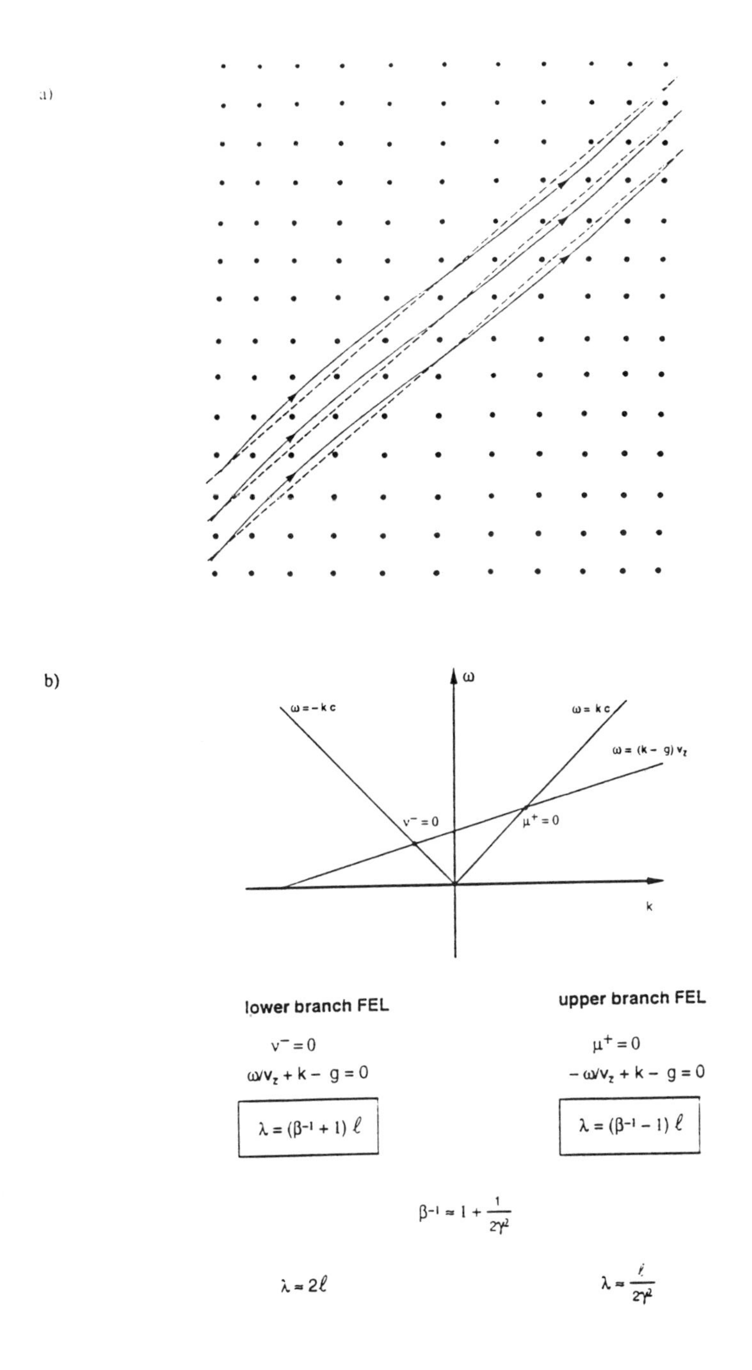

Figure 1. a) Planar channeling paths for positive charged particles in the [110] crystallographic direction. The position of the center of the channeling axis is modulated by the acousti wave periodicity to produce an undulator effect. b) Illustration of acceleration in terms of three-wave mixing.

A beam of relativistic particles while channeling through the crystal follows a well defined trajectory. Figure 1a depicts planar channeling paths for positive charged particles in the [110] crystallographic direction. The center of the channeling axis is modulated by the acoustic wave periodicity to produce an undulator effect. In other words, the particles are periodically accelerated perpendicular to their flight path as they traverse the channel. The micro-undulator wavelength (for a typical acoustic modulation) falls in the range 1000–5000Å, far shorter than those of any macroscopic undulator. Furthermore, the electrostatic crystal fields involve the line averaged nuclear field and can be two or more orders of magnitude larger than the equivalent fields of macroscopic magnetic undulators. Both of these factors hold the promise of greatly enhanced coupling between the beam and the accelerating electromagnetic wave.

The key to collective acceleration via inverse FEL mechanism is a spontaneous bunching of an initially uniform beam channeling through a periodic crystal structure and interacting with the electromagnetic wave. Appropriate phase matching results in energy flow from the wave to the particle beam. This particular kind of particle-density fluctuation has the form of a propagating density wave of the same frequency, ω, as the emitted electromagnetic wave. The phase velocity of the moving bunch matches the velocity of particles in the beam. Therefore, the quantity $k_b \equiv \gamma m\omega/p_z$ represents the wave vector of the propagating particle density bunch. Keeping in mind that the periodicity of the undulator represents a static wave with a wave vector g, and that k is the wave vector of the electromagnetic wave, we can analyze our acceleration mechanism in the language of "three wave mixing" as illustrated in Figure 1b. Here the "lower branch FEL" condition is appropriate for particle acceleration inside a crystal microundulator. It translates into the following acoustic – optical wavelength constraint

$$\lambda = \left(1 + \frac{1}{\beta}\right) l, \qquad \beta = 1. \tag{7}$$

In our numerical example this fixes the accelerating optical wavelength at λ = 1000 nm.

The acceleration efficiency (8), α, is described as the rate of optical amplitude depletion per one particle. Assuming a muon beam of initial energy 250 MeV, a typical value of the beam concentration, $n = 10^{16}\ \mathrm{cm}^{-2}$, acoustic wavelength of l = 5000Å and the initial longitudinal momentum spread of $\frac{\Delta p}{p} = 10^{-2}$, one gets the following value of the acceleration rate (8)

$$\alpha = 2 \times 10^{-3}\ \mathrm{cm}^{-1}. \tag{8}$$

The nominal acceleration efficiency in units of MeV/m will depend on the energy density of the actual optical cavity. The recent advances in high power laser technology based on diode-laser pumped solid-state lasers (9) promise a power of a few MWatts, optically focused to provide energy densities of $E^{max} = 10^{10}$ V/m, where E^{max} is the electric field amplitude of the standing cavity mode. This combined with Eq.(7) sets the final accelerating efficiency equivalent to an accelerating gradient of 2 GeV/m.

We can also test the inverse Cerenkov acceleration mechanism (10), since the index of refraction for silicon is very large, n = 1.5. Matching the phase velocity of the optical mode to the muon velocity requires relatively large crossing angle (between the beam and the laser pulse). This enhances the longitudinal projection of the radial component of the electric field, which in turn yields high accelerating gradient.

EXPERIMENTAL OVERVIEW

The experiment will be done in three stages. The first step, called the transmission experiment, will show channeling of muons in a 4-cm long silicon crystal. The second step will measure the cooling of channeled muons. Finally, the third step will incorporate acceleration.

Phase I - Transmission

Channeling of muons has been observed (11) at 4 MeV. However, channeling of higher energy muons has not been measured conclusively. We require measurements of the critical angle for channeling muons and the ionization energy loss for channeled versus unchanneled muons. We will use the 35 MeV/c momentum surface muons provided by the M13 beamline at TRIUMF, which was selected based on our assessment of the quality of the muon beam. At this energy, the stopping power of amorphous silicon is high, about 4 mm. In this case, only channeled muons will survive the crystal. Unchanneled particles are subject to typical energy loss mechanisms of ionization and bremstrahlung. Whereas, the energy loss of channeled particles is severely reduced (5).

A schematic of the proposed setup for the transmission experiment is illustrated in Figure 2. The beam, incident from the left, passes through a trigger and veto formed by two scintillation counters. The veto rejects muons that do not enter the geometric dimensions of the crystal. The muons then enter the crystal at some incident angle, θ with respect to the axis of the crystal. The crystal's orientation is controlled remotely by a goniometer. Just downstream of the crystal, the exiting muon energy and flux will be measured with a surface-barrier detector. Such detectors are capable of about 20 keV energy resolution at 35 MeV/c. The data will indicate the yield as a function of θ, allowing extraction of the critical angle within which muons are effectively channeled. Furthermore, we can measure the energy spread of the exiting beam. This will indicate the degree of energy straggling we should expect when compared to the incident energy distribution provided by the beamline. It is important to align the beam to hit the crystal within the critical angle for channeling. We will use the tracking information and a goniometer to maximize the number of channeled muons exiting the crystal.

TRIUMF's M13 beamline is the best choice for effective channeling through a 4-cm sample of silicon crystal. It provides surface muons at high intensity – about 1.2 x 10^6 per second (12). They carry momentum of 35 MeV/c with a longitudinal spread of about 4% FWHM. Assuming optimum tuning of the final focus quadrupole doublet in M13 one could achieve a spot size of 2 cm x 2 cm with horizontal and vertical divergences of 25 mrad and 65 mrad respectively. The critical planar channeling angle in a silicon crystal is about 12 mrad for 35 MeV/c positive muons, if one extrapolates critical angle measurements from proton channeling (5). In this case, a sizable fraction of the incident muons will channel through the crystal.

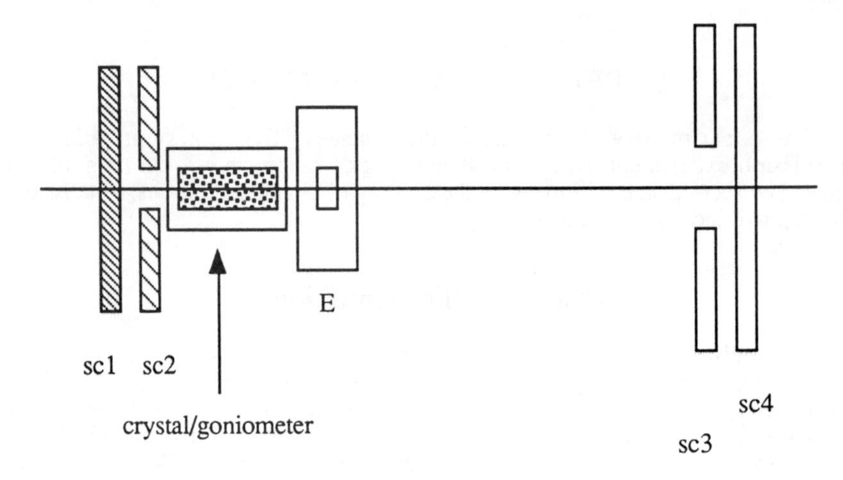

FIGURE 2. A schematic of the Phase I experiment on transmission of muons through the crystal. Sc1-Sc4 are scintillation counters and E is a surface barrier detector measuring the energy of exiting muons. The 4cm x 1cm x 1cm crystal is mounted on a goniometer.

Phase II - Cooling

This phase of the experiment will test the cooling mechanisms summarized in the theory section above. The beam momentum will be about 250 MeV/c as provided by forward decay muons (13) in M11. In this case, both channeled and unchanneled muons will penetrate the 4-cm crystal and the cooling process can be compared for the two. In addition, a higher energy beam tests cooling at the energies considered for realistic collider schemes. The first step of Phase II is to measure initial and final emittances of an unmodified crystal. A schematic of our proposed experimental setup is given in Figure 3. We will track each muon individually using five sets of drift chambers. This way we can identify channeled and unchanneled particles on an event-by-event basis. We can also obtain the exact initial and final emittances for channeled and unchanneled muons separately. It is also possible to separate channeled and unchanneled muons by plotting their energy loss. The trigger will be provided by scintillation counters upstream, combined with a veto counter that rejects muons which do not intersect the crystal. A time-of-flight counter will be placed in a downstream position, to be used in concert with the 1 picosecond timing pulse provided by the M11 beamline. This will aid in particle identification since some positrons and pions will likely contaminate the beam.

To enhance the cooling, we will generate a strain modulation of the planar channels. An acoustic wave of 1 GHz is excited via a piezoelectric transducer. We will also detect predicted channeling radiation by surrounding the crystal with CsI scintillation detectors, which are sensitive to X-rays.

The M11 beamline is presently a source of high energy pions (13). Straightforward modification of the beamline will provide a collimated beam of forward-decay muons at high intensity (14) – about 10^6 per second at 250 MeV/c. The longitudinal momentum spread is about 2% FWHM. Assuming optimum tuning of the final focus quadrupole doublet in M11, we can achieve a spot size of 3 cm x 2 cm with horizontal and vertical divergences of 10 mrad and 16 mrad respectively. The critical angle for planar channeling of μ^+ at 250 MeV/c in silicon is about 7 mrad, extrapolating from proton channeling data. A sizable fraction of the muons should channel through a few centimeters of the crystal.

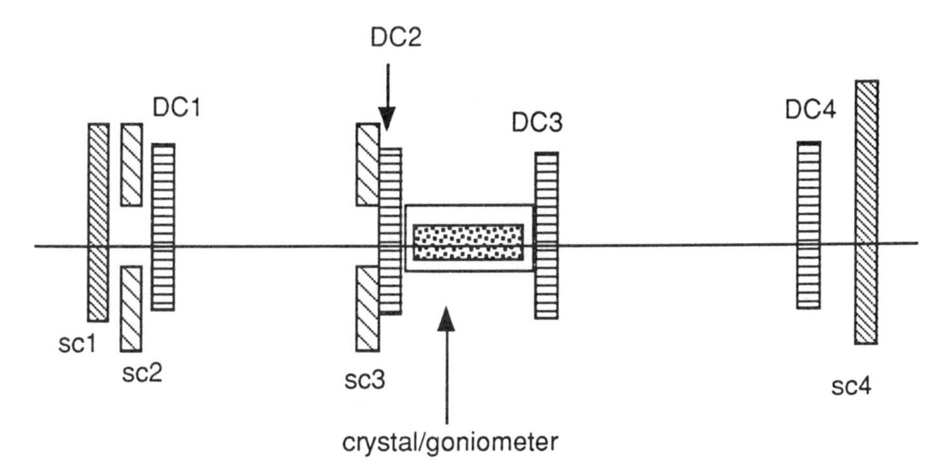

FIGURE 3. A schematic of Phase II: Cooling. Sc1-sc4 are scintillation counters, dc1-dc4 are drift chambers, and the crystal is 4cm x 1cm x 1cm silicon mounted in a goniometer.

Phase III - Acceleration

A high gradient acceleration will be tested in two steps. Initially, an unmodified crystal will be used to test plasma acceleration (7). Then we will apply a laser to test inverse FEL(8) and inverse Cerenkov (10) acceleration of muons. The optical setup is analogous to the Inverse Cerenkov Accelerator Experiment at Brookhaven. It provides a pulse of radially polarized light, which couples energy to the muon beam channeling through a crystal via the inverse FEL mechanism. Here a strain modulation in the crystal imposed by an acoustic wave plays the role of an ultra-short wave undulator. Optical energy will be transferred to the muon beam with an efficiency of GeV per meter. A 4-cm silicon crystal would provide a 40 MeV energy burst. Using a bending magnet in between drift chambers, we will measure the final energy of muons channeling through the crystal. The initial energy of each muons is provided by a spectrometer in the beamline.

References

1 Cline, D.B., "A Crystal μ Cooler for a $\mu^+ \mu^-$ Collider" in *Proceedings of the 1994 Advanced Accelerator Workshop*, Lake Geneva, 1994.

2. Neuffer, D.V., Nuclear Instruments and Methods in Physics Research **A350**, 27 (1994).

3 Cline, D.B., Nuclear Instruments and Methods in Physics Research **A350**, 24 (1994).

4. Z. Hong, P. Chen and R.D. Ruth, "Radiation Reaction in a Continuous Focusing Channel", *Stanford Linear Accelerator Center perprint*, SLAC-PUB-6574, July 1994.

5. Carrigan, R.A. Jr. and Ellison, J.A., *Relativistic Channeling*, New York, Plenum Press, 1986.

6. Chen, P., Private Communication.

7. Chen, P., Cline, D.B., and Gabella, W.E., "Issues Regarding Acceleration in Crystals" in *Proceedings of the Port Jefferson Conference on Advanced Accelerators*, 1993.

8. Bogacz, S.A., Particle Accelerators, **42**, 3-4 (1993).

9. Basu, S. and Byer, R.L., Optic Letters, **13**, 458 (1988).

10. Steinhauer, L.C. and Kimura, W.D., Journal of Applied Physics, **68** ,10 (1990).

11. Maier, K., Hyperfine Interactions, **17**, 19 (1984).

12. Marshall, G.M., Z. Phys. C, **56**, 226 (1992).

13. Oram, C.J., Warren, J.B., Marshall, G.M., and Doornbos, J., Nuclear Instruments and Methods in Physics Research, **A179**, 95 (1981).

14. Doornbos, J., Private Communication.

15. Bazylev, Z.A., and Zhevago, N.K., Sov. Phys. Usp. **33**, 1021 (1990).

A possible ionization cooling experiment at the AGS

R.C. Fernow, J.C. Gallardo & R.B. Palmer
Brookhaven National Laboratory, Upton, NY 11973

D. R. Winn
Fairfield University, Fairfield, CT 06430

D.V. Neuffer
CEBAF, Newport News, VA 23606

ABSTRACT

Ionization cooling may play an important role in reducing the phase space volume of muons for a future muon-muon collider. We describe possible transverse and longitudinal emittance cooling experiments that utilize the capabilities of the AGS at Brookhaven National Laboratory.

INTRODUCTION

There has been considerable recent interest in the development of high energy colliders using muon beams[1]. Most of the proposed designs require a high intensity pion beam resulting from protons interacting in a stationary target. The muons are collected from the decaying pions. However, the resulting muon beam typically has much too large an emittance to give high luminosity in a collider. Barger et al have shown that a luminosity $\sim 10^{34}$ cm^{-2} s^{-1} is required for potential discoveries at 4 TeV in the center of mass[1].

It is possible to reduce the transverse emittance of the muon beam through the process of ionization cooling[2]. Energy loss in a material reduces both the transverse and longitudinal components of the momentum. Multiple scattering in the material is a competing process that effectively heats the emittance. Subsequent acceleration cavities can restore just the longitudinal component, and as a result, the transverse emittance of the muon beam can be reduced.

The longitudinal emittance can also be reduced by placing a wedge-shaped piece of material in a region of the beam with non-zero dispersion. If it is arranged that the higher momentum portion of the beam goes through the thicker side of the wedge, the resulting beam will have a smaller energy spread. Straggling is a competing process that increases the energy spread.

The cooling rates for transverse and longitudinal emittances are coupled[3]. Any practical cooling arrangement must consist of alternating regions of cooling rods, acceleration cavities, and wedges.

We are considering the possibility of developing a prototype cooling rod and testing its performance in a new beam line at the AGS at Brookhaven National Laboratory. A muon momentum of 400 MeV/c would be ideal for the experiment. The proposed rod has to be on the order of 2 m long to obtain a large amount of cooling. Because of the incident beam divergence and multiple scattering, the muon beam would be lost through the sides of the rod unless the rod also acts as a lens. Thus we propose to use a lithium lens[4] as the cooling material.

A 2 m long Li lens has never been constructed or operated. The temporal field profile and its spatial uniformity in the lens need to be measured. Shot to shot variations arising from the power supply, thermal effects, time jitter and stray fields must also be measured. It is important that the monte carlo simulations be fine tuned to accurately describe the action of the lens on the muon beam.

A later refinement would be to build a rod with a tapered end. This would allow matching of the incident beam to a section of rod having the maximum possible surface field. This tapered rod should give the maximum possible emittance reduction per unit length. We would also design the beamline such that a non-zero dispersion could be produced in the experimental area. This would allow the possibility for testing the efficiency of wedge cooling.

MUON PRODUCTION

The muon test beam for the experiment is produced from a 28 GeV/c slow extracted proton beam from the AGS. Table 1 gives the production parameters and Fig. 1 shows the layout of the beam. Pions produced in the target are focused by a triplet of 18" diameter quadrupoles and allowed to decay in the 18 m decay path. The beam profiles and dispersion for the entire beamline are shown in Fig. 1. The beam passes through a 7-interaction length hadron absorber to reduce the pion flux by a factor of 1000 with respect to the muons. Since this causes an energy loss of about 1.35 GeV and the decay muons do not receive all of the pion momentum, the relevant pion momentum at production is around 2 GeV/c.

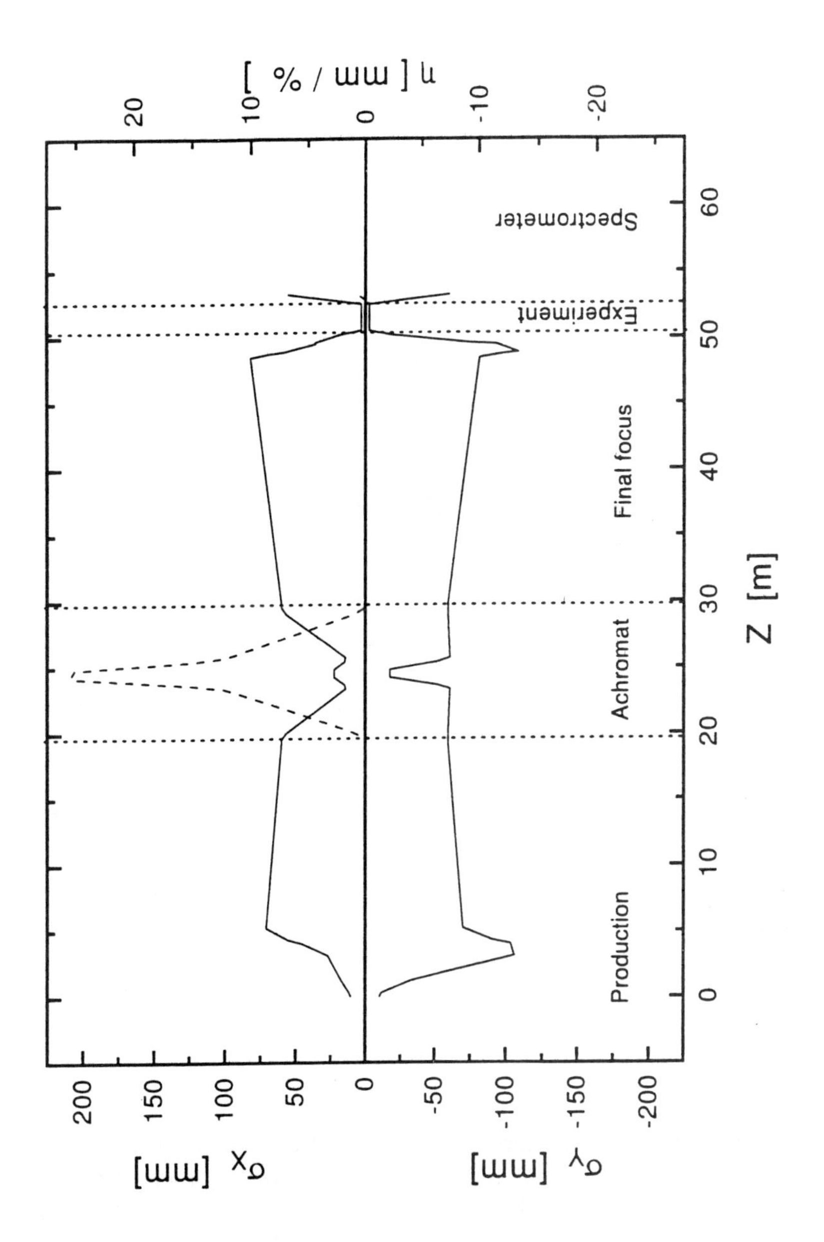

Fig 1. Beam profiles (left axis) and dispersion (right axis) for the cooling rod experiments.

Table 1 Muon production parameters		
proton intensity	$3\ 10^{11}$	p / spill
spill time	1	s
repetition time	2	s
target length	0.1	λ_{int}
π mean momentum	2	GeV/c
collection $\Delta\Omega$	1.0	msr
momentum bite	8	MeV/c
π production rate	$9.3\ 10^5$	π / spill
π decay length	18	m
μ production rate	$1.3\ 10^5$	μ / spill
ΔE in absorber	1.35	GeV
μ mean momentum	400	MeV/c
transport efficiency	75	%
useful μ rate	$1.0\ 10^5$	μ / spill

The pion production yield was estimated using the Wang formula[5]. The accepted momentum bite of 2% is determined by a collimator in the achromat part of the beam line. Taking into account losses in the beam transport, the muon rate at the experiment is about 10^5 per spill. A one second spill is used to stretch out the pulse, so that individual particles can be tracked through the detectors.

BEAM TRANSPORT

The production line optics transports the beam from the target into a waist with $\sigma_X = 60$ mm and $\sigma_{X'} = 3$ mrad at the exit of the hadron absorber. The next section of the beamline shown in Fig. 1 is an achromatic bend, consisting of three quadrupoles and two dipoles. This part of the line bends the experimental beam away from the background behind the absorber dump and provides the opportunity to control the momentum spread in the beam to the experiment. For economic reasons the beam line following the hadron absorber uses 12" diameter quadrupoles. Each of the dipoles bend the 400 MeV/c beam through 20°. The optics transports the beam waist following the hadron absorber into an identical waist at the exit of the achromat. The dispersion near the center of the achromat exceeds 20 mm/% . A collimator at this location could be used to control the momentum spread in the beam.

The final focus part of the beamline shown in Fig. 1 consists of three quadrupoles. The transport optics focus the beam size down to $\sigma_X = 3$ mm and $\sigma_{X'} = 60$ mrad. This is suitable for matching the beam into the rod or wedge for the cooling experiments. An alternative final focus design using a 2.8 T, 1.2 m long solenoid (instead of the three quadrupoles) is also available.

The momentum following the experiments is measured with a dipole in the spectrometer area shown in Fig. 1. In order to guarantee that all tracks leaving the rod are measured, no quadrupoles are used in the spectrometer.

For the rod experiments the currents in the achromat quadrupoles are adjusted so that the beam leaving the achromat has zero dispersion. However, for the wedge experiment, which requires dispersion at the experiment, the currents in the achromat quadrupoles can be adjusted so that the beam leaving the achromat has a finite dispersion. Fig. 2 shows the achromat and final focus portions of the beamline with the quadrupoles adjusted to give a dispersion of 5 mm/% at the wedge experiment.

TRANSVERSE COOLING EXPERIMENTS

The primary goal of this experiment would be to test the performance of a prototype cooling rod. Transverse ionization cooling would be tested by directing the muon beam into a 2 m long lithium lens. (Beryllium would produce more efficient cooling than lithium. However, significant, historical safety concerns would have to be addressed before using beryllium.)

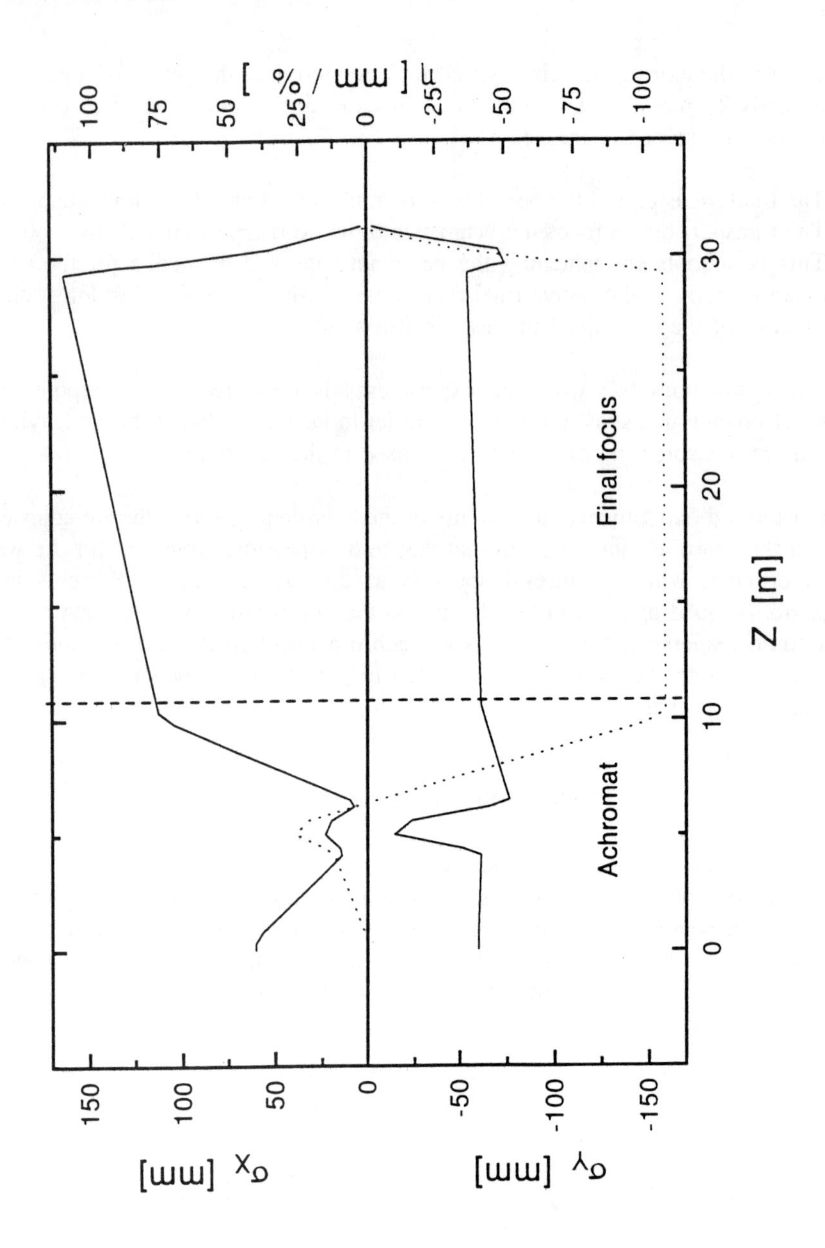

Fig 2. Beam profiles and dispersion in the achromat and final focus portion of the beam line for the wedge experiment.

A large axial current pulse in the lens produces an azimuthal magnetic field. At a certain time in the current pulse, the magnetic field inside the rod varies approximately linearly with r. At this time the muon beam would traverse the rod. With a rod radius R of 1 cm and a current pulse of 0.5 MA, the field B on the surface of the rod reaches 5 T. Under these conditions the muons undergo a number of sinusoidal oscillations before leaving the rod. The betatron focusing parameter $\beta_o = \lambda/2\pi$ in the lens is given by

$$\beta_o = \sqrt{\frac{pR}{eB}} \tag{1}$$

where p is the momentum of the muon and e is the muon charge. The maximum incident beam angle that can be captured by the lens is

$$\theta = \sqrt{\frac{eRB}{p}} \tag{2}$$

The normalized transverse emittance ε_n can be reduced by ionization energy loss dE/dz in the lithium at the rate[3]

$$\frac{d\epsilon_n}{dz} = -\frac{1}{\beta^2}\frac{dE}{dz}\frac{\epsilon_n}{E} \tag{3}$$

where $E = \gamma m_\mu c^2$ is the total energy of the muon. Previous references have often used the relativistic approximation $\beta = 1$ in similar formulas. In the present case, where we are considering momenta down to ≈200 MeV/c ($\beta \approx 0.9$), we are including the β factors. In practical units we can write this as

$$\frac{d\epsilon_n}{dz}\left[\frac{mm-mrad}{cm}\right] = -\frac{1}{\beta^2}\frac{dE}{dz}\left[\frac{MeV}{cm}\right]\frac{\epsilon_n\,[mm-mrad]}{E\,[MeV]} \tag{4}$$

The rod also introduces heating of the normalized transverse emittance due to multiple scattering at a rate[3]

$$\frac{d\epsilon_n}{dz} = \frac{\beta_\perp}{2\,\beta^3}\frac{E_s^2}{E\,m_\mu c^2}\frac{1}{L_r} \tag{5}$$

where $\beta_\perp$ is betatron focusing parameter, β is the velocity of the muon divided by c, E_S = 14 MeV is the characteristic energy for the projected scattering angle from multiple scattering theory, $m_\mu c^2$ is the rest energy of the muon (105.6 MeV), and L_r is the radiation length of the material (155 cm for lithium). In practical units this can be written as

$$\frac{d\epsilon_n\,[mm\text{-}mrad]}{dz\,[cm]} = \frac{60\;\beta_\perp\,[cm]}{\beta^3\;E\,[MeV]} \qquad (6)$$

The results simulated using Eqs. 4 and 6 above for sending the muon beam through a 2 m long, 1 cm radius Li lens with a surface field of 5.33 T are given in Table 2. The surface field value is chosen so that the focusing parameter β_o in the lens matches the focusing parameter $\beta_\perp$ of the incident beam.

Table 2 Uniform rod experiment

	IN	OUT	
p	400	211	MeV/c
δp	8.0	9.9	MeV/c
β	0.967	0.894	
γ	3.91	2.23	
ε_N	681	495	mm-mrad
ε	180	248	mm-mrad
β_o	5.0	3.63	cm
σ_X	3.0	3.00	mm
$\sigma_{X'}$	60	82.7	mrad

The changes in the heating and cooling rates and the resultant normalized emittance are shown in Fig. 3. We see that under the conditions listed in Table 2, the cooling rate is always larger than the heating rate and the normalized emittance of the beam leaving the rod is smaller than the normalized emittance entering it.

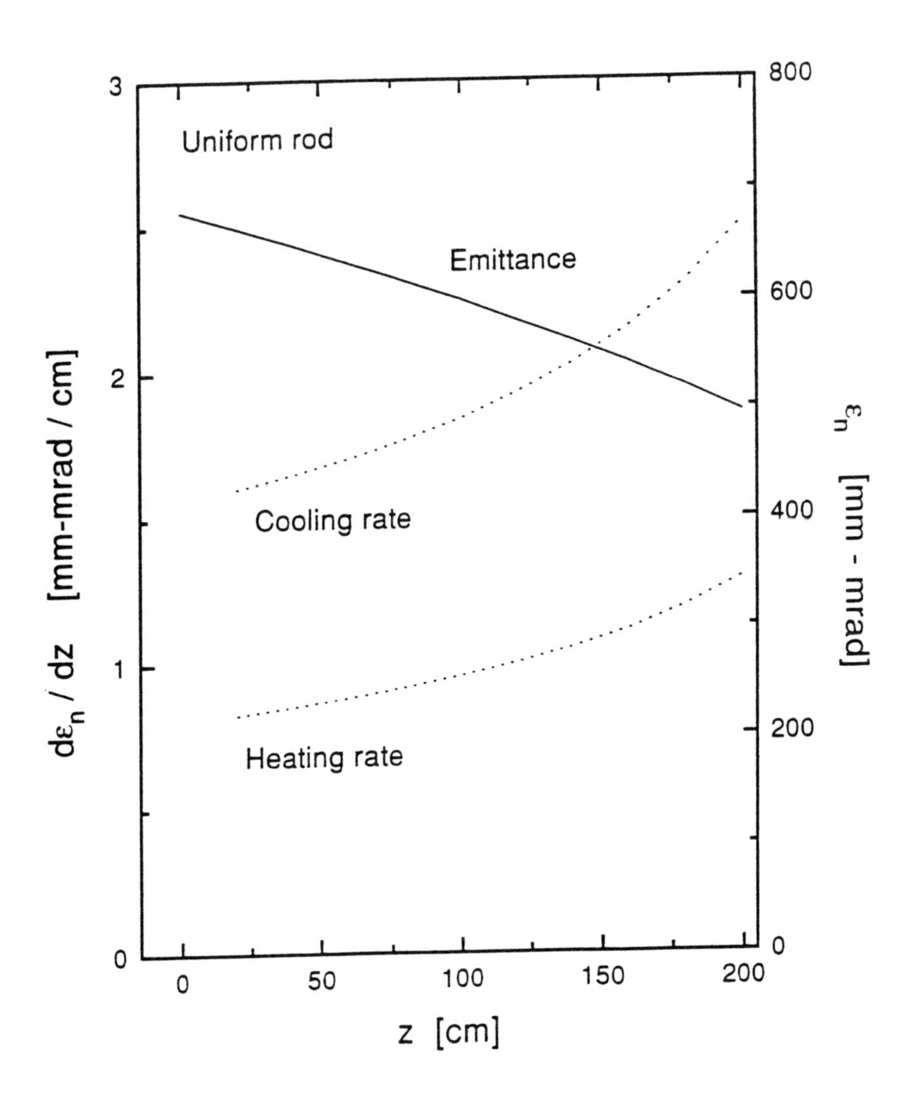

Fig 3. Cooling and heating rates (left axis) and resultant normalized emittance (right axis) for the uniform rod experiment.

One of the design requirements for the experiment is the ability to measure individual muon tracks. This will permit software "control" of the effective cooling rate. Software selection of initial beam tracks will allow us to reconstruct final beam emittance as a function of initial beam emittance. Fig. 4 shows an example of how this might work. The initial normalized emittance for this example is 305 mm-mrad. The heating and cooling rates are almost identical for this case and the final normalized emittance should equal the incident value.

In order to reduce the heating rate as much as possible, we need to make the value of β_o in the rod as small as possible. We see from Eq. 1 that this requires a magnetic field on the surface of the rod that is as large as possible. However, we have just seen that matching the lens to the expected muon beam parameters requires that β_o = 5 cm at the entrance to the rod. One way to accomplish this is to use a tapered entrance end on the rod. Table 3 gives the performance of a possible tapered cooling rod. We assume a lithium rod with a 20 cm region of linear taper, followed by a 2 m length of uniform rod. The taper starts with a radius of 1.37 cm and a 7.3 T field and ends with a 1 cm radius and 10 T field.

Table 3 Tapered rod experiment

	IN	OUT	
p	400	189	MeV/c
δp	8.0	10.0	MeV/c
β	0.967	0.873	
γ	3.91	2.05	
ε_N	681	428	mm-mrad
ε	180	239	mm-mrad
β_o	5.0	2.51	cm
σ_X	3.0	2.44	mm
$\sigma_{X'}$	60	97.6	mrad

The normalized emittance reduction is 37% in this case, as opposed to 27% for the uniform rod in Table 2.

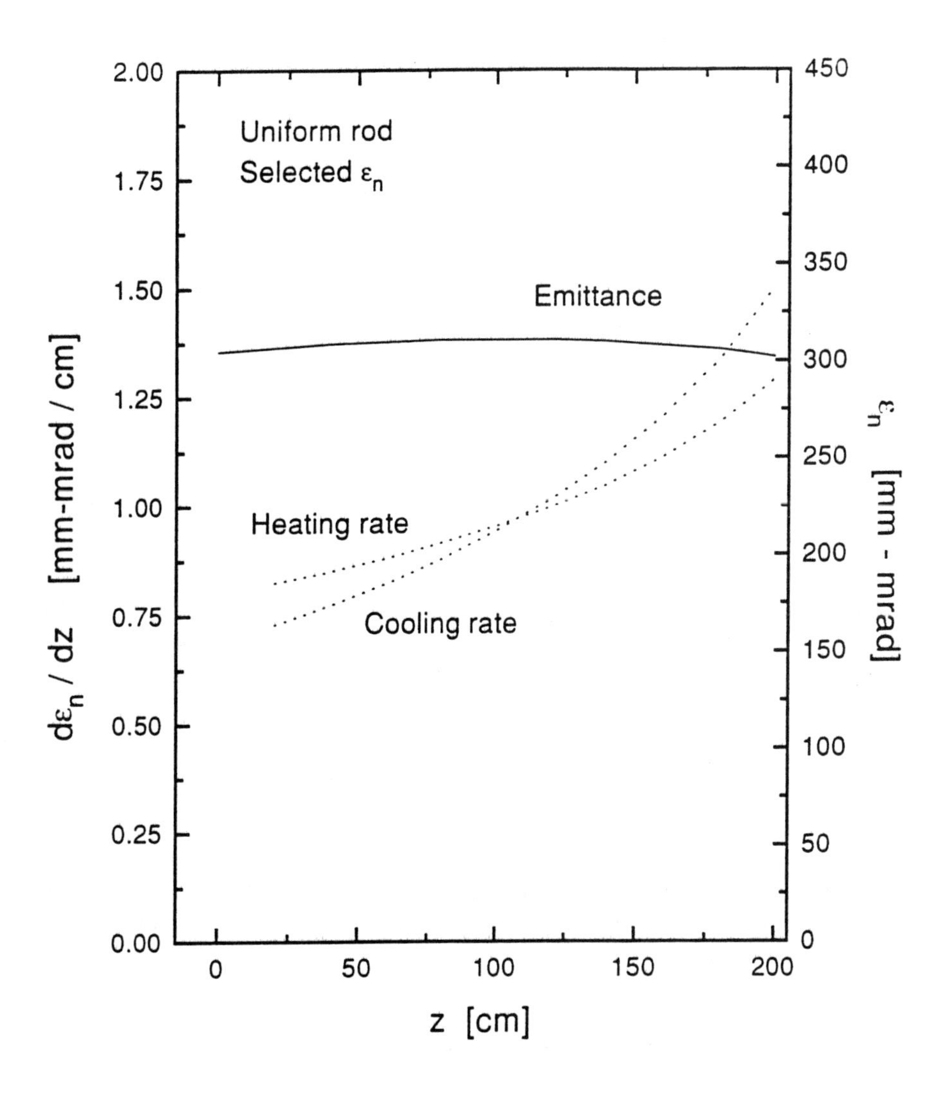

Fig 4. Cooling and heating rates and resultant normalized emittance for selected tracks going into the uniform cooling rod.

LONGITUDINAL COOLING EXPERIMENTS

A later experiment would test the idea of using a wedge shaped absorber to reduce the momentum spread in a muon beam. The incident beam has a $\delta p/p = 2\%$. The quadrupoles are adjusted so that the dispersion at the experiment is 5 mm/%. The high momentum portion of the beam goes through a thicker length of absorber and, conversely, the low momentum portion of the beam goes through a thinner length of wedge. The result is that if the wedge angle is chosen appropriately, all of the beam leaving the wedge has the same mean momentum. However, the beam is spread out horizontally.

Straggling introduces a spread in the mean energy loss in the material given by[3]

$$\frac{d(\sigma_E^2)}{dz} = 4\pi\,(r_e m_e c^2)^2\, n_e\, \gamma^2\,(1 - \tfrac{1}{2}\beta^2) \qquad (7)$$

where r_e is the classical radius of the electron, $m_e c^2$ is the rest energy of the electron, $n_e = ZN_A\rho/A$ is the number of electrons per unit volume of absorber material, and γ and β are the usual relativistic factors. In practical units

$$\frac{d(\sigma_E^2)\ [MeV^2]}{dz\,[cm]} = 0.0363\ \gamma^2\,(1 - \tfrac{1}{2}\beta^2) \qquad (8)$$

for lithium. The amount of momentum spread introduced by straggling changes monotonically with position across the wedge. The experiment would measure the resultant momentum distribution as a function of transverse position and compare with the expected result.

EXPERIMENTAL CONSIDERATIONS

The transverse cooling experiments require measuring the normalized muon beam emittance before and after the lithium rod. The normalized emittance at a beam waist is given by

$$\epsilon_N = \frac{\bar{p}}{mc}\,\sigma_X\,\sigma_\theta \qquad (9)$$

where σ_X and σ_θ are the values for the entrance or the exit to the rod, $\bar{p}$ is the mean momentum for the distribution and m is the mass of the muon. The error on the measurement of ε_N is given by

$$(\delta\epsilon_N)^2 = \left[\frac{\overline{p}^2 \sigma_\theta^2}{m^2} + \frac{2\,\overline{p}^2 \sigma_X^2}{m^2 D^2} + \frac{4\,\overline{p}^2 \sigma_X^2 \sigma_\theta^2}{m^2 \alpha^2 D^2} \right] (\delta\sigma_X)^2 \qquad (10)$$

where D is the distance between chambers, and α is the dipole bend angle for the momentum measurement.

The proposed arrangement of the position detectors is shown schematically in Fig. 5. We have seen in Tables 2 and 3 that we want to measure differences in normalized emittance of about 150 mm-mrad. We determine the required position resolution by demanding that each term in Eq. 10 contribute no more than 10 mm-mrad error. Then, when the three terms are added in quadrature, we will still be able to measure ε_N to an accuracy of 10%. The required properties of the position detectors are given in Table 4. Chambers W1-4 are used to measure the incident momentum, W5-6 measure the incident angle, and W6 measures the incident position. For the outgoing beam W7 measures the position, W7-8 measure the angle, and W7-10 measure the momentum.

Table 4 Position detectors

Detector	Range	Resolution	Channels
	[mm]	[μm]	(x + y)
W1	360	260	2770
W2	360	260	2770
W3	360	260	2770
W4	360	260	2770
W5	72	60	2400
W6	36	45	1600
W7	80	55	2910
W8	160	160	2000
W9	505	410	2465
W10	585	410	2855

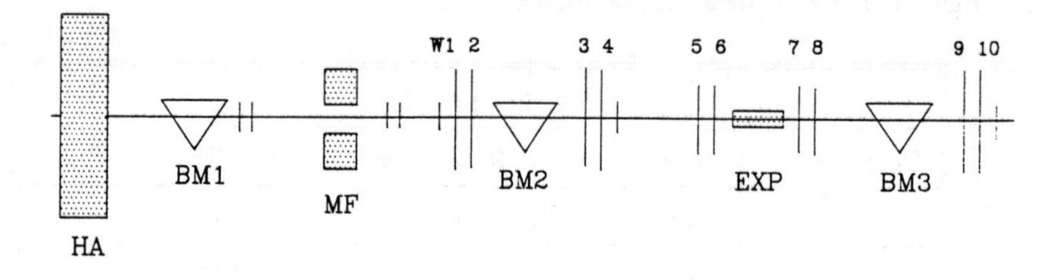

Fig 5. Schematic location of position measuring detectors. HA: hadron absorber, BM: bending magnet, MF: momentum filter, EXP: experimental area.

Detectors W1-4 and W9-10 would probably be wire chambers. The four detectors surrounding the cooling rod might be cathode strip chambers[6]. The detectors for the beam incident on the experiment (W1-6) cover a position range of up to $\pm 3\sigma$. In order to fully understand the performance of the cooling rod, we want to be sure that the chambers after the experiment can detect far into the tails of the angular distribution. For that reason these detectors (W7-10) cover a range of up to $\pm 4\sigma$. Each detector must consist of an x and a y measuring plane. The total number of channels is around 25,000. The instantaneous rate in all the detectors is 10^5 tracks / s.

ACKNOWLEDGEMENTS

This research was supported by the U.S. Department of Energy under Contract No. DE-AC02-76CH00016.

REFERENCES

1. See, for example, Proc. of the Mini-workshop on $\mu^+\mu^-$ colliders: particle physics and design, Napa, CA, Nuc. Instr. Meth. in Phys. Res. A350: 24-56, 1994; see V. Barger, these proceedings.

2. See, for example, D.V. Neuffer, $\mu^+\mu^-$ colliders: possibilities and challenges, Nuc. Instr. Meth. in Phys. Res. A350: 27-35, 1994.

3. D.V. Neuffer & R.B. Palmer, A high-energy high-luminosity $\mu^+\mu^-$ collider, Proc. 1994 European Particle Accelerator Conf.

4. A.J. Lennox, The design parameters for a lithium lens as antiproton collector, IEEE Trans. Nuc. Sci. NS-30:3663-5, 1983.

5. C.L. Wang, Pion production in high energy collisions, Phys. Rev. D10:3876-8, 1974.

6. G. Bencze et al, Position and timing resolution of interpolating cathode strip chambers in a test beam, to be published in Nuc. Instr. Meth. Part. Res. A.

Validity of the differential equations for ionization cooling

R.C. Fernow & J.C. Gallardo

Brookhaven National Laboratory, Upton, NY 11973

We examine the validity of the differential equations used to describe ionization cooling. We find that the simple heating term due to multiple scattering given by D. Neuffer is a good approximation to the expression obtained from a more rigorous derivation.

Ionization cooling of the muon beam emittance is a crucial component of current designs for a muon-muon collider. The idea was first proposed by Skrinsky[1]. Neuffer[2] has continued to champion the idea and has written several articles giving practical equations for the amount of emittance cooling.

1. Neuffer's equations for ionization cooling

Neuffer gives the following expression for the decrease in the normalized emittance ε_N of a muon beam traversing matter due to energy loss

$$\frac{d\epsilon_N}{dz}(cool) = -\frac{1}{\beta^2}\frac{\epsilon_N}{E}\left|\frac{dE}{dz}\right| \tag{1}$$

where β is the usual relativistic factor, E is the total energy of the particles and dE/dz is the ionization energy loss for the particle in the material. There is also an increase in normalized emittance resulting from multiple scattering in the material. Neuffer gives a general expression for the "heating" term

$$\frac{d\epsilon_N}{dz}(heat) = \beta\gamma\,\frac{\beta_\perp}{2}\frac{d}{dz}\langle\theta^2\rangle \tag{2}$$

where γ is the usual relativistic factor, $\beta_\perp$ is the betatron focusing parameter, θ is the angle of the particle trajectory projected onto the y-z plane, and < > represents the mean of the distribution in question over many scatters and over initial conditions. Neuffer also gives a specific formula for the heating term by substituting for the expectation value of θ

$$\frac{d\epsilon_N}{dz}(heat) = \frac{\beta_\perp}{2}\frac{E_S^2}{\beta^3 E m c^2}\frac{1}{L_R} \tag{3}$$

where E_S = 13.6 MeV, mc^2 is the particle's rest energy, and L_R is the radiation length for the scattering medium. Finally, Neuffer defines a "minimum emittance" by equating the cooling derivative in Eq. 1 with the heating derivative of Eq. 3

$$\epsilon_N = \frac{\beta_\perp}{2}\frac{E_S^2}{\beta m c^2 L_R \left|\frac{dE}{dz}\right|} \tag{4}$$

2. General derivation of ionization cooling equations

We start with the definition of normalized transverse emittance

$$\epsilon_N = \beta \gamma \epsilon \tag{5}$$

where ε is the geometric emittance. Consider the change in ε_N as the beam travels along the z direction into the material.

$$\frac{d\epsilon_N}{dz} = \epsilon \frac{d(\beta\gamma)}{dz} + \beta\gamma \frac{d\epsilon}{dz} \tag{6}$$

The first term on the right hand side represents cooling of the normalized emittance and can be written straightforwardly as the expression in Eq. 1. For our case the second term on the right hand side of Eq. 6 takes into account the increase of emittance that results from multiple scattering in the material. It is this second (heating) term, which Neuffer wrote in the form of Eq. 3, that has made some people feel uncomfortable.

In the general case the geometric emittance is defined statistically as

$$\epsilon^2 = \langle y^2\rangle \langle \theta^2\rangle - \langle y\,\theta\rangle^2 \tag{7}$$

where y is a dimension transverse to the particle's direction of motion, θ is the angle of the particle trajectory projected onto the y-z plane, and < > represents the mean of the distribution in question over many scatters and over initial conditions. The change in the geometric emittance as the beam proceeds through the material is given by

$$2\epsilon \frac{d\epsilon}{dz} = \langle y^2\rangle \frac{d}{dz}\langle \theta^2\rangle + \langle \theta^2\rangle \frac{d}{dz}\langle y^2\rangle - 2\langle y\theta\rangle \frac{d}{dz}\langle y\theta\rangle \tag{8}$$

We note here that if we drop the second and third terms in Eq. 8 and use the relation

$$\langle y^2\rangle = \beta_\perp \epsilon \tag{9}$$

from betatron focusing theory, the heating term can be written as the expression used by Neuffer in Eq. 2. There is a natural tendency for a beam of particles scattering their way through some material to spread out laterally and for a correlation to build up between the beam's angle and transverse position. It has been suggested by Palmer that this can be prevented by an external focusing field with sufficient strength to prevent the beams from spreading laterally. We conclude that a necessary condition for dropping the extra terms in Eq. 8, and hence for the validity of Eq. 3, is the presence in the scattering medium of a strong external focusing field.

3. No external focusing

For completeness let us consider first the case of scattering in a material with no external focusing present. In a previous note[3] we have shown that in the gaussian limit, the beam distributions are given by

$$\begin{aligned} \langle y^2\rangle &= \sigma_{yo}^2 + 2z\langle y_o\theta_o\rangle + \sigma_{\theta o}^2 z^2 + \frac{1}{3}\theta_C^2 z^3 \\ \langle \theta^2\rangle &= \sigma_{\theta o}^2 + \theta_C^2 z \\ \langle y\theta\rangle &= \langle y_o\theta_o\rangle + \sigma_{\theta o}^2 z + \tfrac{1}{2}\theta_C^2 z^2 \end{aligned} \tag{10}$$

The characteristic scattering "angle" θ_C is given by

$$\theta_C = \frac{E_S}{pc\beta}\frac{1}{\sqrt{L_R}} \tag{11}$$

where E_S = 13.6 MeV, p is the particle's momentum, and L_R is the radiation length for the scattering medium. The parameter θ_C was considered a constant in the derivation of Eqs. 10. If these expressions are substituted into Eq. 7, the geometric emittance is

$$\epsilon^2 = \left[\sigma_{yo}^2\,\sigma_{\theta o}^2 + \sigma_{yo}^2\,\theta_C^2 z + \theta_C^2 <y_o\theta_o> z^2 + \frac{1}{3}\,\theta_C^2\sigma_{\theta o}^2 z^3 + \frac{1}{12}\,\theta_C^4 z^4\right] \tag{12}$$

If Eqs. 10 are inserted back into Eq. 8, we find that the heating term for the case of no focussing is

$$\frac{d\epsilon}{dz} = \frac{1}{2\epsilon}\left[\sigma_{yo}^2\,\theta_C^2 + 2\theta_C^2 <y_o\theta_o> z + \theta_C^2\sigma_{\theta o}^2 z^2 + \frac{1}{3}\,\theta_C^4 z^3\right] \tag{13}$$

4. Constant external focusing

We consider now the case of scattering in a material in the presence of an external focusing force whose strength is determined by the constant parameter

$$\omega = \sqrt{\frac{eB}{pa}} \tag{14}$$

where e is the particle's charge, B is the focusing magnetic field, and a is the radius of the focusing channel. In the gaussian limit the expectation values of the position and angle are given by [3]

$$\begin{aligned}
\langle y^2\rangle &= \sigma_{yo}^2\cos^2(\omega z) + \frac{\sigma_{\theta o}^2}{\omega^2}\sin^2(\omega z) + z\frac{\theta_C^2}{2\,\omega^2} - \frac{\theta_C^2}{4\,\omega^3}\sin(2\,\omega z) \\
\langle\theta^2\rangle &= \sigma_{\theta o}^2\cos^2(\omega z) + \sigma_{yo}^2\,\omega^2\sin^2(\omega z) + z\frac{\theta_C^2}{2} + \frac{\theta_C^2}{4\,\omega}\sin(2\,\omega z) \\
\langle y\theta\rangle &= \frac{\sigma_{\theta o}^2}{2\,\omega}\sin(2\,\omega z) - \frac{\sigma_{yo}^2\,\omega}{2}\sin(2\,\omega z) + \frac{\theta_C^2}{2\,\omega^2}\sin^2(\omega z)
\end{aligned} \tag{15}$$

In deriving Eq. 15 we have assumed that there is no initial correlation between y_o and θ_o. In other words we have assumed that the initial beam entering the focusing channel is at a waist. In this case

$$\sigma_{\theta o}^2 = \frac{\epsilon}{\beta_\perp} \tag{16}$$

and we have the constraint

$$\frac{\sigma_{yo}}{\sigma_{\theta o}} = \beta_\perp = \frac{1}{\omega} \tag{17}$$

With this constraint we can write Eq. 15 in the simplified form

$$\begin{aligned}
\langle y^2\rangle &= \sigma_{yo}^2 + z\frac{\theta_C^2}{2\,\omega^2} - \frac{\theta_C^2}{4\,\omega^3}\sin(2\,\omega z) \\
\langle\theta^2\rangle &= \sigma_{\theta o}^2 + z\frac{\theta_C^2}{2} + \frac{\theta_C^2}{4\,\omega}\sin(2\,\omega z) \\
\langle y\theta\rangle &= \frac{\theta_C^2}{2\,\omega^2}\sin^2(\omega z)
\end{aligned} \tag{18}$$

The geometric emittance is

$$\epsilon^2 = \sigma_{yo}^2\,\sigma_{\theta o}^2 + \sigma_{yo}^2\,\theta_C^2\,z + \frac{\theta_C^4}{4\,\omega^2}z^2 - \frac{\theta_C^4}{8\,\omega^4} + \frac{\theta_C^4}{8\,\omega^4}\cos(2\,\omega z) \tag{19}$$

and the heating term can be written

$$\frac{d\epsilon}{dz} = \frac{1}{2\epsilon}\left[\sigma_{yo}^2\,\theta_C^2 + \frac{\theta_C^4}{2\,\omega^2}\,z - \frac{\theta_C^4}{4\,\omega^3}\sin(2\,\omega\,z)\right] \tag{20}$$

We can obtain Neuffer's expression for heating, Eq. 3, by keeping only the first term in Eq. 20. Neglecting the second term implies that Neuffer's expression is only valid when

$$\sigma_{yo}^2 \gg \frac{\theta_C^2\,L_{rod}}{2\,\omega^2} \tag{21}$$

where L_{rod} is the length of the cooling rod, while neglecting the third term requires

$$\sigma_{yo}^2 \gg \frac{\theta_C^2}{4\,\omega^3} \tag{22}$$

The approximations in Eqs. 21 and 22 will normally be satisfied if ω is large enough (strong focusing).

5. Effect of energy loss

We have not rigorously derived expectation values for y and θ for the case when energy loss is also present. To estimate the effects of energy loss, we make the (hopefully reasonable) assumption that the basic form of Eqs. 19 and 20 are still correct with the constants θ_C and ω replaced with variables that depend on the local value of the particle energy. The decrease in particle energy is determined from dE/dz in the material. In addition we generalize Neuffer's heating term using Eqs. 14 and 17 as

$$\frac{d\epsilon_N}{dz}(heat) = \sqrt{\frac{p(z)\,a}{e\,B}}\;\frac{E_S^2}{2\,\beta^3(z)\,E(z)\,m\,c^2 L_R} \tag{23}$$

We compare Neuffer's Eq. 23 with the energy dependent prediction of Eqs. 19 and 20 in Fig. 1, using parameters appropriate to a proposed cooling experiment at the AGS[4].

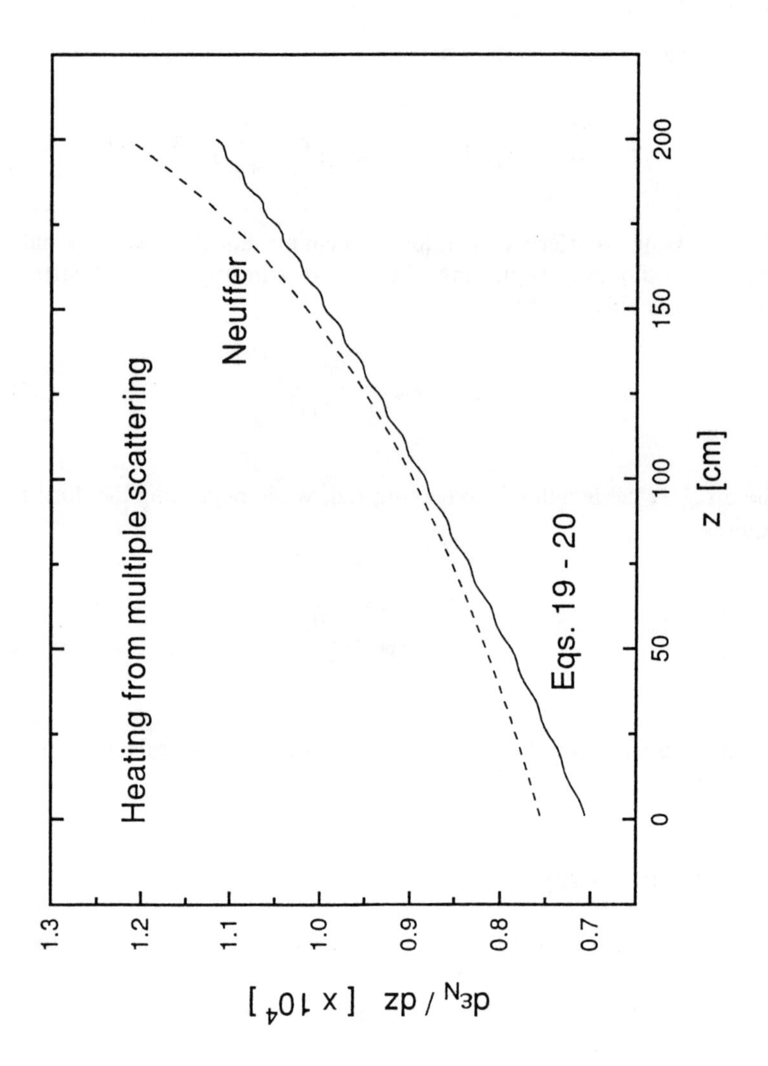

Fig. 1 Comparison of the emittance heating expression of D. Neuffer with the more exact relation derived from Eqs. 19 and 20.

We see that Neuffer's simple expression is a good approximation to the more rigorous result. Eq. 23 overestimates the amount of heating everywhere inside the cooling rod, but the deviation never exceeds about 15%.

Acknowledgements

This research was supported by the U.S. Department of Energy under Contract No. DE-AC02-76CH00016.

Notes and references

[1] E.A. Perevedentsev & A.N. Skrinsky, Proc. 12th Int. Conf. on High Energy Acel. 1983, p.485; A.N. Skrinsky & V.V. Parkhomchuk, Sov. J. Nucl. Phys. 12:3, 1981.

[2] D. Neuffer, Part. Accel. 14:75, 1983; D. Neuffer & R.B. Palmer, A high energy, high luminosity $\mu^+\mu^-$ collider, Proc. 4th European Accel. Conf., 1994; R.B. Palmer, D. Neuffer & J.C. Gallardo, A practical high energy, high luminosity $\mu^+\mu^-$ collider, Proc. Physics Potential & Development of $\mu^+\mu^-$ colliders, 1995; D.V. Neuffer, $\mu^+\mu^-$ colliders: possibilities and challenges, Nuc.Instr.Meth. A350:27-35, 1994.

[3] R.C. Fernow & J.C. Gallardo, Muon transverse ionization cooling: stochastic approach, 1995.

[4] R.C. Fernow, J.C. Gallardo, R.B. Palmer, D.R. Winn & D.V. Neuffer, A possible ionization cooling experiment at the AGS, $\mu^+\mu^-$ collider technical note 01-95, 1995.

Backgrounds and Detector Performance at a 2×2 TeV $\mu^+\mu^-$ Collider

G. William Foster and Nikolai V. Mokhov

Fermi National Accelerator Laboratory

P.O. Box 500, Batavia, Illinois 60510

INTRODUCTION

The rich physics potential of a high-energy high-luminosity $\mu^+\mu^-$ collider and the surprising feasibility of a design [1, 2] attract the attention of many people these days. A few issues define the practicality of such a project, with the enormous particle background levels in a detector due to unavoidable reasons holding first place. In contrast to hadron colliders where particle backgrounds come both from interaction point (IP) and accelerator [3], almost all the backgrounds in the muon collider detectors arise in the machine. The decay length for 2 TeV muons is $\lambda_D^{-1} \sim 10^{-7} m^{-1}$. With 10^{12} muons in a bunch one has 10^5 decays per meter in a single pass through an interaction region, and 10^8 decays per meter per 12 msec store.

This paper examines two major classes of detector backgrounds presented in the muon collider: beam halo backgrounds, and the "direct" backgrounds from electrons from $\mu \rightarrow e \nu \tilde{\nu}$ decay occurring in the beam channel. We describe the nature of both of these backgrounds, provide results of first realistic calculations for both sources and discuss their effects on plausible detectors. Various shielding and collimation geometries have been simulated, and their efficacy and the nature of the surviving background discussed.

DESCRIPTION OF SIMULATIONS

In this paper we study beam muon decays and beam halo interactions in the inner triplet region $\pm 70m$ from the IP. The lattice calculated by B. Palmer and used in our calculations, provides $\beta^* = 3mm$ with $\beta_{peak} = 400km$ at $Q1-B2$ location (Fig. 1). Superconducting dipole magnets B1, B2 and B3 have central field 8 T. Combined function superconducting quadrupoles Q1 and Q2 are with 2 T dipole component and a gradient of 50 T/m. All of the SC components have 8 cm radius aperture.

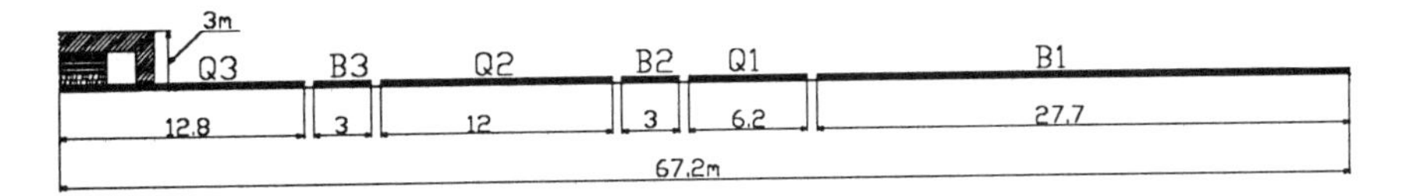

FIGURE 1. Muon collider inner triplet. Dimensions are in meters.

The Q3 quadrupole is resistive with 0.5 T dipole field, except the first 1.3 m near the IP where the dipole component is equal to zero. Its aperture is reduced toward the IP from R=4.5 cm at L=12.8 m to R=0.45 cm at L=1.2 m (see Fig. 2), with the gradient growing from 33.3 T/m to 333.3 T/m, appropriately. Geometries and materials of beam pipes, collars, yokes and cryostats for SC bending dipoles and combined function quadrupoles as well as 2-D POISSON calculated magnetic fields in these components are embedded in the calculational package.

A rather simple model detector is used in this study (Fig. 2): a two-region silicon tracker with volume averaged density $\rho = 0.15 g/cm^3$, a central calorimeter (CH, $\rho = 1.03 g/cm^3$) and a solenoid magnet with 2 Tesla magnetic field. A copper bucking coil is placed on the outside of the Q3 to neutralize the effect of the solenoidal field in the quadrupole.

All the calculations are done with MARS95 code [4], the newest version of the MARS system [5]. Improvements relevant to this problem include better description of muon cross-sections, of muon decay and of the algorithm for electromagnetic fluctuations [6]; improved particle transport algorithm in a magnetic field; synchrotron radiation generation; modified geometry description; extended histogramming and graphical possibilities. The calculation for each case consists of:

- forced $\mu \rightarrow e\nu\tilde{\nu}$ decays in the interaction region (IR) beam pipe;
- tracking of created electrons in the beam pipe under influence of the magnetic field with emission of the synchrotron photons along the track;
- simulation of electromagnetic showers in the triplet and detector components induced by electrons and synchrotron photons hitting the beam pipe;
- simulation of muon interactions (bremsstrahlung, direct e^+e^- pair production, ionization, deep inelastic nuclear interactions and decays) along the tracks in the lattice and detector;
- simulation of electromagnetic showers in the triplet and detector components created at the above muon interaction vertices;
- histogramming and analysis of particle energy spectra, fluences and energy deposition in various detector regions as well as in the whole IR.

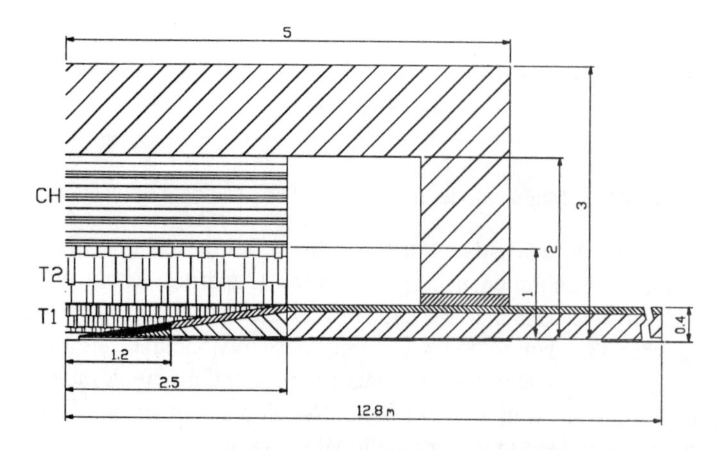

FIGURE 2. Detector, Q3 quadrupole and nozzle configuration used in this study. Dimensions are in meters.

Estimation is done for photo-muon and photo-hadron production including giant resonant neutrons. Energy thresholds are 1 MeV for muons and charged hadrons, 0.1 MeV for electrons and photons, and 0.5 eV for neutrons. In this paper we present results only for electromagnetic component ($e^{\pm}, \gamma$) with $E > 0.5\ MeV$.

In this study we assume that a bunch of 10^{12}, 2 TeV muons pass from right to left toward the IP (Fig. 1), creating showers responsible for backgrounds along its way. Only a single bunch is simulated to study the directionality which is masked in a case of two colliding beams. When studying the integrated effect, we assume 600 turns as a beam lifetime, that gives for two beams about 10^{15} muons per store passing the IR.

DIRECT $\mu \rightarrow e\nu\tilde{\nu}$ DECAY BACKGROUNDS

Muon decays inside the beam channel create decay electrons and result in a high flux of electromagnetic radiation in the beam pipe. A single bunch will release of order 10^8 GeV per meter of path as muons pass through a detector. Thus the direct decays represent a hardly reducible background in the detectors which has the potential of killing the concept of the muon collider, or at least, redirecting the accelerator work towards parameters involving fewer muons per bunch.

Figure 3 shows calculated e^+e^- energy spectrum in the accelerator components. The huge peak sitting around 1 TeV represents the $\mu \rightarrow e\nu\tilde{\nu}$ decay spectrum with a tail at lower energies enriched by electrons and positrons of electromagnetic showers induced in the beam pipe and superconducting coils. Photons emitted due to synchrotron radiation along e^+e^- tracks in a strong 8 T magnetic field have energy around 1 GeV. The number of photons is about 300 times that for electrons and positrons.

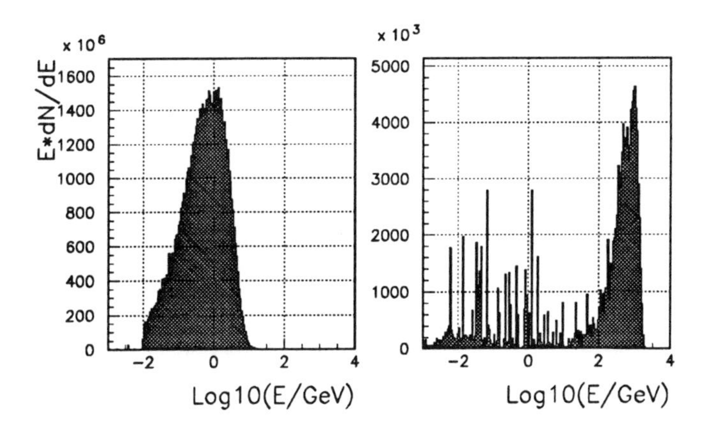

FIGURE 3. Photon (left) and electron/positron (right) energy spectra in the inner triplet accelerator components.

Figure 4 represents energy deposition density in the T1 silicon tracker and first 1.3 meter of the Q3 quadrupole at the either side of the IP. One can see that the energy deposition levels are very high, up to $10^5 - 10^6 GeV/cm^3$ in the innermost parts of the detector. It is clear that one needs to bring these levels down.

To study the source term we have used a tagging techniques in the simulations. Figure 5 shows a distribution of the electromagnetic shower origins: energy deposition in the T1 tracker due to muon decay at distance L from the IP. Due to very high energy of electrons and photons in the large aperture, the whole triplet is a source of backgrounds. There is a strong left(outward)/right(inward) asymmetry in the horizontal plane due to effect of the dipole magnetic field.

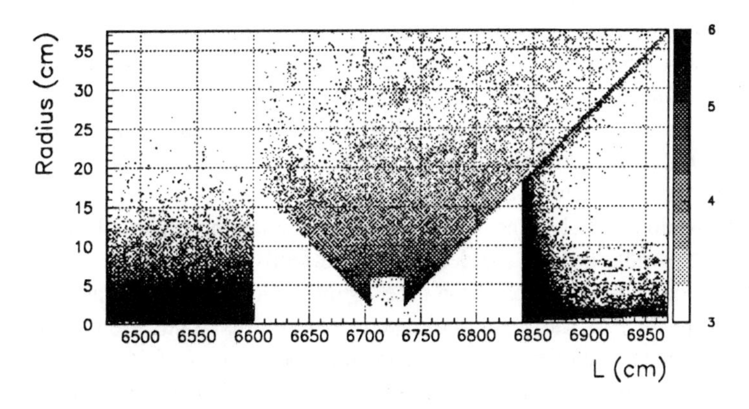

FIGURE 4. Energy deposition in the vicinity of IP, in units of $10^n GeV/cm^3$ per store, where the shade indicates the power n. In this plot muon beam goes from left to right with the IP at L=6720 cm.

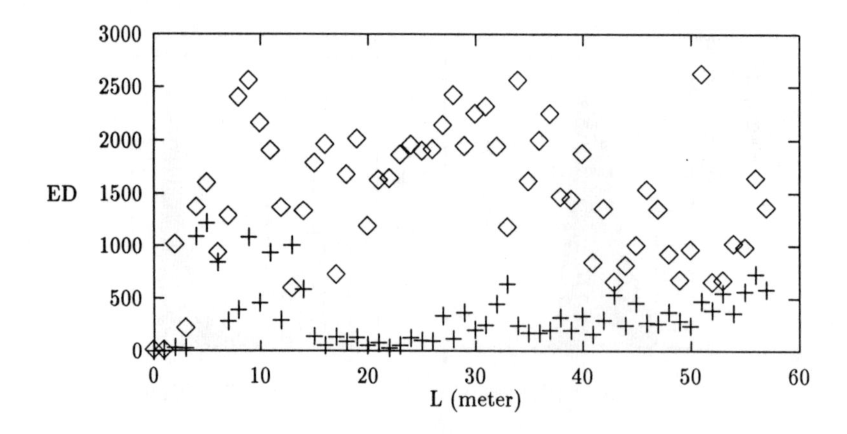

FIGURE 5. Energy deposition (GeV/g per store) in the T1 tracker as a function of distance to the muon decay point. Diamonds are for the outward and pluses are for the inward sides of the tracker.

BACKGROUND REDUCTION WITH COLLIMATORS

The first measure to reduce backgrounds in the detectors was incorporated into the inner triplet lattice from the beginning: a dipole component of the magnetic field in the combined function qudrupoles deflects many of decay electrons into the beam pipe well before the IP. But as it was seen from the previous section, this is not enough. The main job can be done with appropriate collimators.

Several aspects of an effective collimation strategy are clear. The general strategy will involve surrounding all of the the final-focus quads and most of the IR beam pipe with high-Z material. Major variables in this collimator are the inner radius, the distance to which it approaches the interaction point, the angle of the cone which defines the outer surface of the cone (and the angle at which one gives up doing physics). Other variables involve steps and tapers of the inner radius, and covering the collimator with a low-Z cladding.

Since the dominant backgrounds are EM showers with very small transverse momenta, a solenoidal field in the detector will effectively confine all of the charged particles to the vicinity of the beam pipe. Only the low energy, high angle photons at the tails of the EM shower will escape to high radius. Thus the final signal will be an enormous pulse of photons at the critical energy of $\sim 10MeV$.

If the muon decay occurs in the arcs or the superconducting combined-function quad, the decay electron will emit most of its energy as synchrotron photons of energy $\sim 1GeV$ before it hits the beam pipe. If this were the dominant source of background, the most effective collimation strategy might be a "pinhole" collimation very near the IR which limits the solid angle for the synchrotron to emerge from.

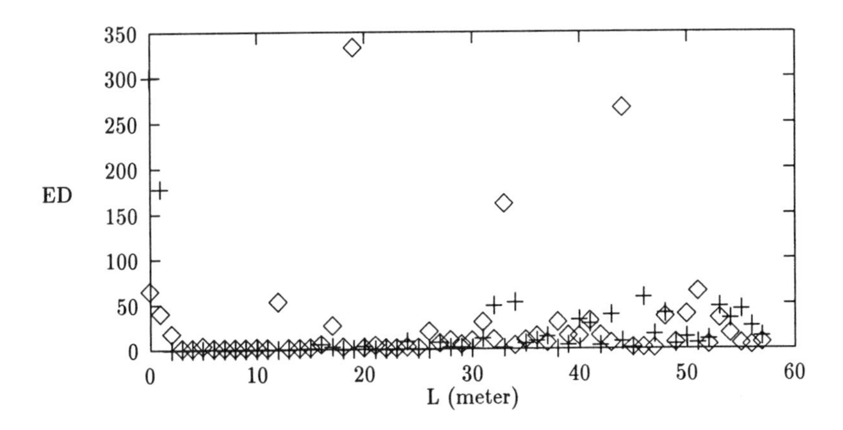

FIGURE 6. Energy deposition (GeV/g per store) in the T1 tracker as a function of distance to the muon decay point with the tungsten nozzle in place. Diamonds are for the outward and pluses are for the inward sides of the tracker.

For muons which decay close to the IR, the decay electron is more likely to survive to initiate a TeV shower when it hits the limiting aperture of the collimator or final focus low-beta quad. For these electrons, additional collimating material is likely to act as an amplifier which causes TeV showers to emerge from the quads right at at shower-maximum. If this were the dominant source of background, the most effective collimation would include a limiting aperture of order 1m from the IR, with an interior conical surface which opens outward as it approaches the IR. These collimators would then have the aspect of two nozzles spraying electromagnetic fire at each other, with the charged component of the EM showers being confined radially by the solenoidal magnetic field and the photons from one nozzle being trapped (to whatever degree possible) by the conical opening in the opposing nozzle.

In this paper we studied collimators of a few configurations occupied the cone $\theta < 150mrad$ in the $15 < L < 120cm$ region on the either side of the IP (Fig. 2). Collimators made of tungsten as well as of a combination of various materials (aluminum, copper and tungsten) with different shapes of the hole have been considered. It turns out that for the main source of the backgrounds, direct muon decays, the best choice is a tungsten nozzle with an aperture radius R=0.45 cm at L=120 cm and R=1 cm at L=15 cm. Figure 6 shows how significant background reduction is in this case. There is a significant difference in the background levels in the left (outward) and in the right (inward) parts of the tracker and calorimeter. This depends on the magnetic field, nozzle and collimator configurations. Calculated absorbed doses in six regions of the detector are given in Table 1.

Muon decays along the whole triplet do contribute into the backgrounds in the detector (Fig. 5 and 6). So, an additional way to suppress further the background levels can be a collimator between Q3 and B3 with a smallest possible aperture. A

Table 1: Dose (mrad/store) absorbed in the left(L) and right(R) parts of the detector for the cases with and without tungsten nozzle and with and without copper collimator.

Nozzle	Collimator	T1 (L)	T1 (R)	T2 (L)	T2 (R)	CH (L)	CH (R)
No	No	1360	326	25	9.89	0.534	0.238
No	Yes	267	108	5.5	3.81	0.075	0.079
Yes	No	22.2	17.6	0.656	0.480	0.074	0.036
Yes	Yes	2.77	7.68	0.050	0.037	0.032	0.006
	K_{max} =	491	43	498	269	17	38

copper collimator 50 cm long with 2.5σ radius aperture in these calculations provided up to a factor of 10 additional dose reduction (see Table 1). The total maximum reduction K_{max} defined as a ratio of background levels in the given detector region without protective measures to that with a tungsten nozzle and copper collimator between Q3 and B3 magnets is given in the last row of the Table 1. It is as high as a factor of 500 on the best, but for the larger radii in calorimeter it is "only" a factor of 20 to 40. Particle spectra in detector are very soft in such a configuration: for all the considered regions background photons and electrons have energy below 100 MeV, being on the average just a few MeV (Fig. 7).

The e^+e^- fluence is given in Table 2 for the crossing of two 10^{12} muon beams. It turns out that the charged particle fluxes calculated as the total track length in the region divided by the region volume, and as the averaged energy deposition density in the region divided by the minimum ionizing power of the region material, coincide within 20-30 percent for small radii and within a few percent for large radii. The behavior in different detector regions is similar to the dose. Maximum reduction is again of about a factor of 500. With an optimal collimation the peak charged

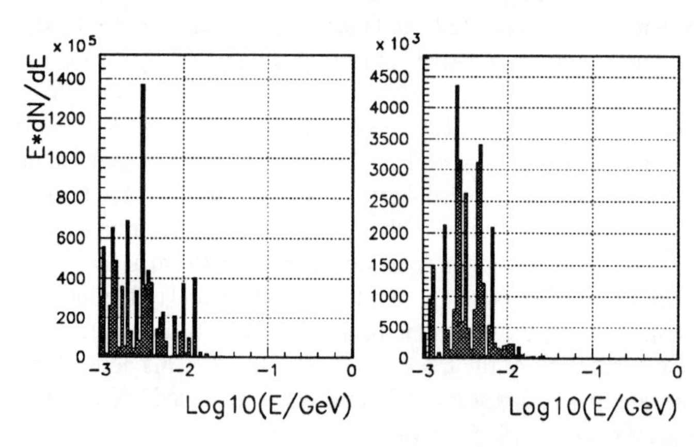

FIGURE 7. Photon (left) and electron/positron (right) energy spectra in the CH calorimeter region with the tungsten nozzle in place.

Table 2: The e^+e^- fluence (cm^{-2} per crossing) in the left(L) and right(R) parts of the detector for the cases with and without tungsten nozzle and with and without copper collimator. Fluence is defined here as a calculated energy deposition density over $(dE/dx)_{min}$ for Si and CH, respectively.

Nozzle	Coll	T1 (L)	T1 (R)	T2 (L)	T2 (R)	CH (L)
No	No	8.54×10^4	2.04×10^4	1.57×10^3	6.21×10^2	2.86×10^1
No	Yes	1.68×10^4	6.74×10^3	3.46×10^2	2.38×10^2	4.00×10^0
Yes	No	1.40×10^3	1.10×10^3	4.11×10^1	3.02×10^1	3.93×10^0
Yes	Yes	1.74×10^2	4.81×10^2	3.15×10^0	2.31×10^0	1.73×10^0

particle flux in the silicon tracker is of the order of 500 cm^{-2} per crossing of two single bunches through the IR in opposite directions. The fluxes fall down very rapidly with radius.

BEAM HALO BACKGROUNDS

These arise from muons which are lost some distance upstream of the detectors. A noninteracting muon which passes through the detector will be of little consequence. However, muons which induce electromagnetic or hadronic showers either upstream or inside of the detector will cause more serious problems. Beam particles injected with large momentum errors or betatron amplitudes will be lost within the first few turns. After this, an equilibrium level of losses will be attained as particles are promoted to larger betatron amplitudes via beam disruption from the collision point, a beam-gas scattering, etc.

In principle the halo background can be calculated fairly accurately in the muon collider. This is due to the small number of turns ($\sim$ 1000) which must be tracked before the muons decay. In contrast, the halo losses in hadron colliders develop over many millions of turns and depend on things like RF noise, magnet power supply ripple moving particles repeatedly over high-order resonances, and other imponderables. These effects do not have time to become significant in the muon collider, and the small number of turns can be tracked numerically to adequate accuracy. However, this simulation will require a detailed knowledge of the machine lattice and a reasonable optimization of the scraping strategy. This will be one of the major projects in the coming months.

At this stage we simulated the beam as entering the IR with a non-truncated Gaussian profile. Muons outside $\pm3\sigma$ will then interact and be scraped by the final arc magnets, low-beta quads, collimators, and detector components. Figure 8 shows a distribution of the muon interaction vertices in the vertical plane in the vicinity of the IP. The distribution is pretty symmetric, but in the horizontal plane there is a strong asymmetry related to the magnetic field. With an energy cut-off for this plot equal to 100 MeV, more than 90% of the vertices are direct e^+e^- pair production,

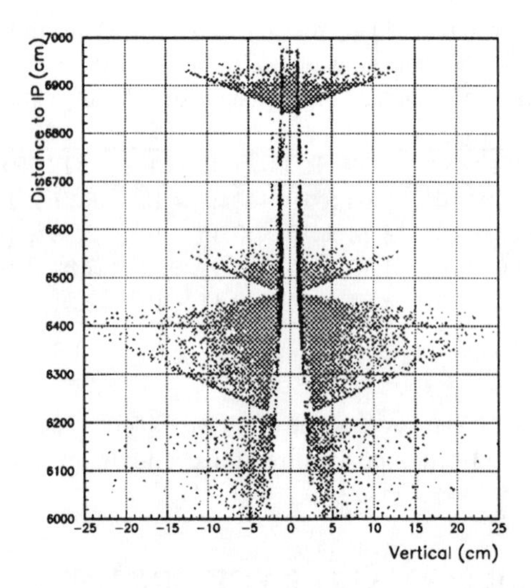

FIGURE 8. Muon beam halo interaction points in the components in the vicinity of the IP with the tungsten nozzle in place. The distribution is given for the vertical plane. In this plot the IP corresponds to L=6720 cm. The muon beam goes from bottom to top.

about 5% are muon bremsstrahlung, and less than a few percent are for deep inelastic nuclear interactions, muon decays and energetic knock-on electron production. For the considered Q3 quadrupole configuration with its aperture decreasing toward the IP (creating the bottleneck for out-of-axis particles), all the contribution to the background in a detector from the beam halo comes from less then about 5 meters upstream of the IP, very different of the direct decay source. So, a copper collimator between Q3 and B3 is useless for this source. Background reduction with a tungsten nozzle is only about a factor of 4. The left(outward)/right(inward) asymmetry is even stronger, with right regions always in a worse condition. With a considered $\pm 3\sigma$ beam loss model background levels are 10 to 100 times higher compared to the best protected configuration in the case of direct decays. So, obviously more work is needed toward a dedicated beam scraper system well upstream of the IR.

EFFECT OF BACKGROUNDS ON DETECTOR PERFORMANCE

The two detector capabilities most at risk from the unavoidable "direct" $\mu \rightarrow e\nu\tilde{\nu}$ decay induced background are the tracking capability (particularly at low radius) and the calorimetry energy measurement. In both cases the background arises from the enormous pulse of $\sim 10MeV$ photons, essentially all which emerge from the unshielded gap between the collimator cones at the interaction point. As hoped, the

Table 3: Averaged occupancy (percent per crossing) T1 and T2 silicon pixels for the direct muon decays of two 10^{12} beams in the left (L) and right(R) parts of the detector for the cases with and without tungsten nozzle and with and without copper collimator.

Nozzle	Collimator	T1 (L)	T1 (R)	T2 (L)	T2 (R)
No	No	34.16	8.16	23.55	9.32
No	Yes	6.72	2.70	5.19	3.57
Yes	No	0.56	0.44	0.62	0.45
Yes	Yes	0.07	0.19	0.05	0.03

combination of the shielding, collimators, and the solenoidal field seems to completely eliminate higher energy gammas and most of charged particles of all momenta.

Silicon Tracker

The inner tracker must be a pixellated silicon detector for reasons that will be evident below. The background in this device will then have the random "salt and pepper" flavor of uncorrelated hits from low energy photons which convert in one plane of silicon, then spiral around with a gyroradius of a few centimeters before coming to rest. With a sufficient tracking redundancy and a low enough density of hits, there should be no mistaking these for high-momentum tracks from physics events. Survival of the tracking measurement relies on preserving a low enough "occupancy" (fraction of hit pixels) to guarantee a high efficiency for the "good" hits from the high momentum tracks from beam-beam interactions. The reconstruction efficiency of a typical tracking device will not suffer as long as the occupancy is less than 1%. Tracking efficiencies typically begin to suffer at occupancies somewhere between 1% and 20% depending on the redundancy and resolution of the tracking system.

Table 3 shows the averaged occupancy calculated from the direct decay source for a silicon tracker with $20\mu m \times 20\mu m$ pixels in the inner tracker (T1) and with $50\mu m \times 300\mu m$ pixels in the outer tracker T2. One can see that the tracking occupancy (fraction of hits per pixel per crossing) is of order 0.1% to 0.2% for the optimized collimation scheme. This should be adequate to ensure the survival and correct function of the inner tracker — a significant conclusion.

Some comments are appropriate about the pixel sizes assumed, which are of course critical to calculating the occupancy. The $20\mu m \times 20\mu m$ dimensions are consistent with current technology for serial readout "CCD" style silicon detectors. The "smart pixels" under development for hadron colliders will be at least an order of magnitude larger. However, for the muon collider there is no need for the "deadtimeless" readout of the smart pixels given the $\sim 20\mu sec$ time between crossings and the small rate of physics events. Thus these pixel sizes, although adequate to the task, represent rather conservative assumptions about what should be possible by the start of the next millennium.

With the best collimation described above, the radiation dose in the innermost layers of pixels is approximately 0.1-0.5 Mrad/yr compared to almost 10 Mrad/yr at the LHC [7], so that electronics survival is not a likely problem.

Calorimeter

The calorimeter energy deposition is approximately 10^6 GeV per beam crossing even for the best-case shielding and collimation we have developed. This is daunting to say the least. However, consider the following argument, which is intended to make it plausible that one could learn to live with this. All of the 1000 TeV of energy appears as 10 MeV photons, so that to first order this appears as more-or-less uniform of "pedestal offset" in the first depth segment of each tower of the electromagnetic (EM) calorimeter. If necessary, the size of this pedestal can be determined on an even-by-event basis and subtracted out from the observed energy in any tower which is suspected of having a physics signal; towers below some threshold corresponding to a few sigma energy fluctuation will be ignored. Now, pedestal sigma from the flood of gammas grows as the square root of the tower area, whereas the signal from a well-localized electron or photon shower is independent of tower size. Thus one wins in signal/noise by reducing the EM tower size until the tower size is reduced to dimensions comparable to the Moliere radius (the transverse width of the EM shower in the calorimeter). These considerations push towards placing the calorimeter at the highest affordable radius. A reasonable system might contain approximately 10^5 electromagnetic towers a few centimeters in size for a calorimeter at a radius of 2 meters. Each of these 10^5 towers will then receive 10 GeV of energy each crossing, in the form of approximately one thousand 10-MeV gammas. Assuming Gaussian statistics on the number of incident photons, the fluctuation in this energy will be approximately $10GeV/\sqrt{(}1000) = 300MeV$. This is less than the pedestal noise in "fast" liquid argon.

Those who might feel that the last argument was a little to glib might be interested in speculations on ways to discriminate further against the background in the calorimeter. A first readily apparent handle is the depth profile in the EM calorimeter: the 10 MeV gammas interact mainly at the front of the EM tower whereas shower maximum for a 100-GeV electron is much deeper. Appropriate weighting and fitting of the shower profile might provide a factor of 3 to 5 more rejection than a simple pedestal subtraction. A second less readily apparent handle is the directionality of the background gammas, which emanate not from the IP but from the tips of the collimators a few centimeters away. If sufficiently directional calorimetry could be obtained (similar to collimators on a gamma-ray telescope) for the first few radiation lengths of the EM calorimeter, then a valuable additional factor might be obtained.

Electron machines have learned to coexist with the flood of keV gammas from synchrotron radiation which accompany each beam crossover, and in the end it does not appreciably interfere with their ability to do physics in the GeV range. Perhaps it is best to view the flood of MeV gammas accompanying the beam crossovers in muon colliders as the "synchrotron radiation" of TeV physics.

CONCLUSIONS

- We have completed a first-pass investigation of the backgrounds from "direct" $\mu \rightarrow e\nu\tilde{\nu}$ decays in a simplified detector geometry. The most effective shielding and collimation geometry that we have devised eliminated all but the low-energy gammas from the tails of electromagnetic showers, and reduced this background by a factor of 50 to 500 compared to the unshielded case.
- This background gives rise to about 500 tracks per cm^{-2} per crossing. This is painful but not fatal to a pixel detector at the innermost radius of the tracker.
- Energy deposited in the calorimeter is about 1000 TeV per crossing, all of it in the first depth segment(s) of the EM calorimeter. Strategies for dealing with this involve segmenting the EM calorimeter as finely as possible and putting it at a large radius.
- We have begun to characterize the beam halo induced backgrounds, but much work remains to be done. Further work on the collider lattice and on a beam scraper system is needed.
- Detector model sophistication and study of photo-hadron and photo-muon contributions to the background levels are in progress.

ACKNOWLEDGMENTS

We thank Robert Palmer and John Peoples for inducing our interest in this problem, Sergei Striganov for his crucial contribution to the muon interaction algorithms, Pat Colestock and David Finley for their support of this work.

References

[1] Cline, D., *Nucl. Instruments and Methods* **A350**, 24-26 (1994).

[2] Neuffer, D., and Palmer, R., "A High-Energy High-Luminosity $\mu^+\mu^-$ Collider", *Preprint BNL-61267*, 1994.

[3] Mokhov, N., "Accelerator-Experiment Interface at Hadron Colliders: Energy Deposition in the IR Components and Machine Related Background to Detectors", *FERMILAB-Pub-94/085*, 1994.

[4] Mokhov, N., "The MARS95 Code System", *Fermilab FN*, 1995.

[5] Mokhov, N., "The MARS10 Code System: Inclusive Simulation of Hadronic and Electromagnetic Cascades and Muon Transport", *Fermilab FN-509*, 1989; Mokhov, N., "The MARS12 Code System", in *Proceedings of the SARE Workshop*, Santa Fe, January 1993.

[6] Striganov, S., *Nucl. Instruments and Methods* **A322**, 225-230 (1992); "Fast Precise Algorithm for Simulation of Ionization Energy Losses", *IHEP 92-80*, Protvino, 1992.

[7] "The Compact Muon Solenoid", *CERN/LHCC 94-38*, 1994.

Detectors and Backgrounds for a Muon-Muon Collider. Working Group Report

David J. Miller*
Physics and Astronomy, University College London,
Gower St., London WC1E 6BT, UK.

Abstract. First estimates of background muon fluxes at a detector vary by more than two orders of magnitude. More realistic calculations are needed. If the lower rates apply then it should be possible to build an experiment with the acceptance and performance required to study the major topics in the physics programme. Some features of a possible solenoid detector have been investigated, with tracking and calorimetry covering angles to within ± 125 milliradians of the beam direction.

INTRODUCTION

This was a small group, but it had members from a range of current e^+e^-, ep and $p\bar{p}$ experiments - as well as veterans of the SSC detector design and simulation campaigns. For many of us it was a first look at the problems of building a detector for a muon collider and we were apprehensive about backgrounds. Willis' introductory talk (1) convinced us that the problems may be tractable, and we examined his assumptions in our discussions. The main verdict is that more detailed lattice and insertion designs are needed before backgrounds can be well enough understood to be sure that a detector can be built to do the important physics at a muon collider.

BACKGROUNDS

Primary backgrounds from muon decay are large and calculable. Decay electrons must be degraded rapidly by synchrotron radiation and absorbed in heavy shielding-collimators. Secondary interactions and beam-halo muons are much more problematical. Willis (1) has given estimates of $10^6/m^2$ per crossing at 10 cm radius, in a vertex tracker, and $10^4/m^2$ at 1m radius in a main tracker. The group agreed that suitably modular pixel detectors could work at such background levels. But are these levels realistic?

As a first step towards a serious prediction, Mokhov presented preliminary simulations of the secondary backgrounds from beam-halo muons interacting in a

---

* Group members: M. Atac (Fermilab), D. Errede (U. of Illinois), W. Foster (Fermilab), V. Kaftanov (CERN/ITEP), N. Mokhov (Fermilab), D. Reeder (U. Wisconsin), W. Willis (Nevis), M. Xie (LBL).
Reporting work done at the workshop *Physics potential and development of $\mu^+\mu^-$ colliders*, Sausolito, California, November 17-19 1994.

specimen detector-insertion. The parameters of the insertion are shown in Table I. The low-beta combined-function quadrupole QF2 is the vital element. It penetrates deep inside the detector to within ±1.2 m of the collision point - giving a β^* of 3 mm. It has a tapered bore, from 45 mm radius at 12.8 m from collision down to 4.5 mm at 1.2 m. Its superimposed bending field degrades decay electrons by synchrotron radiation and sweeps them out into the heavy metal lining of the bore. The other insertion magnets are all superconducting, but this closest one to the detector has a normal coil to cope with the large energy-deposition from showering.

Table I. Parameters of 2x2 TeV Collider Insertion Triplet.
Used in Mokhov's Monte Carlo calculation of backgrounds falling on a detector. All three quadrupole elements (QF, QD) combine focusing and bending. All magnets except the final quadrupole (QF2) are superconducting (SC).

Element	Length, metres	R_{in}, cm	R_{out}, cm	$B_{max.}$, Tesla	Grad, Tesla/m	$B_{deflect}$, Tesla
Pipe	1.2					
Iron QF2	11.6	0.45-4.5	35	1.5	3.33-0.33	0.5
Pipe	0.5	8.0				
SC B3	3.0	8.0	35			8.0
Pipe	0.5	8.0				
SC QD1	12.0	8.0	35	-4.0	-0.5	2.0
Pipe	0.5	8.0				
SC B2	3.0	8.0	35			8.0
Pipe	0.5	8.0				
SC QF1	6.2	7.5	35	4.0	0.533	2.0
Pipe	0.5	8.0				
SC B1	30.0	8.0	35			8.0

The most serious contribution to the background in these first simulation runs comes from muons in the beam halo interacting at a 3σ aperture in QF1 35 metres from the beam-crossing point. They produce electron positron pairs, bremsstrahlung and showers which build up in the magnet materials as they approach the detector region. The dangerous part of the background at the detector comes from tertiary muons which cannot be absorbed by any feasible amount of absorber. This initial result predicts background rates which are 600 times bigger than those estimated by Willis.

The group agreed with Mokhov that there were clear improvements which could be made to his first trial scenario. Apertures in the final insertion should all be at more than 5σ, and the beam halo should be controlled by scrapers in the arcs of the machine, far away from experiments, with toroidal sweepers to get rid of the scraped-off muons.

Soft showers from electron interactions in the tapered bore of the last quadrupole can be shielded from the experiment by the addition of a solid heavy metal "nose" reaching in from the end of the quadrupole to within 15 cm. of the collision point (figures 1 and 3). The inner bore of the nose would match that of the quadrupole, but would flare out gently as it approached the collision point so that most of the shower products from the narrowest beam opening at the opposite side would bury themselves inside the hole rather than hitting the front of the nose and backscattering. As drawn, the nose would present up to 1.05 m of material to showers coming out of the quadrupole yoke; over 240 radiation lengths if it is made of tungsten. Its outer surface may benefit from a thin layer of aluminium to help contain showers due to particles coming from the detector side.

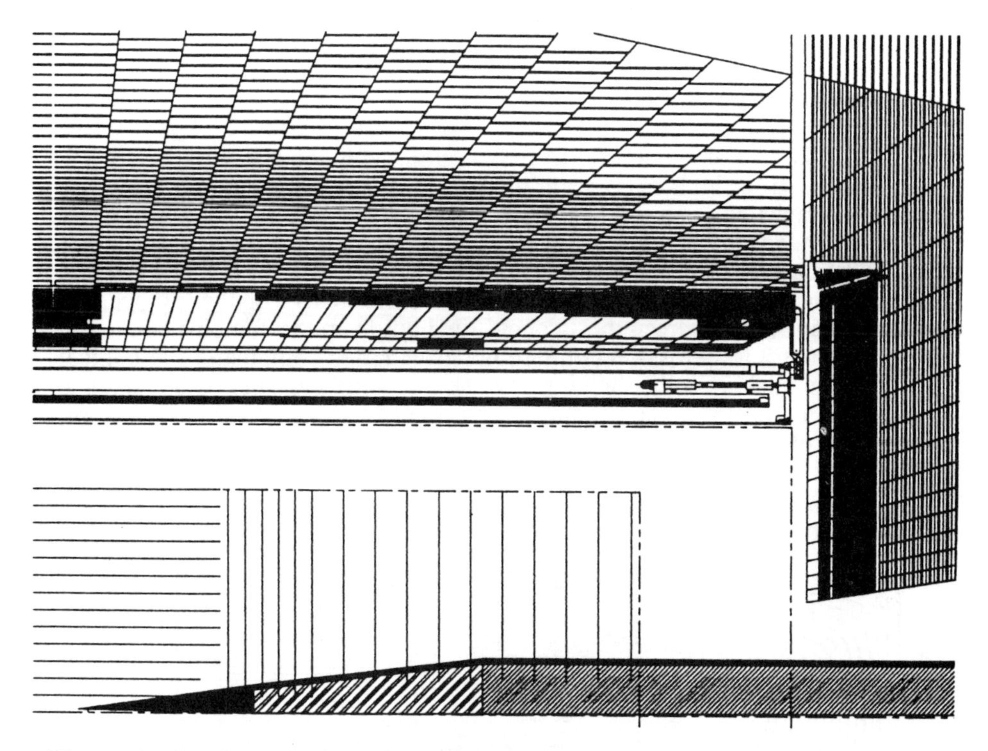

Figure 1. Quadrant-section of a generic detector; (based on the SDC design for SSC, unoptimised) with Mokhov's innermost quadrupole (QF2, Table I) inserted. The tungsten shielding "nose" is heavily shaded. The tapered part of QF2 may be a permanent magnet, with a conventional iron electromagnet for the parallel part. The dark layer around the quadrupole is the "bucking coil", see Figure 2 and text. Tracking is shown to within 125 milliradians of the beam direction.

Two very thorough exercises are needed before a reliable estimate of backgrounds can be made. Mokhov's insertion region design can only be of limited use if it is not matched to a whole machine, so a complete collider lattice design must be put together. Halo muons and scraped muons can then be tracked from wherever they leave the beam envelope. And EGS Monte Carlo simulations have to be made to see how effective such shielding devices as the tungsten nose can be - and the electron-absorbers in the magnet bores. Stumer is setting up EGS to do this.

DETECTOR FIELD CONFIGURATION

Three possibilities were discussed and two of them were soon dismissed. There was no enthusiasm for a toroidal layout. A dipole, or split field dipole, could have minor advantages in keeping the detector field clear of the final combined function quadrupole, but there are at least two serious disadvantages. Dipoles have bad-field directions in which outgoing tracks are not bent, and they can give transverse fields at the interaction point. Such a transverse field would be a double disaster, bending soft electron tracks into the trackers in the horizontal plane with such intensity that they would generate even softer delta rays and other secondaries which spiral along a field lines - filling the whole volume of the detector up with background hits.

A solenoid was seen as the most attractive option, with good acceptance for the high p_T tracks from high mass-scale physics, but capable of trapping low p_T background tracks close to the beam and leading them into the shielding nose. There was concern that the solenoid field would saturate the yoke of Mokhov's normally-conducting final magnet, causing field distortions, but Foster's *Poisson* calculation showed that a suitable configuration of normally-conducting bucking coils could be found to screen the solenoid field out from the combined-function magnet. Figure 2 is a cross-section through a suggested combination of conical and cylindrical bucking coils, showing how the solenoid field-lines could be made to run along the beam direction into the region of the tungsten shielding nose, taking away soft charged-particle backgrounds. The lines are pushed away from the beam axis further from the interaction point. There may be scope for the last metre or so of the magnet to be a permanent quadrupole - if only to ease the space problems of getting current in to the forward cone.

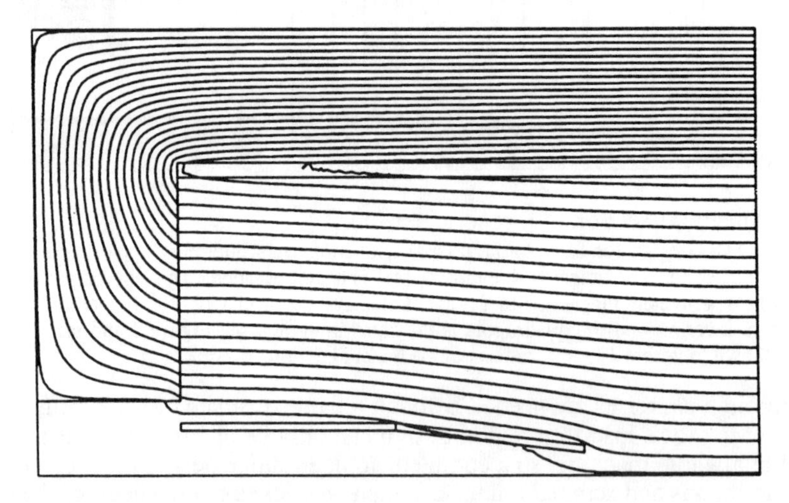

Figure 2. Poisson(2) calculation of solenoid field with bucking coils.
The coil and the conventional iron yoke and pole-pieces of the solenoid can be seen in this single- quadrant section. Conical and cylindrical coils around the beam have been tuned to give a finite field at the interaction point, in order to contain soft background tracks, while eliminating fields in the region of the QF2 quadrupole.

DETECTOR REQUIREMENTS

The main goals are:

- Beauty tagging;
- Sign of charge for full energy e and μ;
- e and μ identification, but not K, π, p;
- Missing Transverse Energy measurement;
- Background toleration.

The group had no disagreements with the Willis approach (1) - so long as the backgrounds are as low as he assumes.

LEP and SLD show that for beauty tagging it will be very important for the innermost layers of silicon pixels to be within a few centimetres of the beam axis. Figure 3 shows how they might be fitted around the tungsten shielding nose. CCD detectors might be able to do this job if they have a common-clear mode which can run continually between untriggered beam-crossings - giving a complete clear within 5 or so crossings. For heavier background rates than Willis assumed, faster-clearing smart pixels would be needed.

Measuring lepton charges needs good resolution over the whole of the main tracker. Willis suggest a multiwire chamber system with pixel-pads to give good background rejection. Assume the pad-readout gives ± 50 microns precision in r-ϕ, with a 1.5 T solenoid field over a tracker with outer radius 1.8 metres. This would give momentum resolution of $\pm 2\%$ for a track with 100 GeV transverse momentum, or $\pm 40\%$ at 2 TeV - just about enough to resolve the sign of the charge.

Electron identification requires good energy resolution in the electromagnetic calorimetry layers so that the deposited energy can be matched with momentum from the tracker. For muon identification Willis' low density calorimetry would be especially attractive if the hadron absorber could be magnetised to give ~3% momentum resolution on muons picked up by an outer tracker.

Muon identification close to the forward and backward directions would be helped by toroidal magnets upstream and downstream of the main detector. Depending on the background fluxes, such toroids might even pick up high energy muons which have been produced within the ± 125 milliradian cone of Figures m and o, and passed through the tungsten nose and the first part of the insertion magnet.

Missing transverse energy measurement will depend upon good hadron calorimetry, with hermetic coverage down to the ± 125 milliradian cone.

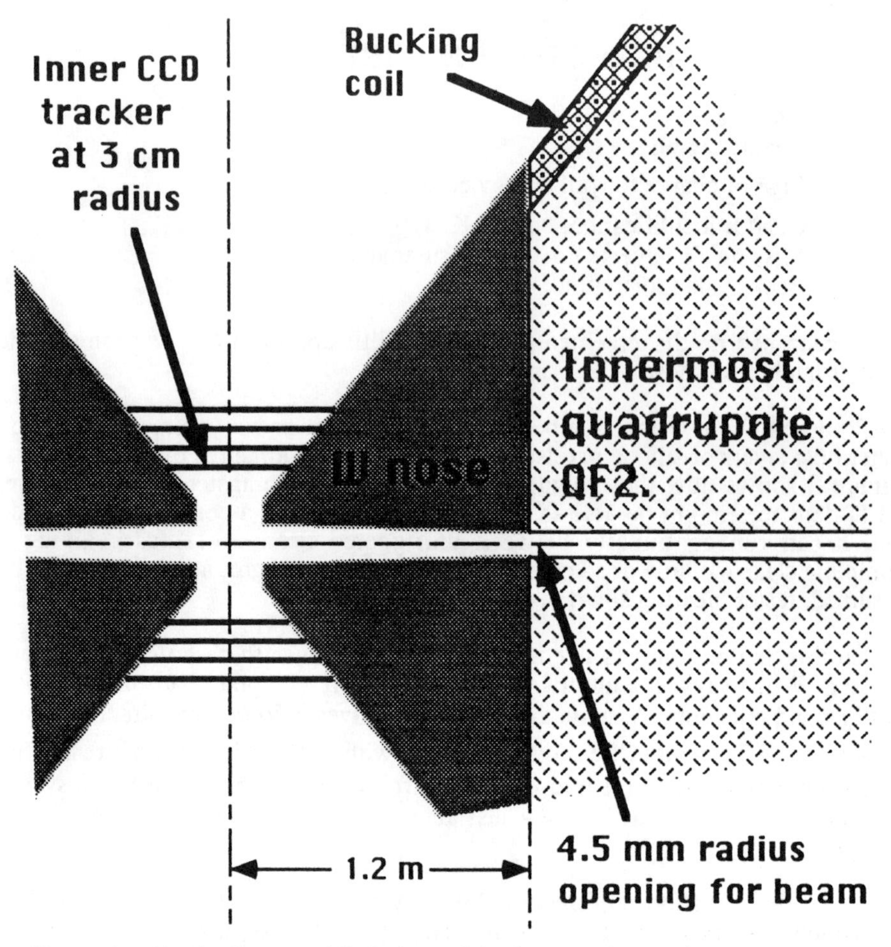

Figure 3. Vertically magnified view of the intersection region.
The vertical scale is 10x the horizontal, for a similar layout to Figure 1. The tungsten shielding nose is shown opening up in aperture from the narrowest opening where the beam leaves the QF2 Quadrupole. A tracker in the position shown will be well shielded from soft electromagnetic background.

LUMINOSITY MONITORING

Because of the small momentum-bite of a circular muon collider, and the low beamsstrahlung, there is no need to monitor the luminosity spectrum in the same way as at an electron-positron linear collider. But the calculations done for Bhabha scattering at the linear collider (3) can be taken over for $\mu\mu$ elastic scattering. The basic pointlike electromagnetic annihilation (s-channel) cross-section at a muon collider sets the unit of R

$$\sigma^{\gamma}_{\mu\mu\to ee} = \sigma^{\gamma}_{ee\to\mu\mu} = \frac{87\,fb}{s(TeV^2)}$$

Other cross sections are measured with respect to this, e.g.

$$\sigma_{t\bar{t}} \to \approx 0.75R \text{ away from threshold,}$$
$$\sigma_{WW} \approx 12R, \text{ depending on acceptance,}$$
$$\sigma_{\gamma\gamma} = 2R, \text{ for two real } \gamma.$$

Elastic μμ scattering has an additional contribution from t-channel γ-exchange, and interference between the s- and t-channels gives significant enhancement even in the barrel region, see Table II.

Table II. Elastic μμ scattering.
(Bhabha scattering rates taken over from (3)).

Angle to beam direction	Rate
180-300 mr.	223R
300-800 mr.	104R
800- 2341 mr.	8R

A few interesting cross sections rise like log s (for instance μνW or μμZ which plateau at ~10 pb; i.e. ~100R for $\sqrt{s} = 1TeV$), but possible Higgs rates are never much more than R.

We conclude that we can monitor the luminosity with sufficient precision for all likely physics channels so long as we can measure the μμ elastic scattering rate for $\theta > 300$ mr. The detector needs to be able to identify collinear muon pairs at beam energy. If it can not do that, what can it do?

VERDICT

A suitable detector can be built; if the background can be brought down to the level Willis assumes (1). Three things are needed to demonstrate that these levels can be achieved:

- a full lattice calculation for the collider;
- EGS simulations of showering, all the way from beam losses in the collider, through collimators, magnets and shielding to the detector;
- agreement between groups checking each other's results.

FNAL is tackling the lattice problem and BNL is running EGS. There is a good chance that firm background predictions will be available for a workshop in November 1995. That will be the time to start on serious detector design.

REFERENCES

1. Willis, W. *Overview of Detectors*, these proceedings.
2. *Users guide to the Poisson/Superfish group of codes*, Los Alamos report LA UR 87 115.
3. Frary, M.N. and Miller, D.J. *Monitoring the Luminosity Spectrum*; in *e^+e^- Collisions at 500 GeV, The Physics Potential*, ed. P.M. Zerwas (1992) DESY 92-123.

MEASUREMENTS OF DYNAMIC APERTURE IN LEP

F. Ruggiero

CERN, SL Division, Geneva, CH 1211

Abstract. We discuss measurement techniques and experimental results relative to the dynamic aperture of LEP. The complicated interplay between radiation damping, machine nonlinearities and resonant beam excitation by the Q-meter is investigated with the help of the Henon map.

History of aperture measurements

A review of aperture measurements in LEP from 1989 to 1992 has been presented in (1) and a summary of the results is shown in Table 1. The measured horizontal aperture $A_x \sim 1.5 \times 10^{-3}\sqrt{\text{m}}$ was nearly a half of the dynamic aperture predicted by particle tracking and most likely due to a physical obstacle in the beam pipe. The subsequent search for a physical aperture limitation gave no result and further measurements in June and in October 1993, with special care about the effect of beta-beating, confirmed the previous small value of A_x.

TABLE 1. Summary of horizontal aperture measurements in LEP from 1989 to 1992: a question mark denotes unpublished or indirect measurements (1). The dynamic aperture predicted by particle tracking with MAD is $A_x^{MAD} \sim 3.5 \times 10^{-3}\sqrt{\text{m}}$ for the injection optics (at 20 GeV) and $A_x^{MAD} \sim 2.5 \times 10^{-3}\sqrt{\text{m}}$ for the low-beta optics (at 45.6 GeV).

DATE	OPTICS	A_x ($10^{-3}\sqrt{\text{m}}$)	METHOD
22/11/1989	n21c20 (60°, 20 GeV)	1.5 ± 0.3	DC injection bump
6/4/1990	n21c20 (60°, 20 GeV)	1.4 ± 0.1	injection kicker
28/5/1990	? (60°, 20 GeV)	1 ?	Q-meter + collimators
31/7/1990	n21h20 (90°, 20 GeV)	0.9 ± 0.03	injection kicker
26/4/1991	? (60°, 20 GeV)	0.9 ?	Q-meter
26/4/1991	? (60°, 20 GeV)	1.1 ?	single corrector bumps
29/4/1991	various (60°, 20 GeV)	$1.4 \div 1.7$	injection kicker
28/5/1990	? (60°, 45.6 GeV)	1.8 ?	Q-meter + collimators
27/8/1990	?05?46 (60°, 45.6 GeV)	≥ 1.5 ?	static arc bumps
24/10/1991	l05g46 (60°, 45.6 GeV)	1.54 ± 0.06 ?	collimator scans
4/8/1992	s05m46 (90°, 45.6 GeV)	1.45	injection kicker
?/?/1992	s06m46 (90°, 45.6 GeV)	1.93 ± 0.14 ?	collimator scans
8/10/1992	s05m46 (90°, 45.6 GeV)	$1.1 \div 1.6$	Q-meter + collimators

The dynamic aperture has mainly been measured by two different methods, namely, resonant beam excitation by the Q-meter plus collimator scans and single kick by the injection or dump kickers. For both methods two possible strategies can be used: a weak excitation option, leading to small current losses and relying on lifetime measurements, and a strong excitation option corresponding, in the case of a single kick, to the loss of half the bunch current.

TABLE 2. Summary of dynamic aperture measurements and predictions in 1993: at 20 GeV (top), at 45.6 GeV (middle) and again at 20 GeV during an experiment to reduce the aperture in a controlled way (bottom). The uncertainties in the results of the single-kick measurements depend on the voltage calibration of the injection kicker IK3E and of the dump kicker.

OPTICS g21p20 (90°/60°, 20 GeV)			
DATE	A_x ($10^{-3}\sqrt{m}$)	A_y ($10^{-3}\sqrt{m}$)	METHOD
18/6/1993	1.4	1.4	Q-meter + collimators: weak excitation
13/10/1993	1.4	1.0	Q-meter + collimators: weak excitation
15/11/1993	$1.8 \div 2.4$	> 1.1	Q-meter + collimators: strong excitation
21/11/1993	3.6 or 2.2	1.7 or 2.1	single kick: 50% current loss
MAD	3.5	2.6	1D tracking
MAD	2.4	1.7	full coupling ($\varepsilon_y = \varepsilon_x/2$)

OPTICS g05p46 (90°/60°, 45.6 GeV): wigglers on to simulate LEP 2 conditions			
DATE	A_x ($10^{-3}\sqrt{m}$)	A_y ($10^{-3}\sqrt{m}$)	METHOD
21/11/1993	3.6 or ?	1.4 or 2	single kick: 50% current loss
MAD	$2.4 \div 2.6$	2.4	1D tracking *without* wigglers

OPTICS g21p20 (90°/60°, 20 GeV): controlled reduction of dynamic aperture			
DATE	A_x measured	A_x MAD (1D)	CONDITIONS
15/11/1993	$1.8 \div 2.4$	3.8	nominal machine, measured quads + sext.
Q-meter	$1.5 \div 1.8$	3.3	4 units of sextupole trim

As shown in Table 2, the measurements by the Q-meter with weak excitation gave results in agreement with those obtained in 1992, while the dynamic aperture measured by the strong excitation option in November 1993 was significantly larger and therefore closer to MAD predictions (2). The results obtained by the injection kicker were comparable or even larger than MAD predictions, although there is an unresolved discrepancy (by almost a factor two) between the old calibration checked in 1989 and the one obtained by measuring the peak-to-peak amplitude of the beam oscillations with the help of the 1000 turn Beam Observation Monitor system (3). However, before 1993 the maximum voltage it was possible to apply at the injection kickers IK3P or IK3E without significant beam losses was around 1 kV at 20 GeV, whereas in November 1993 it was larger than 3 kV for similar values of the horizontal betatron function: this points to the removal of some physical aperture limitation and, if the results of the Q-meter with weak excitation are to be trusted, this event may have taken place between October and November 1993.

A recent revision of the data relative to the single-kick measurements at 45.6 GeV, performed in November 1993 with damping and emittance wigglers at full current to simulate LEP2 conditions (see Table 2), has shown an excellent agreement between the measured horizontal aperture $A_x^{revised} \simeq 2.2 \times 10^{-3}\sqrt{\mathrm{m}}$ and the dynamic aperture $A_x^{MAD} \sim (2.1 \div 2.2) \times 10^{-3}\sqrt{\mathrm{m}}$ obtained by particle tracking with MAD, now including the effect of the wigglers. Together with the single-kick measurements for the injection optics g21p20, also in very good agreement with MAD predictions (if the old kicker calibration is adopted), this seems to indicate that tracking results are reliable and that measurements with the Q-meter, intrinsically more difficult to analyse, tend to underestimate the dynamic aperture.

SINGLE-KICK MEASUREMENTS

A circulating bunch is kicked in the horizontal plane (by the injection kickers IK3E or IK3P) or in the vertical plane by the dump kicker. An example of the relative bunch current loss for different kicker strengths is shown in Figure 1. The kicker strength corresponding to a relative current loss of 50% is directly

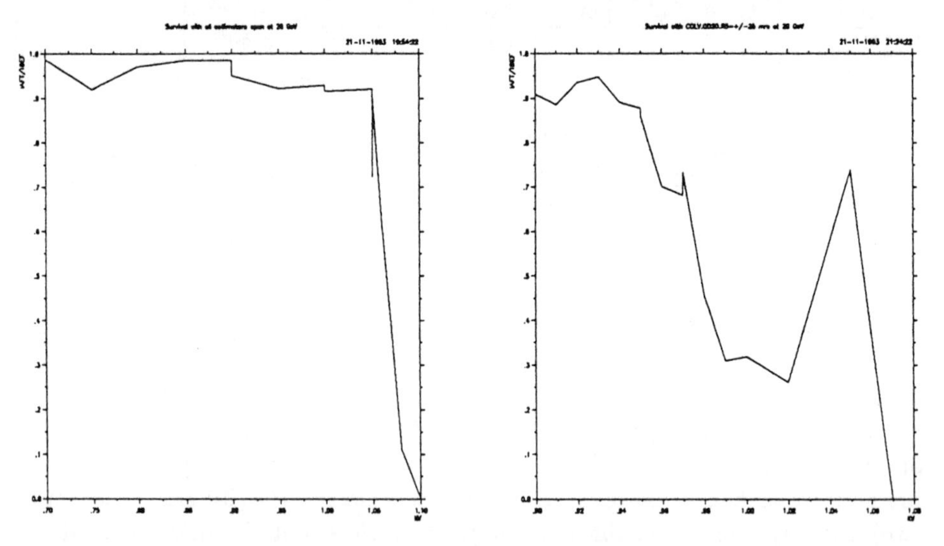

FIGURE 1. Survival plot at 20 GeV corresponding to a vertical kick with the dump kicker. With all collimators open (left) and with collimator COLV.QD20.R5 at ±25 mm (right). In the second case there are three crossings of the 50% level and the situation is not very clear (from Ref. 3).

related to the dynamic aperture. For different current losses, there is a (small) correction due to the finite beam size: see (3) for a more detailed discussion.

MEASUREMENTS BY THE Q-METER

Until November 1993, we adopted the weak excitation method: a bunch is excited in one plane at a time by the Q-shaker, in a frequency range of fixed amplitude around the corresponding betatron frequency, and the beam response at the Q-meter PU is recorded for each excitation level. At the same time, we watch the life-time of the excited bunch and we record its *minimum* value during the frequency sweep. This method is fast, potentially non-destructive and the results are rather reproducible. However, in order to cross calibrate the readings at the Q-meter PU by collimators, we reduce the excitation strength down to a level where the beam lifetime is larger than 10 hours. Since, as shown in Figure 2, the beam response for different excitation levels is very nonlinear, this might significantly reduce the amplitude of the coherent beam oscillations and could be the reason why the measured dynamic apertures are much smaller than the predicted ones. Moreover, the lifetime measurements depend critically on the dynamics of particles in the tails of the distribution and are complicated by the transient beam behaviour during the frequency sweep, which is affected by hysteresis.

To reduce these problems, in the last MD's we adopted a strong excitation method with fixed excitation frequency of the Q-shaker. The excitation strength was increased until a current loss of a few μA was observed over a time interval of some 60 sec. Then we started closing a collimator and took note of the corresponding current losses for the *same* excitation strength: a significant increase of the current loss indicates that the collimator has reached the dynamic aperture and starts scraping the bunch core. Detailed results are given in (2).

In Figure 3, we show the subsequent iterations of the Henon map with damping and resonant excitation for different initial conditions, corresponding to two different limit cycles. Figure 4 refers to the forced beam oscillations observed in LEP under the effect of the Q-shaker and shows a transient rather similar to the evolution towards the inner limit cycle of Figure 3. This data seems to confirm the sensitivity of the Q-meter measurements to initial conditions, such as small dipole beam oscillations or phase of these oscillations relative to the Q-shaker, and thus the difficulty in analysing the results.

REFERENCES

1. F. Ruggiero, "Aperture measurements," in *Proceedings of the Third Workshop on LEP Performance*, Chamonix, 1993, Ed. J. Poole, CERN SL/93-19 (DI), pp. 175–182 (1993).
2. F. Ruggiero, "Measurements of dynamic aperture by resonant beam excitation in 1993," CERN SL-MD Note 115 (1994).
3. L. Ducimetière, E. Keil, M. Lamont, G. Morpurgo, G. Schröeder, A. Verdier, H. Verhagen and E. Vossenberg, "Measurement of the dynamic aperture in both planes with a circulating beam," CERN SL-MD Note 113 (1994).

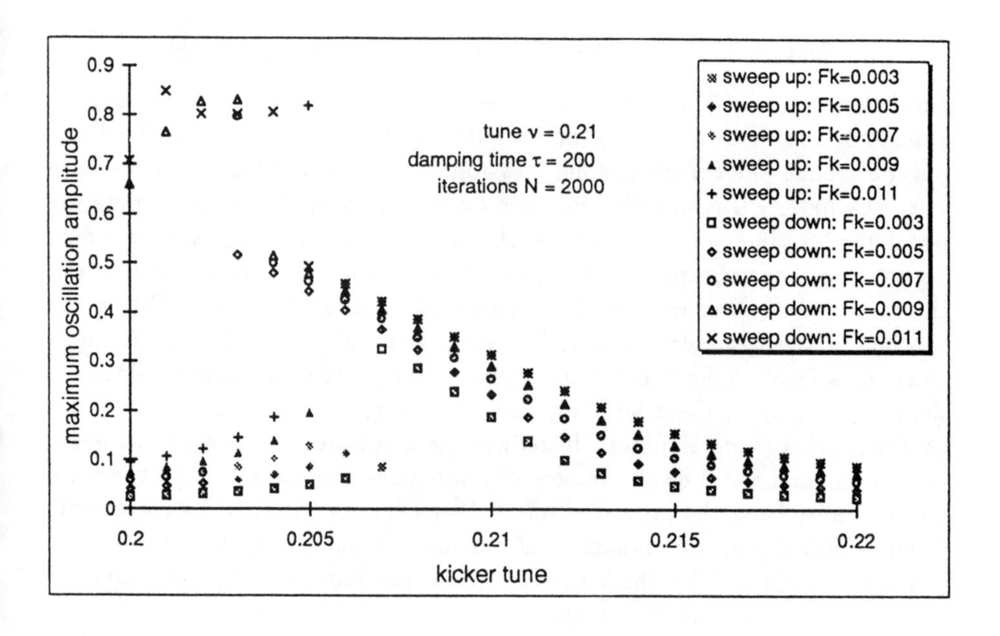

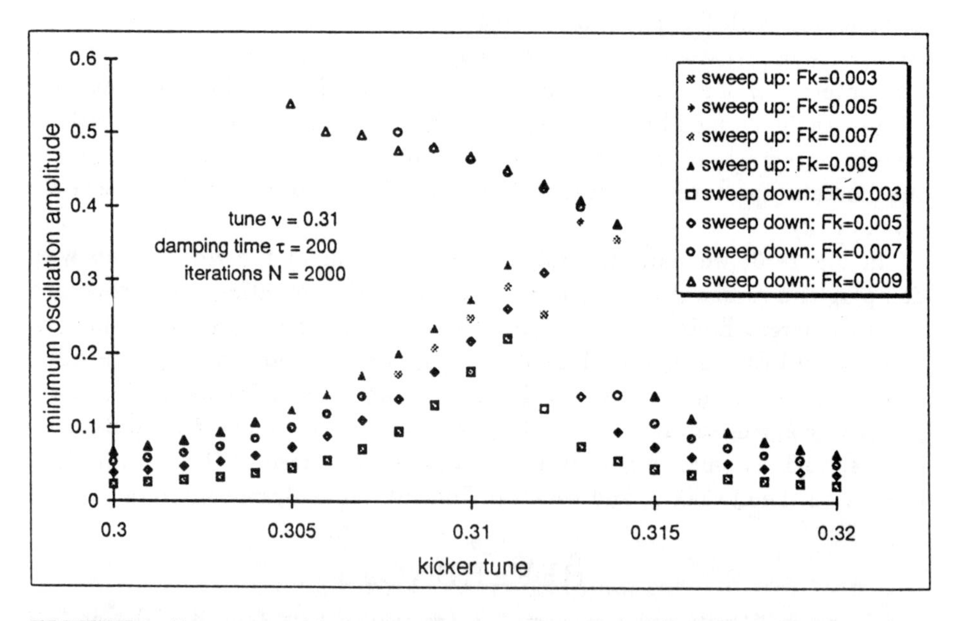

FIGURE 2. Henon map with damping and resonant excitation (tune $\nu = 0.21$ for the upper and $\nu = 0.31$ for the lower plot): maximum (or minimum) oscillation amplitude versus kicker tune for increasing kicker strengths F_k. At the beginning of the frequency sweep, we start tracking at the origin. The corresponding limit cycle is found, after $N = 2000$ iterations, and its maximum (or minimum) oscillation amplitude is evaluated by back-tracking over $N/10$ iterations. Then the frequency is changed and tracking starts from the previous end-point: the system shows histeresis.

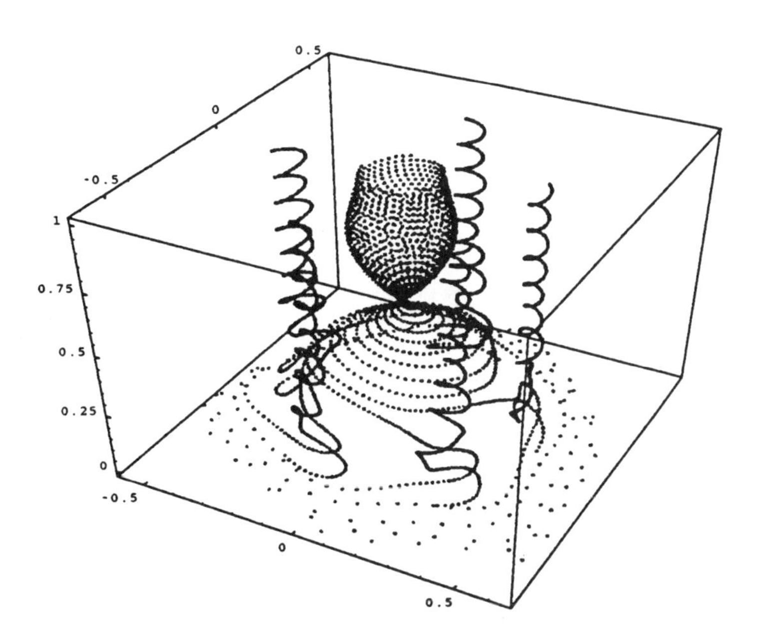

FIGURE 3. Henon map with damping and resonant excitation: tune $\nu = 0.205$, damping time $\tau = 500$ iterations. Limit cycles for weak kicker strength, $F_k = 0.001$ and tune $\nu_k = 0.205$, corresponding to initial conditions $(q,p) = (0.61746, -0.477)$ and $(q,p) = (-0.2, -0.36)$ after 2000 iterations (vertical axis in the figure).

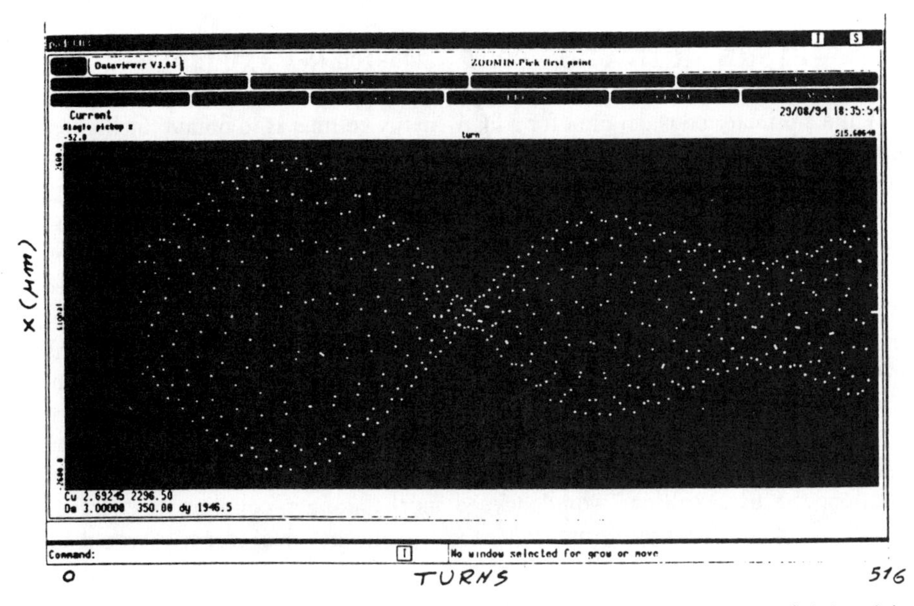

FIGURE 4. Measured beam oscillation at PU.QL1B.R1 (where $\beta_x = 94.2$ m) induced by resonant Q-meter excitation in the horizontal plane ($Q_x = .292$).

Summary of the 200×200 GeV μ^+-μ^- Collider Working Group

D. Neuffer, D. Whittum, P. Bombade, D. Cline, G. Jackson, P. McIntyre, G. Peters, M. Sasaki, D. Summers, K. Yokoya

Abstract. We report the discussions and some preliminary results from the sessions of the 200×200 GeV μ^+-μ^- Collider working group. The physics motivation for such a "medium-energy" collider is discussed. Possible parameters for such a system are described, and compared with higher-energy systems. Relatively-high luminosities (L ~ 10^{33} $cm^{-2}s^{-1}$) appear possible. Modifications of existing facilities to obtain medium-energy μ^+-μ^- collisions are discussed. Other discussion topics (μ-p colliders, low-energy μ sources, etc.) are summarized.

INTRODUCTION

Initial discussions of μ^+-μ^- colliders have concentrated on multi-TeV colliders, because, at that energy, light-lepton (e^+-e^-) colliders appear increasingly impractical, opening a clear opportunity for the use of heavy-lepton (μ^+-μ^-) colliders. However, μ^+-μ^- colliders can explore lower energy physics and may be uniquely suited to some explorations, and a lower energy system may be more readily developed, because of its smaller size and therefore lower cost. A lower-energy collider would be a useful prototype for a future high-enrgy machine, and would test out the concepts and components needed in the high-energy device. Important physics may also occur at lower energies.

PHYSICS MOTIVATION FOR A ~200×200 GeV COLLIDER

While a primary consideration for a high-energy collider is to obtain the highest possible energies, the most important new physics is not necessarily at the highest energies. Specific physics opportunities may occur at lower energies, and a μ^+-μ^- collider may have particular advantages for observing these. The energy region under 100 GeV has been explored very thoroughly by existing hadron and electron machines, and this was set as a lower limit. In the working group, we discussed some of the possible physics opportunities in the 100 to 1000 GeV (center of mass) energy range. This discussion had been initiated by D. Cline in the previous workshop, and is extended by the contributions to this workshop, but further exploration is needed.

A μ^+-μ^- collider would be able to observe all of the physics possible with an e^+-e^- collider at this energy, but without the complications of beamstrahlung. In addition, the μ^+-μ^- collider would have a much stronger coupling to "Higgs" particles. In the working group, it was suggested that this property would be particularly useful in discovering "heavy Higgs", Higgs bosons with masses greater than ~160 GeV, twice the W-boson mass, could be readily observed with a μ^+-μ^- collider.

A relatively clear short-term goal is as a "top factory". Fermilab has seen evidence for the top quark at an energy of ~200 GeV, which means a 200×200 GeV μ^+-μ^- collider could operate as a "top factory" on the t-t* toponium resonance. It would have an advantage over a e^+-e^- collider in that regime, in that it would not be limited by beamstrahlung. The toponium resonance would be clearly resolved and its width could be readily measured, and other toponium properties could be more cleanly observed.

PARAMETERS FOR 200×200 GeV μ^+-μ^- COLLIDERS

We chose 200×200 GeV as a reference collision energy and explored possible parameters for a low-energy collider. Other than the fundamental difficulty of collecting and cooling muons, there are no intrinsic limits to collider performance at these lower energies. The only intrinsic advantage of a higher-energy collider is in adiabatic damping, which reduces the transverse beam size by a factor of E_μ. The longitudinal adiabatic damping may also permit some additional improvements at higher energies. However, provided a high-intensity source of KAON-type class is available, relatively high luminosity is possible. In Table I, we display parameters of a 200×200 collider, based on the use of the same KAON-class source used for the 2×2 TeV scenario,[3] and obtain a luminosity of 10^{33} cm^{-2}s^{-1}.

USE OF EXISTING FACILITIES

We discussed possible adaptations of existing facilities for muon colliders. D. Summers discussed possible adaptations of SLAC and G. Jackson discussed adaptations of Fermilab for muon collider experiments. P. McIntyre discussed use of a π-factory, such as LAMPF or future spallation neutron sources (~1GeV protons) as a $\pi \rightarrow \mu$ source. Since existing sources are at least an order of magnitude less intensity than KAON, a collider based on existing machines should have substantially smaller intensity and therefore a luminosity that is ~2 orders of magnitude smaller. Also most existing machines are not adapted to the high repetition rate of a collider or accommodate large numbers of μ-decays. Certainly existing tunnels and many other facilities could be adapted to a future collider, and further study is needed. Initial concept tests could be run using existing components, and a scenario using the BNL AGS with RHIC was discussed.

μ-p COLLIDERS

A μ^+—μ^- collider can also be operated as a μ^--p collider with both μ and p beams at full energy. High luminosity would be relatively easily obtained because only one beam (μ) is unstable and diffuse. This is a probable initial and debugging operational mode for a storage ring μ^+—μ^- collider. Revolution frequencies of equal energy μ and p beams would be naturally mismatched because of unequal speeds, but they can be rematched by displacing the beams in energy and matching the ring nonisochronicity.[2] A μ-p collider could explore the same physics as an e-p collider. However the center of mass energy of a 200×200 GeV collider is similar to the HERA 30 GeV(e^- beam)× 800 GeV (p-beam) collider, and the 2×2 TeV μ-p collider would explore the same energy scale as LEP × LHC.

COMMENTS AND CONCLUSIONS

Some of the possible motivations and parameters for a 200×200 GeV collider have been presented, and these are discussed in more detail in other contributions to the workshop. A medium energy collider with high-luminosity is possible, and may produce critically important physics. Future studies are needed for clearer development of the physics potential and optimal parameters of such a collider.

References

1. D. Cline, Nucl. Inst. and Meth. **A350**, 24 (1994).
2. D. Neuffer, Particle Accelerators, **14**, 75 (1983).
3. D. Neuffer and R. Palmer, Proc. 1994 European Particle Accelerator Conference, London, p.52 (1994).

Table 1: Parameter list for a 400 GeV μ^+-μ^- Collider

Parameter	Symbol	Value
Energy per beam	E_μ	200 GeV
Luminosity	$L=f_0 n_s n_b N_\mu^2/4\pi\sigma^2$	10^{33} cm^{-2}s^{-1}
Source Parameters		
Proton energy	E_p	30 GeV
Protons/pulse	N_p	$2\times3\times10^{13}$
Pulse rate	f_0	10 Hz
μ-production acceptance	μ/p	.15
μ-survival allowance	N_μ/N_{source}	.33
Collider Parameters		
Number of μ /bunch	$N_{\mu\pm}$	1.5×10^{12}
Number of bunches	n_B	1
Storage turns	n_s	900
Normalized emittance	ε_N	10^{-4} m-rad
μ-beam emittance	$\varepsilon_t=\varepsilon_N/\gamma$	0.5×10^{-7} m-rad
Interaction focus	β_0	0.3 cm
Beam size at interaction	$\sigma = (\varepsilon_t\beta_0)^{1/2}$	12 μm

A Practical High-Energy High-Luminosity μ^+-μ^- Collider

Robert B. Palmer, *SLAC, Stanford University, Stanford CA 94305 and Brookhaven National Laboratory, Box 5000, Upton NY 11973*

David V. Neuffer, *CEBAF, 12000 Jefferson Avenue, Newport News VA 23606*

Juan Gallardo, *Brookhaven National Laboratory, Box 5000, Upton NY 11973*

Abstract. We present a candidate design for a high-energy high-luminosity μ^+-μ^- collider, with E_{cm} = 4 TeV, L = $3\times10^{34}cm^{-2}s^{-1}$, using only existing technology. The design uses a rapid-cycling medium-energy proton synchrotron, which produces proton beam pulses which are focused onto two π-producing targets, with two π-decay transport lines producing μ^+'s and μ^-'s. The μ's are collected, rf-rotated, cooled and compressed into a recirculating linac for acceleration, and then transferred into a storage ring collider. The keys to high luminosity are maximal μ collection and cooling; innovations with these goals are presented, and future plans for collider development are discussed. This example demonstrates a novel high-energy collider type, which will permit exploration of elementary particle physics at energy frontiers beyond the reach of currently existing and proposed electron and hadron colliders.

INTRODUCTION

Lepton(e^+-e^-) colliders have the valuable property of producing simple, single-particle interactions, and this property is essential in the exploration of new particle states. Extension of e^+-e^- colliders to multi-TeV energies is performance-constrained by radiation and "beamstrahlung" effects, which increase as $(E_e/m_e)^4$, and cost-constrained by the need for two full-energy linacs.[1] However, muons (heavy electrons, with m_μ = $200m_e$) have negligible beamstrahlung and can be accelerated and stored in rings. The liabilities of muons are that they decay, with a lifetime of 2.2×10^{-6} E_μ/m_μ s, and that they are created through decay into a diffuse phase space. But that phase space can be reduced by ionization cooling, and the lifetime is sufficient for storage-ring collisions. (At 2 TeV, τ_μ = 0.044 s.) We present the first practical design for a high-energy high-luminosity $\mu^+\mu^-$ collider, with an energy of E_{cm}=$2E_\mu$= 4TeV, and a luminosity of L = $3\times10^{34}cm^{-2}s^{-1}$, which uses only existing technical capabilities.

The possibility of muon (μ^+-μ^-) colliders has been introduced by Skrinsky et al.[2] and Neuffer[3]. More recently, several mini-workshops have greatly increased the level of discussion,[4, 5, 6, 7] stimulating the present developments. In this paper we introduce improvements, and develop a complete scenario for a high-luminosity high-energy collider. Table 1 shows parameters for the candidate design, which is displayed graphically in fig. 1. The design consists of a muon source, a muon collection, cooling and compression system, a recirculating linac system for acceleration, and a full-energy collider with detectors for multiturn high-luminosity collisions.

MUON PRODUCTION

The μ-source driver is a high-intensity rapid-cycling (10Hz) synchrotron at KAON[8] proposal parameters (30 GeV), which produces beam which is formed into two short bunches of 3×10^{13} protons. Combination of the accelerated beam bunches into two short bunches at full energy (possibly in a separate extraction ring) simplifies the subsequent longitudinal phase space manipulations. The two proton bunches are extracted into separate lines for μ^+ and μ^- production. (Separate lines permit use of higher-acceptance, zero-dispersion $\pi \rightarrow \mu$ capture lines.) Each bunch collides into a target, producing π's (~1 π/interacting p) over a broad energy and angular range (E_π = 0—4 GeV, $p_\perp$ < 0.5 GeV/c). The target is followed by Li lenses, which collect the π's into a large-aperture high-acceptance transport line (an r = 0.15m, B= 4T FODO transport with a 0.8m period in a current scenario), designed to accept a large energy width (2±1 GeV) and have a large transverse acceptance($p_\perp$ < 0.4GeV). This array is sufficiently long (~300 m) to insure $\pi \rightarrow \mu$ decay, plus debunching, in which the energy-dependent particle speeds spread the beam longitudinally to a full width of ~6 m, while reducing the local momentum spread.

This is followed by a nonlinear rf system (3 harmonics are sufficient) which flattens the momentum spread. We have conservatively assumed that rf gradients at these frequencies (~10—30 MHz) are limited to less than ~2 MeV/m; this implies a total rf debuncher length of ~1.4 km. (Higher gradients would permit a shorter debuncher. Also, an induction linac rather than resonant-cavity rf could be used, as suggested by Barletta.[9]) The resulting μ-beam is then matched into a beam cooling system. Figure 2 shows a schematic overview of the production and collection system.

A Monte Carlo program (MCM - Monte Carlo Muon) has simulated the muon production and cooling. The generation of π's in the target is calculated using a thermodynamic model or the Wang distribution.[10] The π's are tracked through decay to μ's, and phase-energy rotation, into the cooling system. We obtain ~0.15 captured μ's per inital proton, with ε_N =0.01 m-rad, an rms bunch length of 3m, and energy width of 0.15 GeV with an average energy of 1 GeV.

The μ capture efficiency (0.15μ/p) is larger than estimated in previous scenarios[2, 3, 4], and this is a result of the use of a high acceptance decay transport with a larger momentum acceptance. The transport is followed by a linac-based rf rotation, which reduces the momentum spread to a level acceptable by the subsequent transport and cooling system, while lengthening the beam bunch.[11] Previously, Noble[4] has noted that $\mu^+\mu^-$ collider luminosity increases with the energy spread acceptance ΔE_A by as much as $\Delta E_A{}^4$, with factors accumulating from the production and the decay acceptances of both beams. The high-acceptance transport plus rf rotation has increased that acceptance by almost an order of magnitude above previous estimates. Also, for the first time, the production has been directly calculated in a realistic simulation, and not simply estimated.

BEAM COOLING

For collider intensities, the phase-space volume must be reduced by beam-cooling and the beam size compressed, within the μ lifetime. Much of the needed compression is obtained through adiabatic damping in acceleration from GeV-scale μ collection to TeV-scale collisions. Beam cooling is obtained by "ionization cooling" of muons ("μ-cooling"), in which beam transverse and longitudinal energy losses in passing through a material medium are followed by coherent reacceleration, resulting in beam phase-space cooling.[2,3,12] (Ionization cooling is not practical for protons and electrons because of nuclear scattering (p's) and bremstrahlung (e's) effects, but is for μ's and the necessary energy losses are easily obtained within the μ lifetime.) In this section we present the equations for μ-cooling, use these to deduce optimal cooling conditions, and generate a practical cooling scenario.

The equation for transverse cooling is:

$$\frac{d\epsilon_N}{ds} = -\frac{dE}{ds}\frac{\epsilon_N}{E_\mu} + \frac{\beta_\perp}{2}\frac{(0.014)^2}{E_\mu m_\mu}\frac{1}{L_R} \tag{1}$$

(with energies in GeV), where ε_N is the normalized emittance, $\beta_\perp$ is the betatron function at the absorber, dE/ds is the energy loss, and L_R is the material radiation length. The first term in this equation is the coherent cooling term and the second term is heating due to multiple scattering. This heating term is minimized if $\beta_\perp$ is small (strong-focusing) and L_R is large (a low-Z absorber) .

The equation for energy cooling is:

$$\frac{d\langle(\Delta E)^2\rangle}{ds} \approx -2\frac{\partial \frac{dE_\mu}{ds}}{\partial E_\mu}\langle(\Delta E)^2\rangle + \frac{d\langle\Delta E_\mu^2\rangle}{ds} \qquad (2)$$

Energy-cooling requires that $\partial(dE_\mu/ds)/\partial E) > 0$. The energy loss function, dE_μ/ds, is rapidly decreasing with energy for $E_\mu < 0.2$ GeV (and therefore heating), but is slightly increasing (cooling) for $E_\mu > 0.3$ GeV. This natural cooling is ineffective; but $\partial(dE_\mu/ds)/\partial E$ can be increased by placing a transverse variation in absorber density or a wedge absorber where position is energy-dependent. (This variation is used in two modes: a weak variation to balance cooling rates, or a thick wedge to transfer phase space.) The sum of cooling rates is invariant:

$$\frac{1}{E_{cool,x}} + \frac{1}{E_{cool,y}} + \frac{1}{E_{cool,\Delta E}} = \text{Constant} \approx \frac{2}{E_\mu} \ , \qquad (3)$$

where E_{cool} is the total energy loss needed to obtain an e-folding of cooling and E_μ is the μ energy.

In the long-pathlength Gaussian-distribution limit, the heating term or energy straggling term is given by:[13]

$$\frac{d(\Delta E_\mu)^2}{ds} \approx 4\pi(r_e m_e c^2)^2 N_0 \frac{Z}{A} \rho \gamma^2 (1-\beta^2/2) \ , \qquad (4)$$

where N_0 is Avogadro's number and ρ is the density. Since this increases as γ^2, and the cooling system size scales as γ, cooling at low energies is desired.

To obtain energy cooling and to minimize energy straggling, we require cooling at low relativistic energies ($E_\mu \sim 300$ MeV). For optimum transverse cooling, the ideal absorber is itself a strong focussing lens which maintains small beam size over extended lengths, and a low-Z material. In this design, we use Be(Z=4) or Li(Z=3) current-carrying rods, where the high current provides strong radial focussing. For Be, Z=4, A=9, $dE_\mu/dx = 3$ MeV/cm, and $\rho = 1.85$ gm/cm^3.

The beam cooling system reduces transverse emittances by more than two orders of magnitude (from 0.01 to 3×10^{-5} m-rad), and reduces longitudinal emittance by more than an order of magnitude. This cooling is obtained in a series of cooling cells, with the initial cells reducing the energy toward the cooling optimum of 300 MeV. A typical cooling cell consists of a focusing cooling rod

(~0.7 m long for Be, ~2.1 m for Li) which reduces the central energy by ~200 MeV, followed by a ~200 MeV linac (20—40 m at 10—5 MeV/m), with optical matching sections (~40–60 m total cell length) (see fig. 3). Small angle bends introduce a dispersion (position-dependence on energy), and wedge absorbers (or density gradients) introduce an energy loss dependence on beam energy. Bends are also used to provide path length dependence on momentum, in order to compress the bunch lengths. The cell parameters are adjusted to optimal transverse and longitudinal cooling rates, and cooling by a 6-D factor of ~4 is obtained in each cell. ~15–20 such cells (~800 m) are needed in the complete machine.

From equation 1, we find a limit to transverse cooling when the multiple scattering balances the cooling, at $\varepsilon_N \approx 10^{-2}\ \beta_\perp$ for Be. The value of $\beta_\perp$ in the Be rod is limited by the peak focusing field to $\beta_\perp$~0.01 m, obtaining $\varepsilon_N \sim 10^{-4}$ m-rad. This is a factor of ~3 above the emittance goal of Table 1. The additional factor can be obtained by cooling more than necessary longitudinally and exchanging phase-space with transverse dimensions in a thick wedge absorber. MCM simulations have demonstrated that the desired cooling and phase-space exchange can be obtained.

In the present scenario, we cool only with ionization cooling in conducting Be (or Li) rods, along with phase space exchange, and that is sufficient for high luminosity. However, other techniques (such as ionization cooling in focussing transports or rings, or using plasma lenses, or high-frequency "optical" stochastic cooling[14]) may permit improvements, and are being studied.

ACCELERATION AND COLLISIONS

Following cooling and initial bunch compression to ~.1–.3m bunch lengths, the beams are accelerated to full energy (2 TeV). A full-energy linac would work, but it would be costly and does not use our ability to recirculate μ's. A recirculating linac (RLA) like CEBAF[15] can accelerate beam to full energy in 10—20 recirculations, using only 200—100 GeV of linac, but requiring 20—40 return arcs. The μ-bunches would be compressed on each of the return arcs, to a length of 0.003m at full energy. A cascade of RLAs (i. e., 1–10, 10–100 and 100–2000 GeV), with rf frequency increasing as bunch length decreases, may be used. Rapid-cycling synchrotrons, or hybrid devices, are also possible. (A low-cost scenario requiring only 20 GeV of rf, using an injector and three RLA stages with rapid-cycling in the last stage, has also been developed.) The cooling and acceleration cycle is timed so that less than ~half the initial μ's decay. (In Table 1, we allow a factor of 3 in total losses.)

After acceleration, the μ^+ and μ^- bunches are injected into the 2-TeV superconducting storage ring (~1-km radius), with collisions in one or two low-β^* interaction areas. The beam size at collision is $r = (\varepsilon_N\ \beta_0/\gamma)^{1/2} \sim 2\mu$m, similar to hadron collider values. The bunches circulate for ~300B turns before decay, where B is the mean bending field in T. (This is 150B luminosity turns, a factor of two smaller since both beams decay.) The design is restricted by μ decay within the rings ($\mu \rightarrow e \nu\nu$), which produces 1/3-energy electrons which radiate and travel to the inside of the ring dipoles. This energy could be intercepted by a liner inside the magnets, or specially designed C-dipoles could be used and the electrons intercepted in an external absorber. The design constraints may limit B; we have chosen B = 6T (900 turns) in the present case. μ-decays in the interaction areas will also provide some background levels in detectors. The limitations in detector design are being studied.

COMMENTS AND CONCLUSIONS

We have presented a candidate design for a high-energy high-luminosity μ^+-μ^- collider. The critical features of the scenario (π-collection and decay, phase space compression and cooling) have been modeled with Monte Carlo simulations, as well as by analytical methods. The design uses practical components and concepts within existing technical capabilities, and with requirements within the size and scope of existing facilities (i. e., Fermilab, CERN, HERA). Since we have confined ourselves to existing technology, we can initiate cost estimates, and costs similar to similar-sized facilities are expected.

The 4 TeV energy was set as a benchmark goal for the high-energy frontier. The μ^+-μ^- collider concept naturally increases in luminosity with energy. One factor of E_μ increase results from emittance adiabatic damping. If the injector size/cost is allowed to increase as E_μ, then beam intensities increase as E_μ, and L increases by at least another factor of E_μ. Smaller interaction-region β^* should also be possible through longitudinal adiabatic damping, permitting some further enhancement. This increase of luminosity with energy by at least a factor of E_μ^2 should be followed up to the 100 TeV scale, beyond which μ synchrotron radiation becomes large. Radiation damping would then initially permit further improvements.

Lower energy machines (100+ GeV Higgs, top factories, etc.) are also possible, as proposed by D. Cline.[16] These require specific physics motivations, but could result in important discoveries.

The present scenario is simply a first proof-of-principle calculation of the design concept, using realistic components. Much further optimization and design and concept development is needed. We are forming and recruiting a collaboration for further development of the μ^+-μ^- collider concept. (Interested readers are invited to contact one of the authors for further discussion.) We next describe some of the topics needing further development.

The scenario for production and collection of muons needs further study and improvement. More detailed, and hopefully more accurate, production calculations should be implemented. Many options for targetting, collection and, transport, as well as variations on the rf debunching and compression scenario, should be explored. A critical parameter is the rf-debuncher gradient; a higher gradient would permit a more compact facility. Variations on initial proton and π collection energies should be considered, including the possibility of low-energy stopped π sources, and optimizations should be obtained. (H. Daniel[17] has investigated μ-cooling at low (thermal) energies.)

The ionization cooling scenario has only been developed in broad outline form, and needs to be optimized in full detail, and verified by more complete simulations. The detailed design is strongly dependent on rf gradient and linear and nonlinear optics optimization.

The bunch-compression and acceleration scenario also needs to be optimized and simulated. Variations such as rapid-cycling synchrotrons and fixed-field alternating gradient (FFAG) machines should be considered. Complete lattices are needed. A complete lattice is also needed for the full-energy collider, integrating the interaction-regions with the high-field arcs, and including modifications needed to accomodate μ-decay in the arcs and near the IR's. Tracking of the complete lattice for a muon lifetime should be completed, and any instability limits should be identified.

It would be desirable to collide polarized beams. Muons are naturally polarized when produced in π-decay, however, we do not yet know if that polarization can be maintained through a debunching, cooling, acceleration and collision cycle. Further study is needed.

The luminosity estimate of the present scenario should not be considered as an absolute limit. Luminosity can be readily increased by reducing losses, or by increasing the source repetition rate (increasing the source to above KAON-class intensity).

To initiate practical implementation, some experimental explorations are needed. A first demonstration of ionization cooling is essential. A collaboration is being formed (and recruited; call if interested) to plan an initial cooling experiment. In this experiment, a diffuse muon beam at ~0.5 GeV would be focussed into a Li (or Be) rod, in a configuration similar to the cooling cell of Fig. 3. Since ionization cooling is intensity-independent, only a low-intensity external beam is needed. Beam formation optics, including strong (quad or solenoid) focussing into the Li rod, a high-current power supply for lens activation of the rod, and accurate beam diagnostics are also required.

Other experiments may be needed. More accurate measurements of π-production under the same conditions used in the collider would be useful, and could be important in developing an accurately optimized design scenario. rf acceleration development would also be desirable, both in the low-frequency rf systems needed in the debuncher and cooling and in the high-frequency rf needed in the accelerator and collider.

Much study and development is needed and this initial scenario will be greatly changed and (hopefully) improved before implementation. However, we believe that the improvements reported here have transformed the μ^+-μ^- collider concept into a real possibility, which will permit exploration of elementary particle physics at energy frontiers beyond the reach of currently existing and proposed electron and hadron colliders.

Acknowledgments

We acknowledge extremely important contributions from our colleagues, especially D. Cline, F. Mills, A. Chao, A. Ruggiero, A. Sessler, S. Chattopadhyay, J. D. Bjorken, W. Barletta, D.Douglas, R. Noble, D. Winn, S. O'Day, Y. Y. Lee, I. Stumer and C. Taylor.

References

1. M. Tigner, in *Advanced Accelerator Concepts*, AIP Conf. Proc. **279**, 1 (1993).
2. E. A. Perevedentsev and A. N. Skrinsky, Proc. 12th Int. Conf. on High Energy Accel., 485 (1983), A. N. Skrinsky and V.V. Parkhomchuk, Sov. J. Nucl. Physics **12**, 3 (1981).
3. D. Neuffer, Particle Accelerators, **14**, 75 (1983), D. Neuffer, Proc. 12th Int. Conf. on High Energy Accelerators, 481 (1983), D. Neuffer, in *Advanced Accelerator Concepts*, AIP Conf. Proc. **156**, 201 (1987).
4. R. J. Noble, in *Advanced Accelerator Concepts*, AIP Conf. Proc. **279**, 949 (1993).
5. Proceedings of the Mini-Workshop on μ^+-μ^- Colliders: Particle Physics and Design, Napa CA, to be published in Nucl Inst. and Meth.(1994), D. Cline, ed.
6. Muon Collider Mini-Workshop, Berkeley CA, April 1994.
7. Proceedings of the Muon Collider Workshop, February 22, 1993, Los Alamos National Laboratory Report LA-UR-93-866 H. A. Theissen, ed.(1993)
8. KAON Factory proposal, TRIUMF, Vancouver, Canada (1990)(unpublished).
9. W. Barletta, private communication (1994).
10. C. L. Wang, Phys Rev **D9**, 2609 (1973) and Phys. Rev. **D10**, 3876 (1974).
11. The rf debuncher is similar to one proposed for antiproton acceptance by F. Mills (1982).
12. D. Neuffer, to be published in Nucl. Inst. and Meth. (1994).
13. U. Fano, Ann. Rev. Nucl. Sci. **13**, 1 (1963).
14. A. A. Mikhailichenko and M. S. Zolotorev, Phys. Rev. Lett.**71**, 4146 (1993).
15. CEBAF Design Report, CEBAF, Newport News VA (1986) (unpublished).
16. D. Cline, to be published in Nucl Inst. and Meth., (1994).
17. H. Daniel, Muon Catalyzed Fusion **4**, 425(1989), M. Muhlbauer, H. Daniel, and F. J. Hartmann, Hyperfine Interactions **82**, 459 (1993).

Table 1: Parameter list for 4 TeV μ^+-μ^- Collider

Parameter	Symbol	Value
Energy per beam	E_μ	2 TeV
Luminosity	$L=f_0 n_s n_b N_\mu^2/4\pi\sigma^2$	3×10^{34} cm^{-2}s^{-1}
Source Parameters		
Proton energy	E_p	30 GeV
Protons/pulse	N_p	$2\times3\times10^{13}$
Pulse rate	f_0	10 Hz
μ-production acceptance	μ/p	.15
μ-survival allowance	N_μ/N_{source}	.33
Collider Parameters		
Number of μ /bunch	$N_{\mu\pm}$	1.5×10^{12}
Number of bunches	n_B	1
Storage turns	n_s	900
Normalized emittance	ε_N	3×10^{-5} m-rad
μ-beam emittance	$\varepsilon_t=\varepsilon_N/\gamma$.	1.5×10^{-9} m-rad
Interaction focus	β_0	0.3 cm
Beam size at interaction	$\sigma = (\varepsilon_t\beta_0)^{½}$	2.1 μm

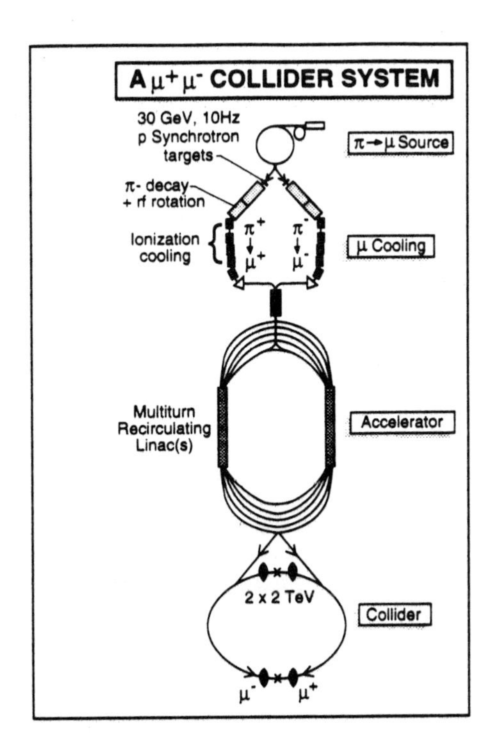

Figure 1: Overview of the μ^+-μ^- collider system, showing a muon (μ) source based on a high-intensity rapid-cycling proton synchrotron, with the protons producing pions (π's) in a target, and the μ's are collected from subsequent π decay. The source is followed by a μ-cooling system, and an accelerating system of recirculating linac(s) and/or rapid-cycling synchrotron(s), feeding μ^+ and μ^- bunches into a superconducting storage-ring collider for multiturn high-energy collisions. The entire process cycles at 10 Hz.

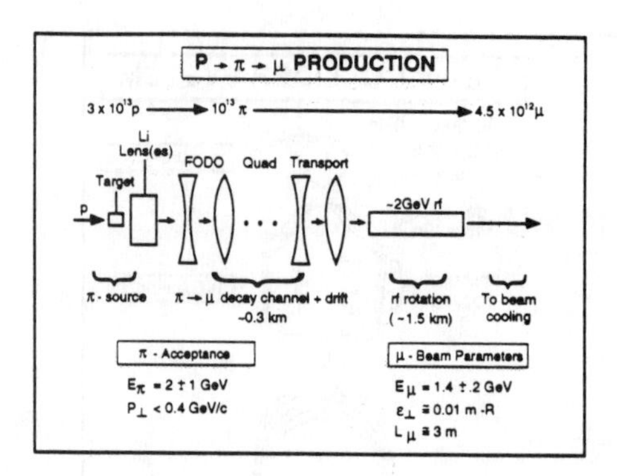

Figure 2: Overview of the $\pi \rightarrow \mu$ production transport line. Each proton bunch collides into a target, producing large numbers of π's (~1 π/interacting p) over a broad energy and angular range (E_π =0-4 GeV, $p_\perp$ < 0.5 GeV/c). The target is followed by a Li lens, which collects the π's into a transport line, which accepts a large energy width (2±1 GeV) and has a large transverse acceptance($p_\perp$< 0.4 GeV/c). This line is sufficiently long to insure $\pi \rightarrow \mu$ decay, plus debunching, in which the energy-dependent particle speeds spread the beam longitudinally, while a nonlinear rf system flattens the momentum spread. The μ-beam is then matched into a beam-cooling system.

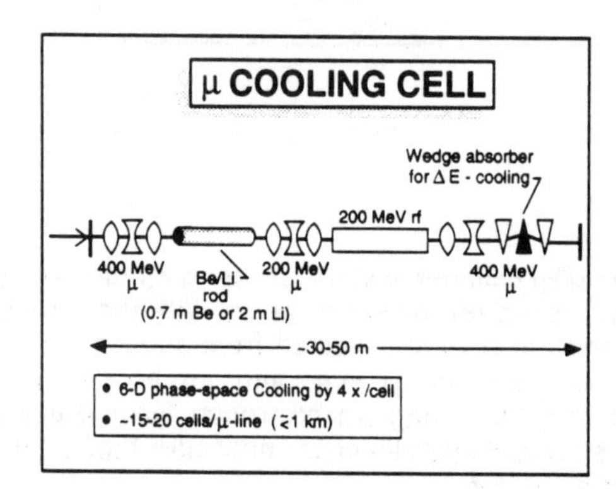

Figure 3: Overview of a μ-cooling (ionization cooling) cell. A typical cooling cell consists of a focusing cooling rod which reduces the central energy from 400 to 200 MeV, followed by a ~200 MeV linac, with appropriate optical matching. Cooling by a 3-D factor of ~4 is obtained in each cell, and 15-20 such cells (~1 km) are needed in the complete machine. For energy cooling, we introduce a dispersion (position-dependence on momentum), and use an absorber with a density gradient or wedge shape. This arrangement permits enhancement of energy-cooling[2, 3], and the parameters are adjusted to obtain optimal transverse and longitudinal cooling.

Physics with Like-Sign Muon Beams in a TeV Muon Collider

Clemens A. Heusch* and Frank Cuypers†

*Santa Cruz Institute for Particle Physics
University of California, Santa Cruz
†Max-Planck-Institut für Physik
Föhringer Ring 6, D-80805 Munich, Germany

Abstract. We point out that both the specific lepton number content and the high energies potentially attainable with muon-muon colliders make it advisable to consider the technical feasibility of including an option of like-sign incoming beams in the studies towards a proposal to build a muon-muon collider with center-of-mass energies in the TeV region. This capability will add some unique physics capabilities to the project. Special attention will have to be paid to polarization retention for the muons.

INTRODUCTION

The prospect of having lepton colliders reach into the TeV range has opened up a slew of fascinating physics perspectives: in a number of workshops through the past several years, electron linear colliders have been investigated in terms of the physics programs that motivate a choice of parameters—such as center-of-mass energies, luminosities, polarization parameters, and the precise particle choice for the initial state - e^+e^-, e^-e^-, $e\gamma$, $\gamma\gamma$. For a number of investigations, particularly those that probe the limits of the Standard Model and reach beyond its boundaries, it has become increasingly clear that it is vitally important to go to the highest attainable energies. Also, a clear definition of incoming polarization states defines the process to be studied and suppresses backgrounds, and is needed for many if not most studies of subtle points in new interactions.

In the framework of machine studies, electron colliders beyond LEP-2 energies are relegated to a linear configuration, thus putting a fiscal limit to the energies that can reasonably be reached (quite apart from the beamstrahlung effects that will exacerbate experimentation at the upper end of the attainable energies). Muon-muon colliders, however, are being proposed well beyond the 2-TeV limit that is practically imposed by economic considerations. This means that interactions that come into their own only well above 1 TeV, and which do not depend on the specific lepton family tag of the incoming beams, are more promising in a muon collider, while some new processes will profit from being studied both in an electron–electron and in a muon–muon system. In this presentation, I will show that both arguments apply for some of the most exciting physics spoils we might expect from the electron-electron mode of a projected NLC.

A LIKE-SIGN MUON COLLIDER

It has been shown [1,2] that, in the electron collider case, an e^-e^- option is a natural ingredient in a broad-based facility for experimentation: It poses no accelerator problems with the possible exception of beam-beam disruption at the interaction point. It has the added advantage that its beams can both be highly polarized - a feature not clearly realizable in the e^+e^- option - certainly not with trivial polarization reversibility. In addition, the e–γ and γ–γ options that are being planned at electron colliders, depend heavily on two highly polarized incoming lepton beams for both spectrum and polarization of the photons. What are the corresponding virtues and difficulties of installing a like-sign option ($\mu^+\mu^+$ or $\mu^-\mu^-$) in the muon collider?

For the argument's sake, I will take D. Neuffer's footprint for a muon collider facility [3] to point out the few essential features that have to be kept in mind for an overall evaluation of the feasibility, at affordable cost, of a like-sign option. Take, e.g., the variant where a muon injector linac feeds the phase-space-compressed muons into a rapid-cycling synchrotron as the principal accelerator: here, like-sign operation would simplify everything except the final collider ring, in which a second beam pipe and guidance system have to be installed. This is not cheap, but presents no technical problem.

We see the main technical challenge in the physics need for highly polarized muon beams—just like the high degree of polarization today available in electron linacs: Muons originate in weak decay with a well-defined helicity—but in the obvious attempt to maximize the muon flux, forward- and backward-emitted muons will be mixed in the cooling and compression process: downstream, the largest fluxes will be a mixture of two helicities. It should not take a great deal of ingenuity, however, to narrow the acceptance of the cooling devices such as to admit only one helicity state, at the expense of a flux loss well below a factor of two. This is a development project that imposes itself, as we will see below from the stress on the optimal definition of helicity states for cogent physics studies of, above all, Beyond-the-Standard-Model processes. It will handsomely pay off in a fuller definition of the chiral couplings that permit or enhance the processes we are trying to study, and thereby have a decisive influence on signal definition and background suppression. These aspects have been fully investigated in the framework of electron–electron scattering, where the additional feature of easy reversal or switching off of a high degree of polarization permits a convincing check on promising but statistically unconvincing signals. Like-sign muon colliders have the advantage of higher energy reach, but will have to strive hard for the added thrust of polarization definition.

In the following, we pass review of a few processes that are sure to benefit greatly from the energy reach of the muon collider as well as from the "exotic" charge, weak isospin, and lepton flavor and number of the incoming channel. While largely based on work performed in more quantitative fashion for the electron-electron version of an NLC, detailed calculations and modelling are straightforward.

STRONG GAUGE BOSON INTERACTIONS

Should the Standard Model go unchallenged by new-physics signals up to the TeV level without producing evidence for an elementary Higgs boson, we expect a new regime of the strong interaction to manifest itself [4]. The attendant new term in the effective Lagrangian, frequently denoted L_5, breaks local gauge symmetry, and becomes observable as the longitudinal components of the W and Z bosons act as Goldstone scalars. An experimental investigation of their interactions then becomes as fundamentally important as that of the pions in the 100 MeV region, and has to be pursued in all J and I channels. e^+e^- annihilation gives evidence only in the J=1, I=0,1 channels; gamma gamma scattering adds data on the J=0,2 channels. The potentially distinctive I=2 channel is accessible through quark–antiquark or gluon fusion in hadron colliders, but with massive attendant backgrounds from heavy-quark decays. While strong WW scattering may not produce easily identifiable signals at e^+e^- colliders, like-sign ee or $\mu\mu$ collisions could unveil a new regime of dynamics that manifests itself, *e.g.*, by the formation of vector ρ-type resonances [5]. Figure 1 shows the basic diagram, where only unobserved escaping neutrinos limit the full reconstruction of the final state. While e^-e^- colliders have the distinctive feature that we can switch polarization parameters essentially at will, like-sign muon colliders with their superior energy reach may well be needed for a full exploration of these new phenomena.

In addition to these dynamical ideas, an extended Higgs sector favors among possible models one version with fundamental doublets and triplets [6] that can lead to easily recognized sharp structure in the s-channel mass plot of final-state W-pair distributions that are emitted isotropically in their center-

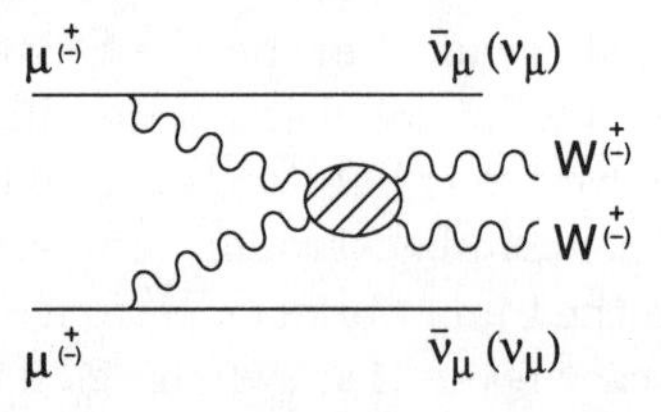

FIGURE 1. Strong WW scattering has a clean signature in $\overset{+}{\mu}{}^{(-)}\overset{+}{\mu}{}^{(-)}$ scattering: only the missing transverse momentum of the escaping neutrinos is absent in the WW final-state observation.

of-mass. This evidence, due to the process

$$\begin{aligned}\overset{+}{\mu}{}^{(-)}\overset{+}{\mu}{}^{(-)} &\rightarrow \overset{(-)}{\nu}_\mu \overset{(-)}{\nu}_\mu \overset{+}{W}{}^{(-)}\overset{+}{W}{}^{(-)} \\ &\rightarrow \overset{(-)}{\nu}_\mu \overset{(-)}{\nu}_\mu \overset{--}{H}{}^{(++)}\end{aligned} \tag{1}$$

where the WW fusion process leads to $\overset{++}{H}{}^{(--)}$ formation with a full unit of R in cross-section (Fig. 2b), could be quite spectacular, as shown in Fig. 3. What increases the interest in performing this search at a muon collider is the possibility that a direct coupling of two like-sign leptons to the $H^{++(--)}$, shown in Fig. 2a, of a theoretically unpredictable strength, might well differ between ee and $\mu\mu$ collisions. This feature adds spice to the search in the $\mu\mu$ case, even if there were prior evidence for such a discovery in e^-e^- experimentation.

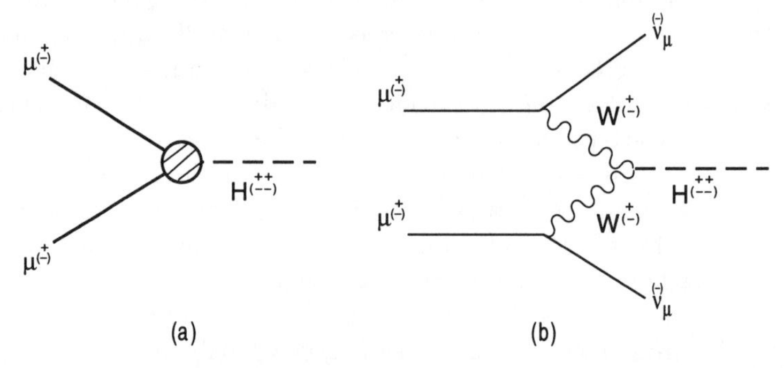

FIGURE 2. For the production of a doubly charged Higgs boson, the $\overset{+}{W}{}^{(-)}\overset{+}{W}{}^{(-)}$ fusion (graph b) will yield a full unit of $R[(=\sigma(e^+e^- \rightarrow \mu^+\mu^-)]$ in cross section. The unknown coupling $\mu\mu H$ in graph (a) may or may not differ from a direct eeH coupling, if it is not hopelessly suppressed.

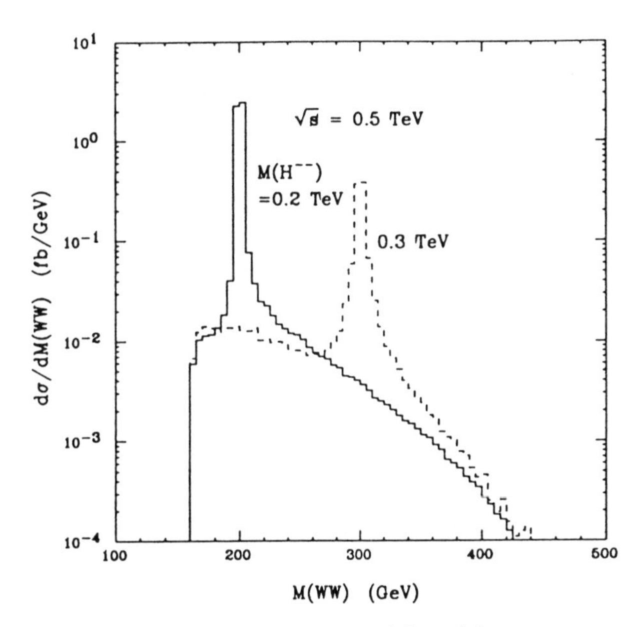

FIGURE 3. Invariant mass distribution for $W^{+(-)}W^{+(-)}$ scattering in the presence of a doubly charged Higgs meson of mass 0.2 or 0.3 TeV (from Ref. [6]).

NEW CONTACT INTERACTIONS

When we enter into a new energy regime of a particle interaction, one question that imposes itself is: Does a simple interaction, like muon-muon Møller scattering, show any signs that the simple QED Lagrangian has to add a new term characterized by the exchange of a heavy gauge boson with a mass well above what would have manifested itself before? Buchmüller and Wyler [7] showed that such a new interaction, with $\Lambda \gg \sqrt{s}$, can be accommodated by a minimal contact term in the Lagrangian

$$\begin{aligned}\mathcal{L}_{\text{eff}} \sim & \frac{g^2}{\Lambda^2}\eta_{LL}\left(\overline{\psi}^1_L\gamma_\mu\psi^1_L\right)\left(\overline{\psi}^2_L\gamma^\mu\psi^2_L\right) + \eta_{RR}\left(\overline{\psi}^1_R\gamma_\mu\psi^1_R\right)\left(\overline{\psi}^2_R\gamma^\mu\psi^2_R\right) \\ & + \eta_{RL}\left(\overline{\psi}^1_R\gamma_\mu\psi^1_R\right)\left(\overline{\psi}^2_L\gamma^\mu\psi^2_L\right) + \eta_{LR}\left(\overline{\psi}^1_L\gamma_\mu\psi^1_L\right)\left(\overline{\psi}^2_R\gamma^\mu\psi^2_R\right),\end{aligned} \tag{2}$$

with ψ_R, ψ_L the standard chirality projections of the electron spinors, $\psi_L = \frac{1}{2}(1-\gamma_5)\psi$, $\psi_R = \frac{1}{2}(1+\gamma_5)\psi$. If we want to test the point-like character of the muon, we can study the appropriate sensitivity of the Møller scattering cross section to a compositeness scale Λ as demonstrated in Fig. 4 [8]. Note that interference of crossing terms makes the Møller cross section more sensitive than the "Bhabha"-type $\mu^+\mu^-$ case, even in the absence of polarization

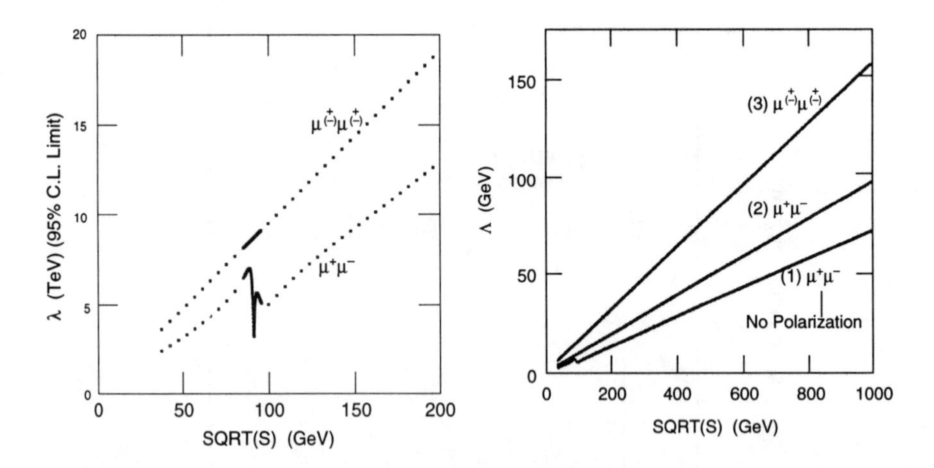

FIGURE 4. a) The sensitivity of Møller vs. Bhabha scattering below and above the Z resonance, in the absence of polarization of the incoming beams, to a compositeness scale Λ [TeV]. b) The corresponding sensitivity comparison in the $\mu^+\mu^-$ case where none (curve 1) or one beam (curve 2) are polarized, and like-sign scattering with both beams polarized (curve 3). The luminosity is chosen such that the statistics at each energy corresponds to the integrated values reached at PETRA for the compositeness limits set there.

(Fig. 4a). At higher energy, and for high degrees of polarizations, the like-sign scattering sensitivity is able to reach impressive new-interaction energy scales for luminosities that are scaled to those reached in the PETRA investigations of present compositeness limits.

Analogous investigations can be performed on the influence of other massive new boson exchanges, like heavy Z' states [9].

HEAVY MAJORANA NEUTRINOS

It has long been pointed out [10–13] that like-sign electron scattering can produce quasi-elastic W pairs, either real or virtual, via Majorana neutrino exchange. While this lepton-number and -flavor violating process (Fig. 5) has been treated mainly in the Left–Right-Symmetric Model, Heusch and Minkowski recently showed [14] that a minimal neutrino mass generating case can be formulated that permits sizeable cross-sections for left-handed WW production in quasi-elastic $e^-e^- \to W^-W^-$ reactions according to Fig. 5: the couplings of the incoming leptons to the exchanged heavy neutrino(s) contain specific mass mixing matrices: clearly, there is a chance that the case where incoming muons have to couple to the exchanged TeV-level Majorana neutrino

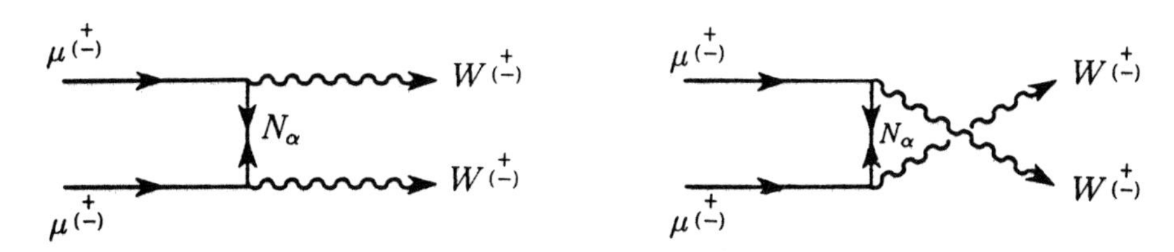

FIGURE 5. Lowest-order graphs for the process $\mu^{\overset{+}{(-)}}\mu^{\overset{+}{(-)}} \rightarrow W^{\overset{+}{(-)}}W^{\overset{+}{(-)}}$, mediated by the exchange of a heavy Majorana neutrino.

may be dominated by matrix elements $\mathsf{U}_{\mu\alpha}$ different from those prevailing in the e^-e^- case.

While this fact in itself is of interest in terms of a novel discovery chance, we stress that the cross section for the process

$$\mu^{\overset{+}{(-)}}\mu^{\overset{+}{(-)}} \xrightarrow{N_\alpha} W^{\overset{+}{(-)}}W^{\overset{+}{(-)}} \tag{3}$$

has a hard-scattering term proportional to s^2,

$$\sigma^{(N)} = \frac{G_F}{16\pi}\frac{s^2}{m_{red}^2}|\eta_N|^2,$$

$$\text{with } \frac{1}{m_{red}} = \sum_\alpha \frac{1}{m_\alpha} \quad (\alpha \text{ is the high} - \text{mass neutrino index}) \tag{4}$$

$$\text{and } \eta_N = \sum_\alpha (\mathsf{U}_{\mu\alpha})^2 \frac{m_{red}}{m_\alpha}.$$

As long as $\sqrt{s} < m(N)$, with N the exchanged heavy neutrino, this cross section will enormously profit from the increased energy range provided by the muon collider when compared with its electron-accelerating analog.

We have not yet explored the details of the change in the theoretical treatment of the active mass matrix elements for the present case. But we emphasize that while electron-initiated reactions (2) are, at least in principle, also present in neutrinoless double beta decay [15], no such equivalence exists in the muon-muon collision case. We therefore believe it to be a valuable discovery aim to look for the potentially spectacular and unmistakable final-state configurations (back-to-back W^- pairs; unlike-flavor, almost back-to-back-emitted negative leptons with missing momenta from escaping light neutrinos, etc. in the otherwise crowded final states of violent muon-muon collisions.

SUPERSYMMETRY

We now turn to the framework of the minimal supersymmetric Standard Model [16]. This implies making use of two *essential* assumptions: (i) *R*-

parity invariance;[1] (ii) the lightest supersymmetric particle is a neutralino, $\tilde{\chi}_1^0$, it is *stable* and interacts *weakly* with matter, *i.e.*, it escapes detection. If these conditions are not fulfilled, the results of the analysis to follow are qualitatively incorrect.

The pair-production of like-sign smuons in $\mu^-\mu^-$ collisions

$$\mu^-\mu^- \to \tilde{\mu}^-\tilde{\mu}^- \tag{5}$$

and their decays into muons and invisible particles

$$\tilde{\mu}^- \to \mu^- \ \tilde{\chi}_1^0 \tag{6}$$

$$\tilde{\mu}^- \to \mu^- \ \tilde{\chi}_2^0 \hookrightarrow \tilde{\chi}_1^0 \ Z^0 \hookrightarrow \nu\bar{\nu} \tag{7}$$

$$\tilde{\mu}^- \to \nu \ \tilde{\chi}_1^- \hookrightarrow \tilde{\chi}_1^0 \ W^- \hookrightarrow \bar{\nu}\mu^- \tag{8}$$

$$\vdots$$

are depicted in the Feynman diagram of Fig. 6. They proceed exactly in the same way as for selectrons in e^-e^- collisions. The whole analysis of Refs. [17,18] can thus be applied here, with higher energies and without polarization.

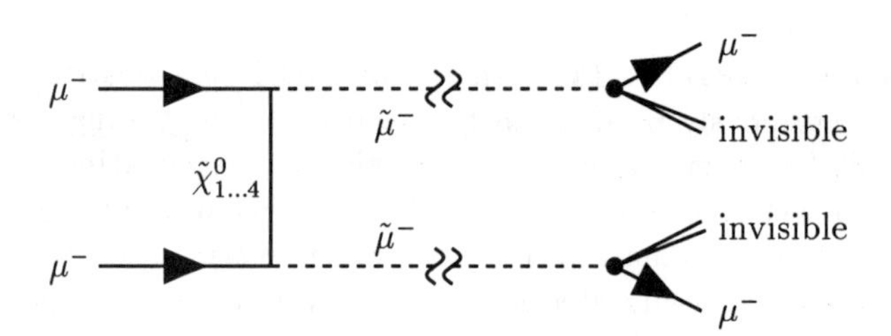

FIGURE 6. Feynman diagram for the production and decay of smuons.

In Fig. 7 we show the energy behavior of the smuon pair production cross section [17] times the branching fractions of these smuons into muons and invisible particles [18] for different smuon masses. For definiteness we have chosen the other supersymmetry parameters to take some "typical" values

$$\tan\beta = 4 \qquad \mu = 500 \text{ GeV} \qquad M_2 = 500 \text{ GeV} \tag{9}$$

[1] If this *ad hoc* symmetry is broken, the supersymmetric phenomenology becomes actually much easier because lepton number violating processes should then show up.

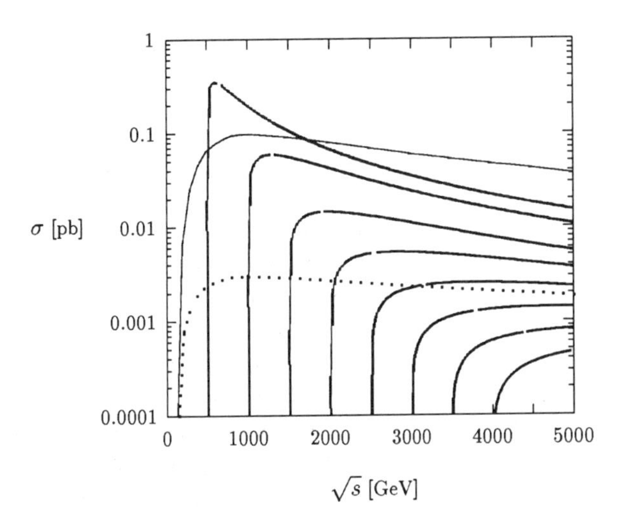

FIGURE 7. Smuon pair-production and decay cross section as a function of the collider energy for smuon masses ranging from 250 GeV to 2000 GeV. The Standard Model background is shown by the thinner line. The dotted line represents the signal cross section needed to exceed three times the Poisson error of the background with 100 fb^{-1} of data.

and have assumed the soft supersymmetry breaking terms to have a common value at the grand unified scale. The corresponding mass of the lightest neutralino is 244 GeV; it is predominantly a mixture of photino and zino. The other neutralinos and charginos have masses around 500 GeV. The asymptotic behavior of the cross section is

$$\sigma \sim \frac{\pi\alpha^2}{2s} \ln \frac{s}{\Lambda^2_{\mathrm{SUSY}}} \, . \tag{10}$$

The expected Standard Model background, mainly from W^- bremsstrahlung, is also shown in Fig. 7. To evaluate this background, we have imposed the mild detector acceptance cuts $|\eta_\mu| < 2$ and $E_\mu > \sqrt{s}/10$ on the emerging muons. Close to threshold, *i.e.*, for maximum signal, these cuts do not affect the smuon cross sections.

Operating the collider at 1 TeV, smuons up to almost 500 GeV can be produced with a signal-to-noise ratio of one. Insisting that the supersymmetric signal exceeds the Standard Model background by at least three standard deviations, we find, with a luminosity of 100 fb^{-1} and a collider energy of 3 TeV, that smuons up to 1250 GeV can be discovered just by comparing total rates. For heavier smuons, a more subtle analysis will be required. It must be noted, however, that such heavy smuons are not favored at all by low energy supersymmetry.

ANOMALOUS QUARTIC GAUGE COUPLINGS

The non-abelian part of the electroweak sector remains up to now a little-explored area of the Standard Model. Although existing LEP I data are already testing the trilinear Yang-Mills couplings through loops [19], no effects of the quartic gauge interactions have yet been observed.

New physics at the TeV scale is bound to induce some anomalies in the gauge sector. The exact form of these deviations from the Standard Model expectations depends of course on the precise nature of unknown phenomena, and it is customary to parametrize our ignorance in terms of effective lagrangians. There are infinitely many operators which can induce non-standard vector boson couplings. However, those which have the lowest dimension and which do not induce any trilinear couplings are expected to yield the largest contributions, because they can result from tree-level exchanges of heavy particles. There are only two such so-called "genuine" quartic operators of dimension four. They modify the W^4 and W^2Z^2 vertices and induce a novel Z^4 vertex, by adding the following pieces to the Standard Model lagrangian [20]:

$$\begin{aligned}\mathcal{L}_0 &= g_0 g_W^2 \Bigg(W^{+\mu} W^-_\mu W^{+\nu} W^-_\nu \\ &\quad + \frac{1}{\cos^2\theta_w} W^{+\mu} W^-_\mu Z^\nu Z_\nu + \frac{1}{4\cos^4\theta_w} Z^\mu Z_\mu Z^\nu Z_\nu \Bigg) , \\ \mathcal{L}_c &= g_c g_W^2 \Bigg[\frac{1}{2} (W^{+\mu} W^-_\mu W^{+\nu} W^-_\nu + W^{+\mu} W^+_\mu W^{-\nu} W^-_\nu) \\ &\quad + \frac{1}{\cos^2\theta_w} Z^\mu W^+_\mu Z^\nu W^-_\nu + \frac{1}{4\cos^4\theta_w} Z^\mu Z_\mu Z^\nu Z_\nu \Bigg] ,\end{aligned} \tag{11}$$

where θ_w is the weak mixing angle, g_W is the usual W coupling and g_0 and g_c parametrize the strength of the anomalies. Note that no anomalous quartic operators of dimension four involve the photon.

As in e^-e^- scattering [21], such anomalous couplings would show up dramatically in high energy $\mu^-\mu^-$ reactions such as

$$\mu^-\mu^- \to \nu_\mu \nu_\mu W^- W^- , \tag{12}$$

$$\mu^-\mu^- \to \mu^-\mu^- W^+ W^- , \tag{13}$$

$$\mu^-\mu^- \to \mu^- \nu_\mu Z^0 W^- , \tag{14}$$

$$\mu^-\mu^- \to \mu^-\mu^- Z^0 Z^0 , \tag{15}$$

because they spoil the normally very effective unitarity cancelations between the diagrams involving the quartic vertices and the diagrams. Even tiny values

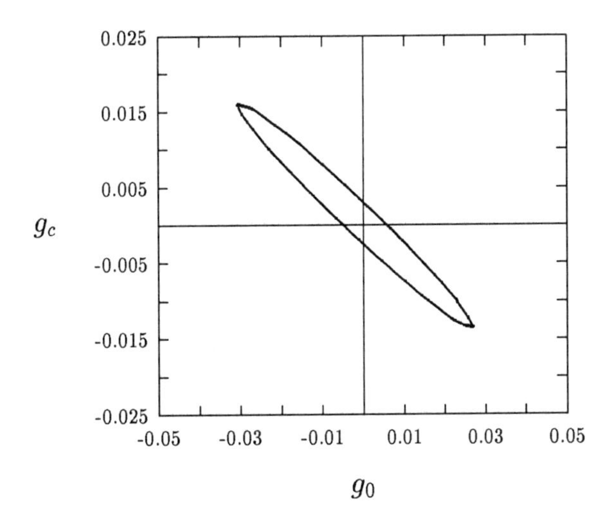

FIGURE 8. Contours of observability at 95% confidence level of the anomalous quartic gauge couplings g_0 and g_c, for the reaction (11) at 2 TeV and with 100 fb^{-1} of data.

of g_0 or g_c provoke significant increases of the total cross sections, the more so at higher energies.

Assuming 100 fb^{-1} of accumulated luminosity at a 2 TeV $\mu^-\mu^-$ facility, and concentrating solely on the hadronic decays of the W's, we show in Fig. 8 the area in the (g_0, g_c) plane beyond which these anomalies would be observed in reaction (11) with better than 95% confidence. The strong correlation can be lifted by also studying reactions (12-14) [21]. The anomalous parameters should then be constrainable down to a few tenths of a percent.

CONCLUSION

In the preceding brief discussion, we have shown how muon-muon scattering in the TeV region can be considerably enriched by making like-sign beams available as the input channel. The list we give is indicative rather than complete. It becomes evident that a major effort towards making these beams well-defined in their helicity states is easily motivated by the physics promise, and we urge an early feasibility study for implementation of a like-sign version, including a well-defined high degree of polarization.

ACKNOWLEDGMENTS

We thank David Cline for encouraging this contribution to the $\mu\mu$ Workshop, and the Workshop participants for many instructive discussions about muon colliders. It is a pleasure for F.C. to thank Karol Kołodziej and Geert Jan van Oldenborgh for many fruitful discussions and their collaboration on the topics of the last two sections.

REFERENCES

1. Heusch, C.A., *Nucl. Instrum. Methods* **A355**, 75 (1995).

2. Heusch, C.A., presented at the First Arctic Workshop on Future Physics and Accelerators, Saariselka, Finland, August 21–27, 1994, SCIPP 95/02, to be published.

3. Neuffer, D., *Nucl. Instrum. Methods* **A350**, 27 (1994).

4. Chanowitz, M., and Gaillard, M.K., *Nucl. Phys.* **B261**, 379 (1985)

5. T. Han, in *Proceedings of the 2nd International Workshop on Physics and Experiments with Linear e^+e^- Colliders*, Waikoloa, HI, 26-30 April 1993, F.A. Harris *et al.*, editors (World Scientific, Singapore, 1993), p. 270.

6. Barger, V., *et al.*, *Phys. Rev.* **D50**, 6704 (1994); Gunion, J., in *Proceedings of the 2nd International Workshop on Physics and Experiments with Linear e^+e^- Colliders*, Waikoloa, HI, 26-30 April 1993, F.A. Harris *et al.*, editors (World Scientific, Singapore, 1993).

7. Buchmüller, W. and Wyler, D. *Nucl. Phys.* **B268**, 621 (1986); Schrempp, B., et al., *Nucl. Phys.* **B296**, 1 (1988).

8. Barklow, T., private communication (1994), to be published.

9. Choudhury, D. *et al.*, *Phys. Lett.* **B333**, 531 (1994).

10. Rizzo, T. *Phys. Lett.* **116B**, 23 (1982); in *Proceedings of the 2nd International Workshop on Physics and Experiments with Linear e^+e^- Colliders*, Waikoloa, HI, 26-30 April 1993, F.A. Harris *et al.*, editors (World Scientific, Singapore, 1993), p. ?.

11. Maalampi, J., Pietilä, A., and Vuori, J., *Phys. Lett.* **B297**, 327 (1992); *Nucl. Phys.* **B381**, 544 (92).

12. London, D., Belanger, G., and Ng, J.N., *Phys. Lett.* **B188**, 155 (1987).

13. Gluza, J., and Zralek, M., SU-TP-1/95 (to be published).

14. Heusch, C.A., and Minkowski, P. *Nucl. Phys.* **B416**, 3 (1994).

15. Heusch, C.A., Nucl. Phys. (Proc. Suppl.) **B38**, 313 (1995); Heusch, C.A., and Minkowski, P., SCIPP 95/07 or BUTP 95/11, to be published.

16. H.P. Nilles, *Phys. Rep.* **110**, 1 (1984); H.E. Haber and G.L. Kane, *Phys. Rep.* **117**, 75 (1985).

17. W.-Y. Keung and L. Littenberg, *Phys. Rev.* **D28**, 1067 (1983); F. Cuypers, G.J. van Oldenborgh and R. Rückl, *Nucl. Phys.* **B 409**, 128 (1992).

18. F. Cuypers, *Phys. At. Nucl.* **56**, 1515 (1993) [*Yad. Fis.* **56**, 23 (1993)].

19. S. Dittmaier, D. Schildknecht, K. Kołodziej, M. Kuroda, *Nucl. Phys.* **B426**, 249 (1994).

20. G. Bélanger, F. Boudjema, *Phys. Lett.* **B288**, 210 (1992).

21. F. Cuypers and K. Kołodziej, *Phys. Lett.* **B344**, 365 (1995).

Overview of $\mu^+\mu^-$ Collider Options

D. B. Cline

Center for Advanced Accelerators, Physics Department, University of California Los Angeles, Los Angeles, CA 90095-1547 USA

Abstract. Two dedicated workshops have been held in California on $\mu^+\mu^-$ colliders: Napa in 1992 and Sausalito in 1994. As a result of this and a new round of simulations, as well as $\mu^\pm$ cooling experimental plans, it is possible to consider the feasibility of constructing a $\mu^+\mu^-$ collider. However, the energy range and required luminosity must be specified by the particle physics goals before proceeding. Also, the unique features of the physics potential vs LHC must be shown. We will describe the status of this field and the key problems that must be solved before a definite proposal can be made.

INTRODUCTION

At Port Jefferson Advanced Accelerator Workshop in the Summer of 1992, a group investigating new concepts of colliders studied anew the possibility of a $\mu^+\mu^-$ collider since e^+e^- colliders will be very difficult, in the several TeV range[1]. A small group also discussed the possibility of a $\mu^+\mu^-$ collider[2]. A special workshop was then held in Napa, California, in the fall of 1992, for this study. There are new accelerator possibilities for the development of such a machine, possibly at an existing or soon to exist storage ring[3]. For the purpose of the discussion here, a $\mu^+\mu^-$ collider is schematically shown in Figure 1. In this brief note we study one of the most interesting goals of a $\mu^+\mu^-$ collider: the discovery of a Higgs Boson in the mass range beyond that to be covered by LEP I & II (~80-90GeV) and the natural range of the Super Colliders $\geq 2M_Z$ [4,5,6]. In this mass range, as far as we know, the dominant decay mode of the h^0 will be

$$h^0 \rightarrow b\bar{b} \tag{1}$$

whereas the Higgs will be produced by the direct channel

$$\mu^+\mu^- \rightarrow h^0 \tag{2}$$

which has a cross section enhanced by the ratio

$$\left[\frac{M_\mu}{M_e}\right]^2 \sim (200)^2 = 4\times 10^4 \qquad (3)$$

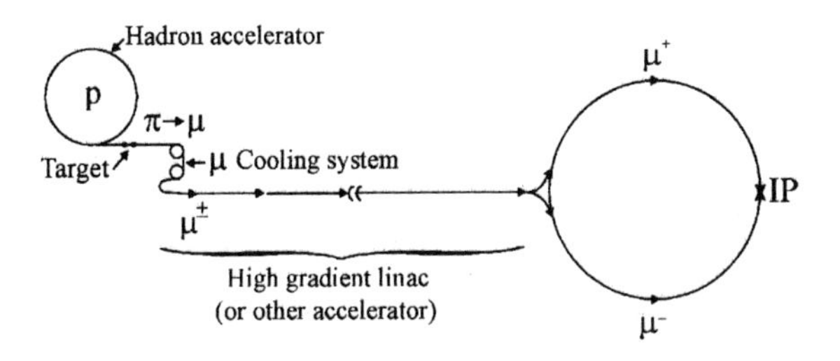

FIGURE 1. Schematic of a possible $\mu^+\mu^-$ collider scheme - few hundred GeV to few TeV.

much larger than the corresponding direct product at an $e^+ e^-$ collider. However, the narrow width of the Higgs partially reduces this enhancement. Recent results suggest that the low mass Higgs is preferred[5](Figure 2a).

In the low mass region the Higgs is also expected to be a fairly narrow resonance and thus the signal should stand out clearly from the background from[3]

$$\begin{aligned} \mu^+\mu^- &\rightarrow \gamma \rightarrow bb \\ &\rightarrow Z_{tail} \rightarrow bb \end{aligned} \qquad (4)$$

If the resolution requirements can be met, the machine luminosity of $\sim 10^{32}$ $cm^{-2}sec^{-1}$ could be adequate to facilitate the discovery of the Higgs in the mass range of 100-180GeV.

Finally, another possibility is to use the polarization of the $\mu^+\mu^-$ particles orientated so that only scalar interactions are possible. However, there would be a trade-off with luminosity and thus a strategy would have to be devised to maximize the possibility of success in the energy sweep through the resonance (see Figure 2b for other physics issues)

RESULTS FROM THE NAPA AND SAUSALITO $\mu^+\mu^-$ COLLIDER WORKSHOP

At the Napa meeting (1992) a small group of excellent accelerator physicists struggled with the major concepts of a $\mu^+\mu^-$ collider; some results are published in NIM, Oct. 1994[3]. At the Sausalito (1994) meeting a larger group of accelerator, particle and detector physicists were involved. The proceedings will be published by AIP press in 1995.

The major issue confronting this collider development is the possible luminosity that is achievable. Two collider energies were considered: 200x200GeV and 2x2TeV. The major particle physics goals are the detection of the Higgs Boson(s) in the *s* channel for the low energy collider and

TABLE 1. Parameter list for 400GeV μ+μ- collider, Sausalito, 1994, 200x200GeV working group report.

Parameter	Symbol	Value
Collision energy	E_cm	400GeV
Energy/Beam	E_μ	200GeV
Luminosity	$L = \frac{f_n n_s n_b N_\mu^2}{4\pi \sigma^2}$	$1x10^{31}cm^{-2}sec^{-1}$
	Source Parameters	
Proton Energy	E_p	30GeV
Protons/Pulse	$N - p$	$2x3x10^{13}$
Pulse Rate	$f - o$	10Hz
μ (prod./accept.)	μ/p	0.03
μ survival	N_μ/N_{source}	0.33
	Collider Parameters	
Number of μ/bunch	$N_{\mu\pm}$	$3x10^{11}$
Number of bunches	n_B	1
Storage turns	$2n_s$	1500 (B=5T)
Normalized emittance	ε_N	10^{-4} m-rad
μ-beam emittance	$\varepsilon_t = \varepsilon_N / \gamma$	$5x10^{-8}$ m-rad
Interaction focus	β_0	1cm
Beam size at interaction	$\sigma = (\varepsilon_t \beta_0)^{1/2}$	2.μm

WW scattering as well as supersymmetric particle discovery[3].

The workshop goal was to see if a luminosity of 10^{32} to $10^{34}cm^{-2}sec^{-1}$ for the two colliders might be achievable and useable by a detector. There were five working groups on the topics of (1) Physics, (2) 200x200GeV Collider, (3)

2x2TeV Collider, (4) Detector Design and Backgrounds, and (5) μ Cooling and production methods.

Table 1 gives the parameters and luminosity for the low energy collider. Table 2 gives the somewhat more optimistic parameter list for a 2x2TeV collider from the work of Neuffer and Palmer[7]!

The $\mu^+\mu^-$ collider has a powerful physics reach, especially if the $\mu^\pm$ polarization can be maintained. One interesting possibility is the observation of the super symmetric Higgs Boson(s) in the direct channel (see the report by V. Barger et al., from Sausalito meeting). The detector backgrounds will be considerable due to high energy μ decays upstream of the detector.

TABLE 2. Parameter list for 4TeV μ+μ- Colliders (Neuffer and Palmer)[7].

Parameter	Symbol	Value
Collision energy	$E_c m$	2TeV
Energy per beam	E_μ	200GeV
Luminosity	$L = f_n n_s n_b N_\mu^2 / 4\pi\sigma^2$	10^{34}cm^{-2}sec^{-1}
	Source Parameters	
Proton energy	E_p	30GeV
Protons/pulse	N_p	2x3x10^{13}
Pulse rate	f_0	10Hz
μ (prod./accept.)	μ/p	15
μ survival	N_μ/N_{source}	25
	Collider Parameters	
Number of μ/bunch	$N_{\mu\pm}$	10^{12}
Number of bunches	n_B	1
Storage turns	$2n_s$	500
Normalized emittance	ε_N	3x10^{-5} m-rad
μ-beam emittance	$\varepsilon_t = \varepsilon_N/\gamma$	1.5x10^{-9} m-rad
Interaction focus	β_0	0.3cm
Beam size at interaction	$\sigma = (\varepsilon_t\beta_0)^{1/2}$	2.1μm

In the summary of working group 4 it was concluded that these backgrounds might be manageable. One key to achieving a high luminosity collider is the collection of the $\mu^\pm$ from $\pi^\pm$ decays over nearly the full phase space over which they are produced. This is far from trivial and leads to conclusions from groups 2 and 3 that the present uncertainty in luminosity is of order $10^{2\pm1}$ (at most 4 orders of magnitude, but perhaps, realistically, 2 orders of magnitude, which, unfortunately, spans the range from being uninteresting to being very interesting). Hopefully, before the next meeting this uncertainty can be reduced. Perhaps the most interesting aspect of a $\mu^+\mu^-$ collider is the need to cool the $\mu^\pm$ beams over a very large dynamics range. Three experimental programs were discussed and are being initiated to study μ cooling: at BNL, FNAL, and a UCLA group is proposing to study cooling and acceleration in crystals at TRIUMF[8]. One major conclusion of the meeting is that $\mu^+\mu^-$ colliders are complimentary to both *pp* (LHC)[6] and e^+e^- (NLC) colliders, especially for the Higgs Sector and for the study of supersymmetric particles.

THE POSSIBLE LUMINOSITY OF A $\mu^+\mu^-$ COLLIDER

The luminosity is given by

$$L = M \frac{N_{\mu^+} N_{\mu^-} f}{4\pi \, \varepsilon_N \, \beta^*} \gamma \tag{5}$$

The $N_{\mu\pm}$ depend directly on the $\mu^\pm$ production and capture rate (μ/p), f is related to the magnetic field of the collider, ε_N the final μ invariant emittance from the final stage of cooling, and β^* will depend on the bunch length (and the longitudinal cooling of the $\mu^\pm$ beams), as well as the collider lattice.

We can rewrite the luminosity as

$$L \propto \frac{(\mu / p)^2 B_{\text{collider}} \gamma}{\varepsilon_N \, (\text{final } \beta^*)} \tag{6}$$

In order to increase L we must increase (μ/p) and B and decrease ε_N and β^*. At the Napa meeting the best judgment of the group was that $(\mu/p) \sim 10^{-3}$ and $\varepsilon_N \sim 1 \times 10^{-5}$ πm-rad, $\beta^* \sim 1$cm. For the case of $\gamma = 200$, $L = 2\text{x}10^{30}\text{cm}^{-2}\text{sec}^{-1}$. If, on the other hand, we use the optimistic values of $(\mu/p) = 0.2$[7], $\beta^* = 1/3$cm, and $\varepsilon_N = 3\text{x}10^{-5}$m-rad, we find

$$\begin{aligned} L &= \left(2 \times \left(4 \times 10^4\right) \times 3\right) L_0 \\ &= 4.8 \times 10^{35}\,\text{cm}^{-2}\text{sec}^{-1} \end{aligned} \tag{7}$$

a very large luminosity, never before achieved by any collider! Clearly this must be far too optimistic. It is clear that the (μ/p) ratio is the key parameter of the machine. Table 1 and 2 give some parameters of low energy and high energy colliders.

a)

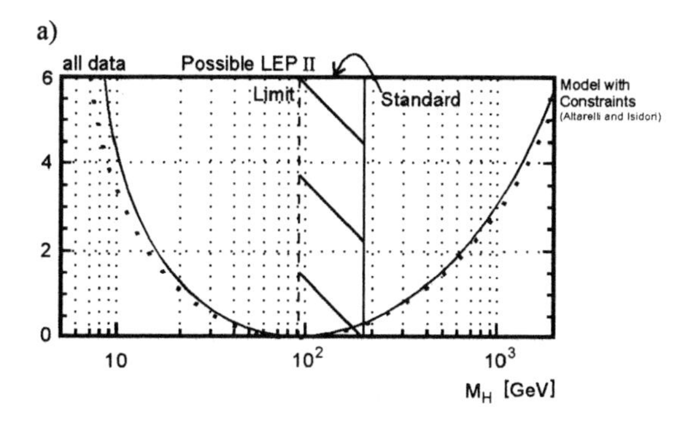

b)

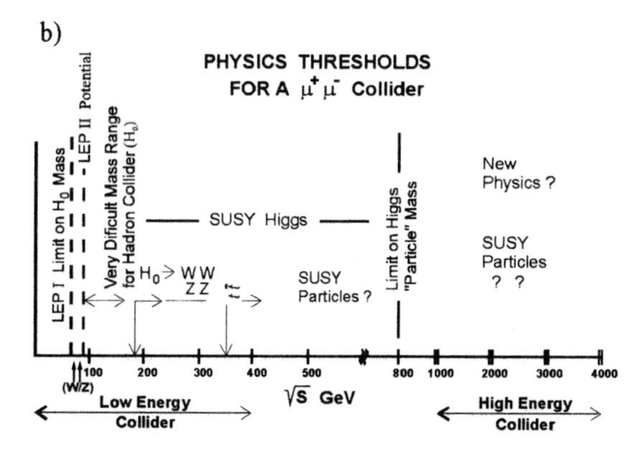

FIGURE 2. a) Upper and lower bound s on m_{ϕ_0} as a function of m_t, coming from the requirement of a perturbative theory[4]. b) Physics threshold for a μμ collider.

MUON COOLING EXPERIMENT AT TRIUMF

This phase, the first phase of the experiment TRIUMF, will test the cooling mechanisms summarized in the proposal to TRIUMF. The beam momentum will be about 250MeV/c unchanneled muons will penetrate the 4cm crystal and the cooling process can be compared for the two. In addition, a higher energy beam tests cooling at the energies considered for realistic collider schemes. The first step of Phase II is to measure initial and final emittances of an unmodified crystal.

To enhance the cooling, we will generate a strain modulation of the planar channels. An acoustic wave of 1GHz is excited via a piezoelectric transducer. We will also detect predicted channeling radiation by surrounding the crystal with CsI scintillation detectors, which are sensitive to X-rays. Recent ideas on crystals and beams are very interesting.[9,10]

The M11 beamline is presently a source of high energy pions[11,12]. Straightforward modification of the beamline will provide a collimated beam of forward-decay muons at high intensity – about 10^6 per second at 250MeV/c. The longitudinal momentum spread is about 2% FWHM. Assuming optimum tuning of the final focus quadrupole doublet in M11, we can achieve a spot size of 3x2cm with horizontal and vertical divergences of 10m-rad and 16m-rad respectively. The critical angle for planar channeling of μ^+ at 250MeV/c in silicon is about 7m-rad, extrapolating from proton channeling data. A sizable fraction of the muons should channel through a few centimeters of the crystal.

TABLE 3.

	For a $\mu^+\mu^-$ Collider Development	Example in Past $p\bar{p}$ Collider (1976-1987)[14] (Cline,McIntyre, Rubbia proposal)
1)	Strong Physics Motivation Higgs, SUSY, etc. etc. (Higgs mass unknown - but it may be at low mass)	W/Z Discovery (M_W, Z known)
2)	Parameters Study Are they realistic? How can we make a convincing argument	FNAL/CERN Studies (1976-1981)
3)	Beam Manipulation and Cooling Rapid Acceleration Possibility? (μ lifetime constraint)	AA Ring and Beams ($\bar{p}$ production yield)
4)	Demonstration of $\mu^\pm$ Cooling (Experiments) (New Ideas)	$p/\bar{p}$ Cooling ICE Ring Novosibirsk, FNAL (1976-1981)
5)	Detector Concepts and Feasibility Study	UA1/UA2 CDF/D0 Designs 1977-1987

HOW TO GET A $\mu^+\mu^-$ COLLIDER STARTED

There are many problems and also possibilities to start a $\mu^+\mu^-$ collider in the USA. For example, if crystal cooling could be used a collider of the type shown in

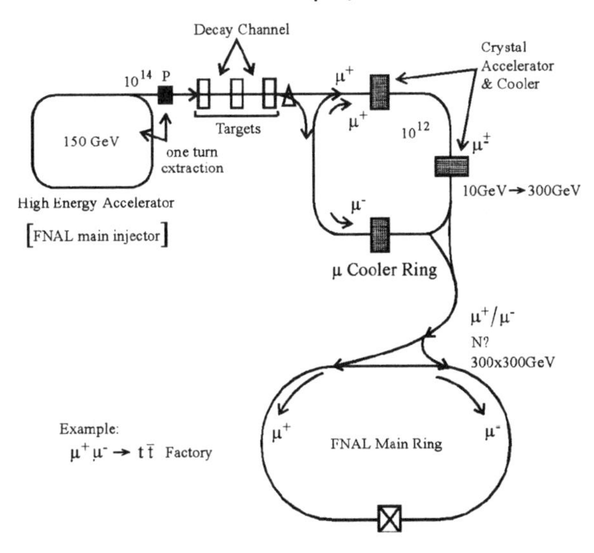

FIGURE 3. A scheme for a $\mu^+\mu^-$ collider using crystal cooling.

Figure 3 might be constructed[13]. We list the major issues in the development of a collider in Table 3. In the past the only example of such an innovative machine is the $p\bar{p}$ collider initiated by Cline, McIntyre and Rubbia in 1976[14]. In Table 3 we attempt to make a comparison between these two projects! In my opinion the key problem is comparison with the NLC and LHC.

I wish to thank W. Barletta, D. Neuffer, A. Sessler, R. Palmer, F. Mills, and the other participants of the Napa and Sausalito workshops for may interesting discussions. Thanks also to A. Bogacz and P. Sandler for help with this paper.

REFERENCES

1. M. Tigner, presented at the 3rd Int. Workshop on Adv. Accel. Concepts, Port Jefferson, NY (1992); other paper to be published in AIP, (1992), ed. J. Wurtle.
2. Early references for μμ colliders: E.A. Perevedentsev and A.N. Skrinksy, Proc. 12th Int. Conf. on H.E. Accel., eds. R.T. Cole and R. Donaldson, (1983), p. 481; D. Neuffer, Part. Accel. 14 (1984) 75; D. Neuffer in Adv. Accel. Concepts, AIP Conf. Proc. 156 (1987) 201.
3. D. Cline, NIM, A350, 24 (1994) and 4 following papers constitute a mini-conf. proc. of the Napa meeting.
4. See for primary references to the theoretical estimates here, S. Dawson, J.F. Gunion, H.E. Haber and G.L. Kane, The Physics of the Higgs Bosons: Higgs Hunter's Guide (Addison Wesley, Menlo Park, 1989).
5. Most recently limits on Higgs mas were reported in a talk by D. Schaile at the WW meeting at UCLA, Feb. 1995.
6. For example, the Compact Muon Solenoid Detector at the LHC, CERN reports and proposal.
7. Presentation of D. Neuffer and R. Palmer at the Sausalito workshop (to be published in the proc. of the conf.).
8. Proposal to TRIUMF for μ Cooling Experiment, A. Bogacz, et al., unpublished.
9. Z. Hong, P. Chen and R.D. Ruth, Radiation Reaction in a Continuous Focusing Channel, SLAC prep. SLAC-PUB-6574, July 1994.
10. S. Basu and R.L. Byer, Optic Letts., 13, 458 (1988).
11. Z.A. Bazylev and N.K. Zhevago, Sov. Phys. Usp., 33, 1021 (1990).
12. C.J. Oram, J.B. Warren, G.M. Marshall and J. Dornobos, J. Nucl. Instr. & Meth. in Phys. Res., A179, 95 (1981).
13. D. B. Cline, A Crystal μ Cooler for a $\mu^+\mu^-$ Collider, in Proc. of the 1994 Adv. Accel. Workshop, Lake Geneva, 1994.
14. C. Rubbia, P. McIntyre and D. Cline, Proc. of 1976 Aachen Neutr. Conf., p. 683 (eds. H. Faissner, H. Reithler and P. Zerwas) (Viewag Braunschweig, 1977).

Attendees List

Atac, Muzaffer: Fermi National Accelerator Laboratory, MS 223, P.O. Box 500, Batavia, IL 60510, USA
email: atac@fnal, phone: 708 840 3960

Bambade, Philip S.: LAL, Department of Physics, Bat. 200, F-91405 Orsay Cedex, France
email: bambade@frcpn11.in2p3.fr

Barger, Vernon D.: University of Wisconsin, Madison, Physics Department, 1150 University Avenue, Madison, WI 53706, USA
email: barger@phenxi.physics.wisc.edu or ldolan@phenxd.physics.wisc.edu, fax: (608) 262-8628, phone: (608) 262-8908

Barletta, William: Lawrence Berkeley Laboratory, Accelerator and Fusion Research Div., 1 Cyclotron Road, Berkeley, CA 94720, USA
email: barletta@uclahep, fax: (510) 486-6003, phone: (510) 486-5408

Berger, Michael S.: Indiana University, Department of Physics, Swain Hall West 117, Bloomington, IN 47405-4201, USA
email: berger@gluon.physics.indiana.edu, fax: (812) 855-5533

Bogacz, S. Alex: University of California Los Angeles, Physics Department, 405 Hilgard Avenue, Los Angeles, CA 90095-1547, USA
email: bogacz@calvin.fnal.gov, fax: 708 840 4552, phone: 708 840 3873

Bross, Alan: Fermi National Accelerator Laboratory, Department of Physics, P.O.Box 500, Batavia, IL. 60510, USA
email: bross%rdiv@fnal.gov, phone: (708) 840-2950

Chattopadhyay, Swapan: Lawrence Berkeley Lab., Accel. & Fusion Research Div., 1 Cyclotron Road, Berkeley, CA 94720, USA
email: chapon@lbl, phone: (510) 486-7217

Chen, Pisen: Stanford Linear Accelerator Center, Bin 26, P.O. Box 4349, Stanford, CA 94305, USA
email: chen@slacvm, phone: (415) 926-3384

Cline, David B.: University of California Los Angeles, Physics Department, 405 Hilgard Avenue, Los Angeles, CA 90024-1547, USA
email: laraneta@physics.ucla.edu, fax: (310) 825-2673, phone: (310) 206-1091

Diebold, Robert E.: United States Department of Energy, High Energy Physics Division, ER 221 GTN, Washington, DC 20545, USA
fax: (303) 903-5079, phone: (303) 903-5490

Errede, Debbie: University of Illinois, Loomis Laboratory of Physics, 1110 West Green Street, Urbana, IL 61801, USA
email: derrede@uihepa.hep.uiuc.edu

Fernow, Richard: Brookhaven National Laboratory, Dept. of Physics, Bldg. 901A, Upton, NY 11973-5000, USA
fax: (516) 282 3248

Foster, Bill: FNAL, M.S. 223, P.O. Box 500, Batavia, IL 60510.

Fujii, Keisuke: KEK, Physics Division, 1-1 Oho, Tsukuba, Ibaraki, Japan
email: fujiik@jlcux1.kek.jp, fax: 02 98 64 2580

Gallardo, Juan C.: Brookhaven National Laboratory, Department of Physics, 901-A, Upton, NY 11973, USA
email: jcg@bnlcl6.bnl.gov, fax: 516 282 3248, phone: 516 282 3523

Gunion, John F.: University of California Davis, Physics Department, High Energy Division, Davis, CA 95616, USA
email: bsmiv@ucdhep.ucdavis.edu, fax: (916) 752-4717, phone: (916) 752-1134

Han, Tao: University of California, Davis, Department of Physics, Geology Building, Davis, CA 95616, USA
fax: 916 752 4717, phone: 916 752 9855

Heusch, Clemens: University of California Santa Cruz, Physics Department, Institute for Particle Physics, Santa Cruz, CA 95064, USA
fax: 408 459 3043, phone: 408 459 2635

Hong, Woopyo: University of California Los Angeles, Department of Physics, 405 Hilgard Avenue, Los Angeles, CA 90024-1547, USA
email: hong@uclahep, fax: (310) 206-1091, phone: (310) 825-1214

Jackson, Gerry: Fermilab, P.O. Box 500, Batavia, IL 60510, USA
email: gpj@almond.fnal.gov

Kaftanov, Vitali: CERN, PPE Division, CH-1211, Geneva 23, Switzerland
email: kaftanov@cernvm.cern.ch

Matsui, Takayuki: KEK. , 1-1 Oho, Tsukuba, Ibaraki, 305, Japan
email: matsui@kekvax.kek.jp

McIntyre, Peter: Texas A & M University, Department of Physics, High Energy Division, College Station, TX 77843, USA

Miller, David J.: University College London, Dept. of Physics & Astronomy, Gower St., London, U.K. WCIEGBT
email: uclva::djm@v1.ph.ucl.ac.uk, fax: 71 380 7145, phone: 71 380 7152

Mills, Frederick: 14491 N. Alamo Canyon Dr., Tucson, AZ 85737, USA
email: fredmills@aol.com

Mokhov, Nikolai: Fermilab, MS 345, P.O. Box 500, Batavia, IL 60510, USA
email: mokhov@fnalv.fnal.gov, fax: (708) 840-4552, phone: (708) 840-4409

Mori, Yoshiharu: KEK, National Laboratory for High Energy Physics, Oho 1-1, Tsukuba-shi, Ibaraki-ken 305, Japan
email: moriy@kekvax.kek.jp, fax: 81-298-64-3182, phone: 81-298-64-5209

Nagamine, Kanetada: University of Tokyo, Meson Science Lab., Faculty of Science, Hongo, Tokyo 113, Tsukuba, Ibaraki 305, Japan
fax: 81 02986 4 4979

Nakajima, Kazuhisa: KEK, Natl. Lab. High Energy Physics, Accel. Division, 1-1 Oho, Tsukuba-shi, Ibaraki-ken, 305, Japan
email: nakajima@kekvax.kek.jp, fax: 81 298 64 3182, phone: 81 298 64 5248

Neuffer, David: CEBAF, Particle Physics, 12000 Jefferson Avenue, Newport New, VA 23606, USA
email: neuffer%micro@cebaf.gov

Noble, Robert J.: Fermi National Accelerator Laboratory, MS 307, P.O. Box 500, Batavia, IL 60510, USA
email: noble@almond.fnal.gov, fax: 708 840 8590

Norem, Jim: Argonne National Laboratory, HEP Division, , Argonne, IL 60439, USA
email: norem@hep.anl.gov, fax: 708 252 5076, phone: 708 252 6548

O'Day, Steve: Fermi National Accelerator Laboratory, Department of Physics, MS 341, P.O.Box 500, Batavia, IL. 60510, USA

Oh, Sunkun: Kunkuk University, Department of Physics, Deoung-Dong Gu Mojin-Dong, Seoul, South Korea

Palmer, Robert B.: Brookhaven National Laboratory, Physics Department, Bldg. 510D, Upton, NY 11973, USA
email: palmer@slacvm, fax: 415 926 4999, phone: 415 926 2190

Parsa, Zohreh: Brookhaven National Laboratory, Physics, Bldg. 510D, Upton, NY 11973, USA
email: parsa@bnl.gov, fax: (516) 282-3248, phone: (516) 282-2085

Peters, Gerald J.: United States Department of Energy, High Energy Physics Division, ER-224, GTN, Washington, DC 20585, USA
email: jerry.peters@mailgw.er.doe.gov, fax: 301 903 2597, phone: 301 903 5228

Rajpoot, Subhash: California State University Long Beach, Department of Physics, High Energy Division, Long Beach, CA 90840, USA
email: rajpoots@beach1.csulb.edu, fax: (310) 985-2315, phone: (310) 985-4924

Reeder, Don D.: University of Wisconsin, Madison, Department of Physics, 1150 University Avenue, Madison, WI 53706, USA
email: reeder@wishep.physics.wisc.edu, phone: (608) 262-8798

Ronan, Michael T.: Lawrence Berkeley Laboratory, 50B-5239, 1 Cyclotron Road, Berkeley, CA 94720, USA
email: ronan@csa2.lbl.gov, phone: (510) 486-4396

Ruggiero, Alessandro G.: Brookhaven National Laboratory, Building 475B, P.O. Box 5000, Upton, NY 11973, USA
email: agr@bnlcl1.bnl.gov, fax: 516 282 7650, phone: 516 282 4997

Sandler, Pamela: University of California Los Angeles, Physics Department, 405 Hilgard Avenue, Los Angeles, CA 90024-1547, USA
email: sandler@physics.ucla.edu, fax: (310) 206-1091

Sasaki, Makoto: CERN, PPE/LE, CH-1211, Geneva 23, Switzerland
email: sasaki@cernvm.cern.ch, fax: 41 22 782 1928, phone: 41 22 767 4524

Sessler, Andrew M.: Lawrence Berkeley Laboratory, MS 71-259, 1 Cyclotron Road, Berkeley, CA 94720 , USA
email: Andrew_Sessler@macmail.lbl.gov

Stumer, Iuliu: Brookhaven National Laboratory, Department of Physics, 510-A, Upton, NY 11973, USA
email: stumer@bnlux1.bnl.gov

Summers, Don: University of Mississippi, Department of Physics, Oxford, MS 38677, USA
email: summers@fnalv.fnal.gov

Sutter, David F.: United States Department of Energy, Division of High Energy Physics, ER 224, GTN, Washington, DC 20585, USA

Takayama, Ken: KEK, , Oho 1-1, Tsukuba, Ibaraki, 305, Japan
email: takayama@kekvax.kek.jp, fax: 81 298 64 3182, phone: 81 298 64 5240

Thun, Rudi: University of Michigan, Physics Department, , Ann Arbor, MI 48109, USA
email: thun@mich19.physics.lsa.umich.edu, fax: (313) 936-1817

Tollestrup, Alvin: Fermilab, CDF Group, Wilson Road; P.O. Box 500, Batavia, IL 60510, USA
email: alvin@fnald.fnal.gov

Whittum, David: Stanford University, SLAC, MS-26, 2575 Sand Hill Road, Stanford, CA 94309, USA
email: whittum@yeoman.slac.stanford.edu, fax: (415) 926-4999, phone: (415) 926-2302

Willis, Bill: Columbia University, Department of Physics, 538 West 120th Street, New York, NY 10027, USA
email: willis@nevis.nevis.columbia.edu, fax: (212) 854-3379

Winn, David: Fairfield University, Department of Physics, , Fairfield, CT 06430-75424, USA

Yokoya, Kaoru: KEK, , Oho, Tsukuba-shi, Ibaraki-ken, 305, Japan
email: yokoya@kekvax.kek.jp

Zhao, Yong Xiang: Brookhaven National Laboratory, Physics Department, Bldg. 510D, Upton, NY 11973, USA

Author Index

T

W

Y